The Gut-Brain Axis in Neurological Disorders

The human gut microbiome is a diverse population of microorganisms. The development of native microbiome starts at birth and is highly adaptable to changes with age, food habits, lifestyle, and physiological conditions (ailments). Advanced "omics-based" techniques are aiding in identifying and establishing the quantitative and qualitative metabolic profiles, activity profiles, and complex interactions between gut microbiota and many physiological ailments, including obesity, intestinal bowel diseases, type 2 diabetes, and several types of cancers. Several studies have identified a direct relation between gut microbiota composition and subject's health including in aging and neurological ailments. Therefore, internal microbiota composition might become one of the critical factors in maintaining good health as well as a crucial indicator of disease theranostics. This book explores all the dimensions for microbiota composition, in relation with geriatric health issues, and provides insight into the available contemporary medicine and traditional or alternative strategies for the treatment and prevention of diseases.

Neuroscience of Neurodegenerative Disorders
Series Editor: Wael Mohamed
Wael Mohamed Yousef Mohamed
BBCH (Egypt), MMSc (Egypt), MD/PhD (USA)

Dr. Mohamed earned his PhD from Penn State University (USA) and currently is an Associate Professor at IIUM [International Islamic University Malaysia] Medical School in Malaysia. He serves as an editor for *Frontiers in Neurology* and *PLOS-One* Journals. His research is well supported by extramural funding. He is the founder and chair of the AfrAbia-PD-Genomic Consortium.

ABOUT THIS SERIES

This book series aims to publish volumes focusing on both basic and clinical research in the field of Neurodegenerative Disorders with a focus on delineating the recent advances in various disorders including Parkinson's Disease, Alzheimer's Disease, ALS, Huntington's Disease, among other common Neurodegenerative disorders. The books will examine different disorders aspects including pathophysiology, diagnostic, treatment, and precision medicine. Volumes in the series will also highlight the recent advances in the management of Neurodegenerative disorders especially in OMICS and Genomic era and precision medicine. Volumes will be useful for neuroscientists, neurologists, psychiatrists, psychologists, neurosurgeons, and those interested in preventive medicine.

Integrating Neuroimaging, Computational Neuroscience, and Artificial Intelligence
Edited by Indranath Chatterjee and Nasrallah Moradikor

Non-Pharmacological Interventions for Dementia: A Comprehensive Approach
Edited by Abraham Olufemi Asuku, Ebele Joan Ajagun, and Gbonjubola Oyinlola Ogungbangbe

Burn-Induced Neurodegeneration: Subtitle Therapies for Neuromotor Complications
Edited by Amany Atiaa and Shehab E-Kader

Neurological Disorders: Insights into Disease Biology
Edited by Jitendra Kumar Sinha and Shampa Ghosh

The Gut-Brain Axis in Neurological Disorders: Clinical Implications and Holistic Management
Edited by Vishal Ahuja, Preeti Solanki, and Harmail Singh

The Gut-Brain Axis in Neurological Disorders

Clinical Implications and Holistic Management

Edited by
Vishal Ahuja, Preeti Solanki
and Harmail Singh

CRC Press
Taylor & Francis Group
Boca Raton London New York

CRC Press is an imprint of the
Taylor & Francis Group, an **informa** business

Designed cover image: Shutterstock

First edition published 2027
by CRC Press
2385 NW Executive Center Drive, Suite 320, Boca Raton FL 33431

and by CRC Press
4 Park Square, Milton Park, Abingdon, Oxon, OX14 4RN

CRC Press is an imprint of Taylor & Francis Group, LLC

ISBN: 978-1-041-00522-3
ISBN: 978-1-041-00523-0
ISBN: 978-1-003-61027-4

DOI: 10.1201/9781003610274

Typeset in Times
by codeMantra

Aging is more psychological than physical. The acceptance of changes is the best way to adapt.

We dedicate our efforts

"To all the therapists and counselors

For creating awareness and a healthy environment for old age."

Smiles have a better impression than wrinkles on faces.

Contents

Foreword

Aging is the natural process that, in simpler terms, is defined as getting older. However, older is associated with physical appearance, while aging is a type of physical, physiological, and molecular remodeling of biological systems, which is clearly reflected in gut microbiome composition variation. Aging has shown a close relationship with microbiome composition and activity, the production of metabolites, and signaling molecules that establish a bidirectional communication system between the gut and the brain, the "Gut–Brain Axis" (GBA).

This book, *The Gut-Brain Axis in Neurological Disorders: Clinical Implications and Holistic Management*, has been planned to summarize the close association between the gut and the brain with a special emphasis on neurological disorders. The authors have made commendable efforts in preparing quality manuscripts for the volume. Each chapter of the volume is a rigorous exploration of molecular mechanisms, metabolite production, activity, and interactions.

The editors and reviewers also tried to present information in an order from foundation to advanced approach for management. Furthermore, this volume also tried to synergize the contribution of indigenous knowledge and traditional food with scientific evidence that can be employed to prevent or lower the impact of neurological disorders and maintain gut microbiome.

Preface

This book, *The Gut-Brain Axis in Neurological Disorders: Clinical Implications and Holistic Management*, has been planned to explore the foundation concepts and root causes responsible for microbiome disturbance, metabolic-neurological disorders, aging, and possible treatment approaches. This book represents a crucial advancement regarding the central axis (gut and brain) and their intercommunications. The editors have tried to arrange the book chapters to represent all the information in an order to make it understandable and readable. This book has been categorized into three sections and 15 chapters, written by active researchers and teachers. This book also includes professionals and medical practitioners from hospitals as well as research centers to have the best possible information in a single volume.

Section A encompasses the structure and functional concepts for gut–brain axis in five chapters. Chapter 1 (Development and Stabilization of Microbiome in Human) is written by a group, led by Dr. Gohil. It represents the development and establishment of microbiome in human from childhood to adulthood and then old age. It also summarizes the crucial changes with age. Chapter 2 (Health and Disorders: Modulation of Microbiome) is contributed by professional practitioners Dr. Kaur, Dr. Singla, and Dr. Behl. The chapter compiles the health ailments associated with the microbiome. Chapter 3 (Microbiome Metabolites: Direct and Indirect Communication and Relation with Host) by Dr. Singh and colleagues represents the intercommunication pattern for microbiome. Dr. Gupta et al. emphasize the role and importance of the gut–brain axis in Chapter 4 (Current status, Impact, and Significance of Gut–Brain Axis). Chapter 5 (Brain and Neural System: Clinical Implications and Mental Health) by Aarti and Kamal described the clinical significance of the neural system and mental health.

Section B advances the progression of aging and impact of microbiome disturbance in the form of aging. A group led by Dr. Bhalchim and Dr. Sangu has summarized the core mechanisms for aging with molecular-level analysis in Chapter 6 (Aging: Overview and In-Depth Analysis). Chapter 7 (Aging, Neural Communication, and Neurodegeneration) by Dr. Kumar depicts the mechanisms of neurodegeneration and its link with aging. Chapter 8 (Gut Microbiota, Composition, and Relation with Age-Related Neurological Disorders) by a group of professional practitioners (Dr. Bansal et al.) correlates the neurological disorders with the microbiome. Chapter 9 (Clinical Management of Geriatric and Psychiatric Cases: Implications of Surgery and Hospitalization on Gut–Brain Axis) by another group of professional practitioners (Ms. Saini and group) analyzes the possible negative impact of surgery and other treatments on gut microbiome. Chapter 10 (Epigenetics, Gene, and Microbiome Engineering) by Dr. Kumar and colleagues evaluates the genetic and epigenetic makeup of microbiome and its engineering.

Section C, comprising five chapters, is designed to correlate the environmental impact with gut microbiome as well as aging. Chapter 11 (Built-Up Environment Microbes and Mental Health) by health practitioners led by Dr. Trivedi compiles the role of built-up environment on mental health and identifies the critical parameters that can affect the biological system functioning. Chapter 12 (Traditional Foods: Global Coverage and Processing Methods) by research groups of Dr. Kumar and Chapter 13 (Traditional Foods: Components and Biological Effects) by the group of Dr. Patel represent the common traditional foods around the globe and how it is associated with health. Chapter 14 (Evolution of Functional Foods for Brain Health: Practices and Interventions) by Dr. Jariwala extended the discussion of health and food and its association with the brain. Chapter 15 (Adaptations in Lifestyles, Herbal Care, Physical and Psychosocial Therapies) by Dr. Gill summarizes the role of herbal practices, psychosocial therapies, pranayama, and lifestyle in balancing the gut microbiome and aging.

The overall focus of this book is to depict the foundation of aging, gut and brain function, and neurological disorders and acknowledge the traditional practices along with correlating with the current scenario.

Acknowledgments

Aging is a natural phenomenon associated with mental and physical changes in appearance, function, and communication. It is not only associated with age but it is more related to lifestyle and well-being of the individual. The compilation of vast information is a tedious task, which is possible due to continuous support. The editors extended their gratitude to our authors, contributors, and collaborators for their crucial contributions and sincere efforts.

Personally, Dr. Vishal Ahuja would like to share love and gratitude to his parents Mr. Ashok Kumar Arora and Mrs. Sunita Arora for sharing the unconditional support, love, and trust. He also offers special thanks to his mentor Professor (Dr.) Arvind Kumar Bhatt for nurturing and enlightening him.

Dr. Solanki also conveyed her sincere thanks to all the contributors of this book. As the editor, she is deeply indebted to the chapter contributors whose scholarly insights, rigorous research, and timely cooperation have significantly enriched the depth and quality of this volume. She warmly acknowledges the dedicated team at the publishing house for their professional support, meticulous production work, and unwavering commitment to maintaining high academic and production standards throughout the publication process. Her heartfelt thanks are also due to my esteemed co-editors, whose intellectual contributions, critical feedback, and collaborative spirit have been invaluable at every stage of conceptualization, review, and refinement.

Dr. Harmail Singh is grateful to his family, especially his wife Dr. Mandeep Kaur, for her constant encouragement, the publishing team for this great venture, a special thanks to co-editor Dr. Preeti Solanki who initiated the whole idea of this book, the co-authors especially Dr. Kanwarbir Singh and Monica Saini, and his mentors and the critique partners who helped him to polish this work. It is a collective effort of dedication and journey of hard work with passion, so I would also like to appreciate every individual who contributed their time, trust, and belief in this book writing.

Abbreviations

ACTHs	Adrenocorticotropic hormones
AD	Alzheimer's disease
AhR	Aryl hydrocarbon receptor
AI	Artificial intelligence
ALA	Alpha-linolenic acid
ASD	Autism spectrum disorder
BAs	Bile acids
BBB	Blood–brain barrier
BGCs	Biosynthetic gene clusters
CD	Crohn's disease
CFS	Chronic fatigue syndrome
CNS	Central nervous system
CRC	Colorectal cancer
CRS	Chronic rhinosinusitis
CVDs	Cardiovascular diseases
EBF	Exclusively breastfed
ENS	Enteric nervous system
FMT	Fecal microbiota transplantation
FOS	Fructo-oligosaccharides
FOSHU	Foods for specified health use
FXR	Farnesoid X receptor
GABA	Gamma aminobutyric acid
GBA	Gut–brain axis
GLP-1	Glucagon-like peptide
GOS	Galacto-oligosaccharides
GPCRs	G-protein-coupled receptors
HDAC	Histone deacetylase
HMOs	Human milk oligosaccharides
HPA	Hypothalamic–pituitary–adrenal
IAld	Indole-3-aldehyde
IBD	Inflammatory bowel disease
IBS	Irritable bowel syndrome
IECs	Intestinal epithelial cells
MDCs	Microbiota-driven components
mGWAS	Microbiome genome-wide association studies
NCDs	Noncommunicable diseases
OF	Oligofructose
PAT	Pulsed antibiotic therapy
PIDs	Pelvic inflammatory diseases
PTSD	Posttraumatic stress disorder
RBANS	Repeatable battery for the assessment of neuropsychological status
ROS	Reactive oxygen species
SASP	Senescence-associated secretory phenotype

SCFAs	Short-chain fatty acids
TFPs	Traditional food products
TMAO	Trimethylamine N-oxide
UC	Ulcerative colitis
UTIs	Urinary tract infections

About Editors

Vishal Ahuja has received his Ph.D. and Master's degree in Biotechnology from the Department of Biotechnology, Himachal Pradesh University. He has worked at the CSIR-Indian Institute of Petroleum, Dehradun, Uttarakhand, India, in the FTT project and is currently working as an Assistant Professor of Research and Development at Chandigarh University, India. He has more than 8 years of research experience in microbial fermentation, healthcare, and the production or extraction of bioactive compounds like phytochemicals and microbial metabolites from natural biomass. He has more than 65 publications in the form of research-review articles, chapters, and edited books, with >1,200 citations and an h-index of 17. Based on his research contribution, he was awarded the "Young Scientist Award" in 2023 by the Society of Agriculture Research and Social Development, New Delhi, India. He is a reviewer and editorial board member of many top-reputed journals.

Preeti Solanki has more than 11 years of teaching and research experience in the fields of Environmental Science, Biotechnology, Aging Research, Neurobiology, and Neurophysiology. She has worked with government institutions of national repute like DRDO, University of Delhi, and PGIMS (Rohtak) besides private institutions like LPU, DAV, and Chandigarh University. She has published more than 20 international publications and received fellowships, research grants, and international travel grants from national and international bodies like the American Association of Pharmaceutical Sciences (USA), International Brain Research Organization, Society for Psychophysiological Research (USA), DRDO (India), UGC (India), and AICTE (India).

Harmail Singh has been working as a medical officer and anesthesiologist under the Directorate of Health Services, Punjab, India, since April 2015, and is presently posted at Civil Hospital, Banur, Punjab. He completed his MBBS from Government Medical College, Amritsar, Punjab, in 2008, after which he pursued his junior residency at Fortis, Mohali, and in Transfusion Medicine at GMCH, Chandigarh. After completing his Diploma in Anesthesia and Critical Care, he worked as a senior resident at GMCH, Chandigarh, during 2012–2015. He is a young and dynamic health professional with more than 10 years of experience in critical care medicine.

Contributors

Aarti
Department of Anatomy
Pt. B. D. Sharma PGIMS
Rohtak, Haryana, India

Wafa Aziz
Heart and Vascular Institute
Penn State College of Medicine
Hershey, Pennsylvania

Yashik Bansal
Department of Microbiology, Manipal Tata Medical College Jamshedpur
Manipal Academy of Higher Education
Manipal, Karnataka, India

Preeti Behl
Department of Pathology
Yosemite Pathology Medical Group
Modesto, California

Vrushali Bhalchim
Amity Institute of Pharmacy
Amity University Maharashtra
Mumbai, Maharashtra, India

Shamba Chatterjee
Department of Biotechnology
Haldia Institute of Technology
Haldia, West Bengal, India

Jaibeer Singh Chhina
General Practice
Wessex Deanery
London, UK

Sucheta Das
Department of Biotechnology
Haldia Institute of Technology
Haldia, West Bengal, India

Mantaz Kaur Dhillon
Psychiatry
Apollo Hospital
Chandigarh, India

Kanwarbir Singh Gill
Department of Health and Family Welfare
Civil Hospital Rupnagar
Rupnagar, Punjab, India

Nikitasinh Gohil
Department of Microbiology, Faculty of Science
Marwadi University
Rajkot, India

Namrata Gupta
Department of Biotechnology, University Institute of Biotechnology
Chandigarh University
Mohali, Punjab, India

Ruchi Jariwala
Department of Microbiology
Vanita Vishram Women's University, VNSGU
Surat, Gujarat, India

Jillwin Joseph
Department of Microbiology and Immunology
American University of Antigua College of Medicine
Osbourn, Antigua and Barbuda

Kashish Julka
University Institute of Biotechnology
Chandigarh University
Mohali, Punjab, India

Mandeep Kaur
Department of Microbiology
Government Medical College
Patiala, Punjab, India

Simran Kaur
Psychiatry Rotatory Internship
Gian Sagar Medical College
Punjab, India

Sajpreet Kour
Department of Biotechnology, University Institute of Biotechnology
Chandigarh University
Mohali, Punjab, India

Dimple Kothari
Department of Microbiology
Marwadi University
Rajkot, India

Anil Kumar
Food Water Waste Research Group
University of Nottingham
University Park, Nottingham, UK

Deepak Kumar
Department of Biotechnology
Chandigarh University
Mohali, India

Ritik Kumar
Department of Biotechnology
Thapar Institute of Engineering and Technology
Patiala, Punjab, India

Sachin Kumar
Department of Biotechnology, University Institute of Biotechnology
Chandigarh University
Mohali, Punjab, India

Santosh Kumar
Department of Biology
McMurry University
Abilene, Texas

Vijay Kumar
Centre for Medicines Discovery
University of Oxford
Oxford, UK

Vinay Kumar
College of Medicine
Pennsylvania State University, Hershey Medical Center
Hershey, Pennsylvania

Shilpi Kumari
Bioserve Biotechnologies (India) Private Limited
Hyderabad, Telangana, India

Nivedita Mama
Department of Microbiology, Faculty of Science
Marwadi University
Rajkot, India

Divya Mori
Department of Microbiology, Faculty of Science
Marwadi University
Rajkot, India

Vrushali Neve
Department of Pharmacology
Dr. D. Y. Patil Institute of Pharmaceutical Sciences and Research
Pune, Maharashtra, India

Apurva Patel
Department of Pathology
Roy J. and Lucille A. Carver College of Medicine
University of Iowa
Iowa City, Iowa

Niralee Patel
Department of Microbiology
Marwadi University
Rajkot, Gujarat, India

Sachin Pahal
Department of Biotechnology, University Institute of Biotechnology
Chandigarh University
Mohali, Punjab, India

Mohini Rana
Department of Biotechnology, University Institute of Biotechnology
Chandigarh University
Mohali, Punjab, India

Monica Saini
Psychology Department
Chandigarh University
Mohali, Punjab, India
and
Civil Hospital
Rupnagar, Punjab, India

Sakshi
Department of Biotechnology
Thapar Institute of Engineering and Technology
Patiala, Punjab, India

Muthuraju Sangu
Radiology
Academic Institute Houston Methodist
Houston, Texas

Namrata Sarma
Department of Biotechnology, University Institute of Biotechnology
Chandigarh University
Mohali, Punjab, India

Harnoor Kaur Sibia
Department of Internal Medicine
Desert Valley Hospital
Victorville, California

Kamal Singh
Department of Anatomy
Pt. B. D. Sharma PGIMS
Rohtak, Haryana, India

Varshika Singh
University Institute of Biotechnology
Chandigarh University
Mohali, Punjab, India

Nidhi Singla
Department of Microbiology
Government Medical College and Hospital
Chandigarh, India

Preeti Solanki
Department of Biotechnology, University Institute of Biotechnology, University Centre for Research and Development
Chandigarh University
Mohali, Punjab, India

Sonali
Centre for Stem Cell Tissue Engineering and Biomedical Excellence
Panjab University
Chandigarh, India

Abhijit Swain
Genomiki Solutions Private Limited
Greater Noida, Uttar Pradesh, India

Subrata Trivedi
Department of Biotechnology
Haldia Institute of Technology
Haldia, West Bengal, India

Anuradha Tyagi
School of Medicine
University of Missouri
Columbia, Missouri

Section A

Microbiome Development and Gut–Brain Axis

1 Development and Stabilization of Microbiome in Human

Divya Mori, Nivedita Mama, Anuradha Tyagi, and Nikitasinh Gohil

1.1 INTRODUCTION

Over the past few decades, microbiome research has grown rapidly, attracting the attention of both scientists and the public. The field's initial focus was environmental studies, but it has now evolved to consider how eukaryotic organisms, including humans, are closely related to the microbial communities with which they inhabit. As a sophisticated system, the human body is home to trillions of microorganisms that interact with their host in various ways. Microorganisms that inhabit nearly every part of the body are referred to as the "microbiome" because of their genes (Shreiner, Kao, and Young 2014). Because of its significant impact on human health, scientists frequently refer to it as a "second genome." Phenotypes that improve health and survival in various ecological niches have been shaped by this co-evolution (Dominguez-Bello et al. 2019). This relationship can be neutral, damaging, or even helpful depending on the situation. With an estimated 50–100 times more genes than the human genome, the microbiome greatly expands the body's genetic and functional potential (Sender, Fuchs, and Milo 2016). Researchers have shifted their focus from examining the interactions between individual microbes and their hosts in isolation across time. The holobiont theory provides a broader perspective. This theory acknowledges that the equilibrium among a host and its associated microbiota is crucial for health (Simon et al. 2019). When this balance is disturbed, a disorder known as dysbiosis often occurs. An interesting fact is that just a small percentage of bacteria can cause hazardous diseases. These pathogens are frequently the result of changes in the microbiome's makeup rather than being the only cause of illness (Theis et al. 2016). Conversely, the great majority of microorganisms play a crucial role in encouraging positive interactions within microbial communities as well as with their hosts and preserving the stability of ecosystems. Opportunistic infections highlight the complex relationships that exist between microorganisms and their hosts. Their behavior is influenced by both the host and microbial population (Aggarwal et al. 2022). In infancy, the microbiota experiences substantial growth and is essential in influencing the immune system and overall well-being. The makeup of the microbiome is influenced by several factors, such as the method of delivery, dietary habits, and environmental influences.

The microbiological composition of various body organs varies significantly and is unique to each individual. Although there is no clear definition of what constitutes a "healthy microbiome," research suggests that probiotics, prebiotics, and synbiotics help maintain a balanced microbial community. To understand the significance of bacteria in health, it is essential to differentiate between the core and transient microbiota (Berg et al. 2020). The core microbiota is made up of microbial communities that consistently associate with a particular host or environment, while transient microbiota fluctuates throughout time due to external factors (Figure 1.1A). Identifying these differences and improving their statistical analysis and experimental techniques can enhance the study of microbiomes for medicinal or therapeutic purposes.

DOI: 10.1201/9781003610274-2

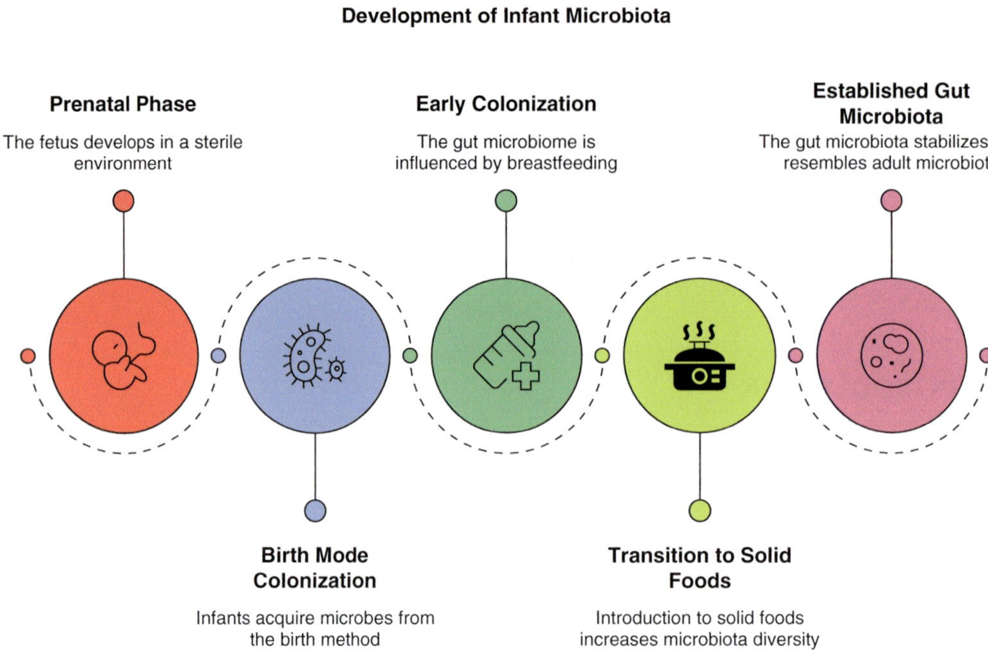

FIGURE 1.1A Development of microbiome in infant and adult human body.

1.2 FACTORS SHAPING THE HUMAN MICROBIOTA

Neonates and children have different patterns of microbiota colonization, which affects how immune responses are regulated later in life and the likelihood of illness. Numerous variables might affect a child's intestinal microbiota's composition and structure during the establishing period (ages 1–4), such as the mother's prenatal health, the method of birth, breastfeeding or formula feeding, early nutrition, and postnatal medication treatments.

1.2.1 Delivery Mode

The manner of delivery is the primary determinant for the establishment of a robust microbiome during the early stages of existence. Maternal cutaneous bacteria are the first to colonize newborns born via Cesarean section, and the microbial communities of these infants closely resemble their mothers' skin microbiota (Aggarwal et al. 2022).

The genital area communities of the mothers of vaginally delivered infants are reflected in the bacterial communities of the various regions. However, the microbiota maturation of newborns born via Cesarean section is significantly differently plateaued in the subsequent 6 months, in contrast to their vaginally born counterparts. The microbiota diversity of babies delivered by Cesarean section progressively developed and became similar to that of vaginally born infants (Bokulich et al. 2016). The main microbiota of newborns differs depending on the delivery method and shows significant modifications. Various microbiota of newborns delivered via Cesarean section steadily developed gut microbiota that of newborns born vaginally (Kuhle, Tong, and Woolcott 2015). Compared to those delivered vaginally, who were mostly colonized by *Bifidobacterium, Bacteroides, Clostridium,* and *Lactobacillus*, infants born by C-section showed lower levels of Actinobacteria and Bacteroidetes in the first 3 months of life, together with increased levels of Firmicutes. Studies indicate that Cesarean-born infants have a higher prevalence of type 1 diabetes, obesity, asthma, celiac disease, and childhood mortality relative to vaginally delivered infants (Black et al. 2015). Consequently, the mode of delivery significantly influences gut microbiota in the first year of life.

1.2.2 Feeding Practice

Feeding methods significantly influence the development of an infant's gut microbiome. *Bifidobacterium* present in milk provides essential beneficial bacteria from the mother and increases the development of beneficial bacteria (Bäckhed et al. 2015). Human milk oligosaccharides function as natural prebiotics, which accounts for this phenomenon. Infants who are breastfed exhibit a unique microbial composition in contrast to those who receive formula feeding. Breast milk nutrients contribute significantly to the maintenance of a balanced and healthy gut environment. Conversely, formula feeding results in variations in bacterial populations. *Bifidobacterium* levels are significantly elevated in infants who are breastfed. This advantageous bacterium facilitates digestion and enhances the immune system. Initial variations in the microbiome can influence long-term health outcomes. Breastfeeding is essential for the development of gut flora during infancy. During the initial 3 months of life, the development of microbiota, bacterial richness growth rate, and phylogenetic diversity in formula-fed infants is significantly lower compared to that of breastfed infants. In this period, *Staphylococcus, Lactobacillus, Actinobacteria*, and *Megasphaera* are more prevalent in breastfed infants, while *Clostridiales* and *Proteobacteria* genera are more prevalent in formula-fed infants (Savage et al. 2018). Research indicates a positive correlation between the prevalence of beneficial *Clostridium, Lactobacillus,* and *Bifidobacterium,* which are still underexplored microbes in breastfeeding. Additionally, the presence of functional *Lachnospiraceae* has been observed to diminish in breastfed infants (Baumann-Dudenhoeffer et al. 2018). The microbiome of the intestinal tract of most infants who were breastfed exhibited enhanced capacity for synthesizing branched-chain amino acids such as arginine, methionine, and threonine (Thompson et al. 2015).

There is a significant difference in the microbiota of the intestines of babies who had been exclusively breastfed (EBF) immediately before the introduction of solid diets compared to non-EBF infants, with a greater number of Bifidobacterium and a decreased number of *Clostridiales* and *Bacteroidetes* (Lv et al. 2021). In addition to breastfeeding, the introduction of solid food resulted in significant alterations in microbial composition, which is distinguished by a transition to comparatively stable communities, and also an increase in the production of vitamins, short-chain fatty acids, biosynthesis, and carbohydrates (Leong et al. 2018). The gut microbiota is also influenced by patterns of supplemental food consumption. The bacterial diversity and *Roseburia* abundance of infants in a 12-month-old baby-led introduction to solids group (an improved version of baby-led weaning, where solid foods are introduced at 6 months of age and then the family self-feeds the meals) were lower than those in a conventional intake-style group (i.e., the parents used a spoon to feed mushy food) (Chelimo et al. 2020).

1.2.3 Environmental Factors

The development of the microbiota in one's gut may be hampered by environmental factors, especially the use of antibiotics throughout early life. For example, compared to newborns not exposed to antibiotics, those exposed to them showed delayed development, which was most noticeable between the 6th and 12th months, mostly attributable to the deficiency of *Lachnospiraceae* and *Erysipelotrichaceae* (Korpela et al. 2016). Additionally, the host intestinal microbiota's metabolism may be impacted by early pulsed antibiotic therapy (PAT) with ab-lactamormacrolide; the impacts included lowering the increased cycle of citric acids, nucleoside synthesis, and amino acid synthesis, as well as promoting the glycolysis and gluconeogenesis pathways (Yatsunenko et al. 2012).

Moreover, in mice given with modest doses of penicillin in PAT, the quantity of *Allobaculum, Candidatus*, the expression of genes related to *Lactobacillus* and *Arthromitus* was refined, resulting in reduced differentiation, activation, adherence, recruitment, and the number of immune cells. These effects were observed to contribute to obesity and to have a lasting impact on the host throughout its lifespan (Leong et al. 2018). Moreover, changes induced by early exposure to antibiotics made the host vulnerable to various diseases, including type 1 diabetes, obesity, and asthma.

The human body is a natural home to various internal and external microorganisms that thrive in environments that support their growth. With the help of variables like pH, nutrient availability, temperature, oxygen concentration, pressure, and osmolarity, the human body establishes an environment akin to its indigenous habitat (Kim et al. 2016). Besides internal locations such as the lungs, gastrointestinal tract, the bladder, kidneys, and enlarged vagina, microorganisms also inhabit external sites like the eyes, skin, and exposed regions under the nails. Microorganisms can also enter through entry sites such as the respiratory system, oral cavity, and urogenital tract. The diversity of microbiota is influenced by a number of extrinsic factors in addition to these intrinsic ones, such as seasonal fluctuations, geographic location, climate, medication, nutrition, and lifestyle. Additionally, microbial diversity in various body areas is influenced by host-specific factors like age, gender, ethnicity, and genetics. While some secretory fluids, such as sebum on the skin, not only change pH but also act as a carbon source, encouraging microbial growth, and the human body temperature also encourages microbial colonization. Similar to this, the intestinal mucus layer aids in colonization by supporting microbial attachment and adhesion in addition to providing nutrients (Herath et al. 2020; Wang et al. 2022).

1.3 PHYSIOLOGY AND REGULATORY ROLE OF MUTUALISTIC AND COMMUNALISTIC MICROBES

Mutualistic microorganisms participate in ecological interactions that provide a net advantage to each species involved. These interactions are essential inside the human body, especially in the gut and vaginal microbiota, where they play a vital role in host health. The human gut has a significant and diverse microbiota of symbiotic microorganisms that provide sustenance and protection to the host. Development and regulation of the immune system, depends on stability of gut microbiota. These microbes exhibit a co-evolutionary relationship with the human immune system, playing roles in metabolic function, pathogen defense, and the provision of nutrients and energy. The gut microbiota plays a crucial role in energy and nutrient extraction from food due to its different metabolic genes, which enable the production of distinctive enzymes and biochemical routes. It enables the synthesis of bioactive compounds, such as amino acids, vitamins, and lipids (Berg et al. 2020).

A variety of compounds are included in microbiome-related products, such as microbiota-driven components, microbiota-derived metabolites, and microbiota-secreted proteins. A vibrant microbial population within the intricate ecosystem of the digestive tract generates a wide range of metabolic enzymes, including glycosidases, proteases, and biogenic enzymes, which improve the body's capacity for digestion and metabolism. These amazing beings have shown their significant influence on our well-being by becoming important players in immune response, metabolism, brain function, tissue development, and behavioral regulation (Table 1.1). Therefore, while microorganisms in other regions of the body play a key role in physiologic control through mechanisms such as immune modulation, colonization resistance, and maternal transmission, the oral and gastrointestinal microbiomes are crucial for the digestion and absorption of nutrients and substances (Hill and Round 2021).

1.4 INTERACTION OF HUMAN MICROBIOME, HEALTH, AND ILLNESS

Microbiomes, which are diverse and intricate microbial ecosystems, are present in the human body and consist of innumerable microbial species that have a significant impact on host physiology. Numerous and intricate interactions exist between humans and their associated microbiota (Figure 1.1B). These microorganisms have an enormous effect on our physiology in the context of both health and illness, having a broad spectrum of functions in various regions of the human body, spanning from symbiotic to pathogenic (Aggarwal et al. 2022).

TABLE 1.1
The Regulatory Application and Physiology of Symbiotic Microorganisms

Microflora	Location	Type of Symbiosis	Recent Example of Application	References
Lactobacillus spp.	Vagina	Mutualism	*Lactobacillus* species contribute to vaginal health by generating lactic acid, which lowers pH and suppresses harmful bacteria.	Zafar and Saier (2021)
Bacteroides fragilis	Gut	Mutualism	Facilitates the digestion of complex carbohydrates and contributes to immune system development, thereby reducing inflammation.	Lee et al. (2021)
Bifidobacterium spp.	Gut	Mutualism	Associated with decreased inflammation and enhanced gut barrier function.	Dominguez-Bello et al. (2019)
Faecalibacterium prausnitzii	Gastrointestinal tract	Mutualism	Associated with decreased inflammation and enhanced gut barrier integrity.	Rinninella et al. (2019)
Roseburia spp.	Gut	Mutualism	Enhancing gut health and metabolic functions.	Maioli et al. (2021)
Faecalibacterium prausnitzii	Colons	Commensalism	Support colon health by generating butyrate through fiber fermentation. Main energy source for colonocytes, improves intestinal barrier function, and exhibits anti-inflammatory properties.	Sakanaka et al. (2022)
Fusobacterium nucleatum	Oral cavity	Commensalism	While associated with periodontal disease, it also plays a role in oral health by participating in biofilm formation that protects against more harmful pathogen.	Dekaboruah et al. (2020)

1.4.1 Oral Microbiome

These microbes live in the connective tissues of the oral mucosa, and they are particularly prevalent on the hard surfaces of teeth, especially dentures. Among the common bacterial species found in the mouth are *Streptococcus, Granulicatella, Gemella, Actinomyces, Corynebacterium, Rothia, Veillonella, Fusobacterium, Prevotella, Porphyromonas, Capnocytophaga, Neisseria, Haemophilus, Treponema, Eikenella, Leptotrichia, Lactobacillus, Peptostreptococcus, Staphylococcus, Eubacterium*, and *Propionibacterium*. Fungi such as *Aspergillus, Candida, Saccharomyces, Fusarium, Cladosporium*, and *Cryptococcus* are also present (Jensen et al. 2021). Some oral bacteria, including *Streptococcus salivarius* strain K12, actively promote the body's defense response,

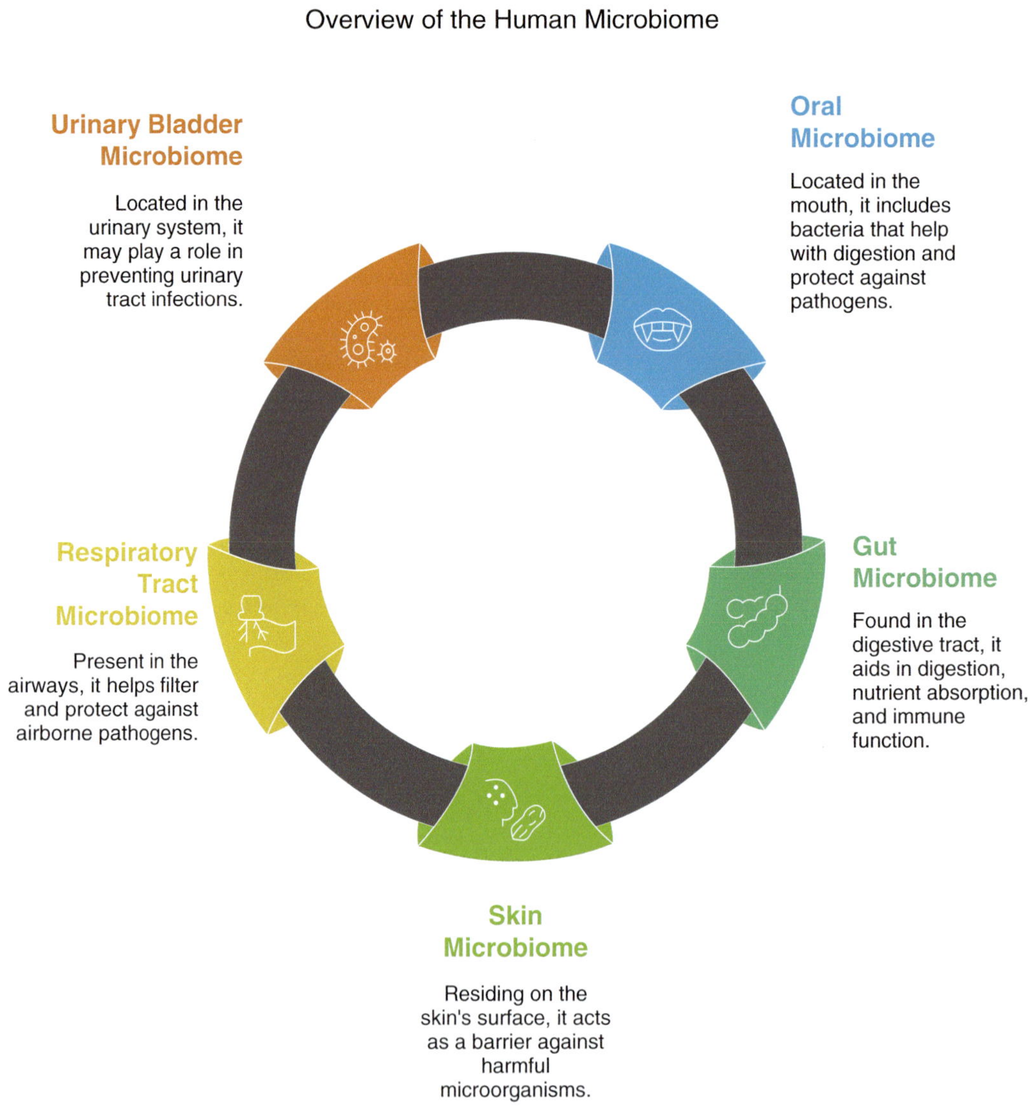

FIGURE 1.1B Development of microbiome in infant and adult human body.

which generates bacteriocin, by creating an environment that prevents the growth of Gram-negative bacteria that lead to gum disease and other issues with oral health. The most common oral health issue is tooth decay, also referred to as dental caries. Tooth decay is mostly caused by three bacteria: *Lactobacillus acidophilus, Streptococcus mutans,* and *Streptococcus sobrinus*. Other bacteria, including *Veillonella, Bifidobacterium, Propionibacterium, Actinomyces, Atopobium*, and *Scardovia*, have also been linked to cavities. Furthermore, research has shown that oral health can influence how noncommunicable diseases develop and progress, neuropsychiatric disorders, diabetes, obesity, and cancer. This has led to the theory that an individual's mouth microbiota could be a helpful tool for assessing their risk of contracting specific diseases. Future studies are expected to provide insight into how to restore a healthy oral microbiome, which could improve overall health and reduce the risk of illness (Craig et al. 2018; Willis and Gabaldón 2020).

1.4.2 Gut Microbiome

The colon has more bacteria than the stomach and small intestine. 90% of the gut microbiota is made up of Bacteroides and Firmicutes. *Actinobacteria, Proteobacteria, Fusobacteria,* and *Verrucomicrobia* are other intestinal bacteria. *Bacillus, Lactobacillus, Enterococcus, Clostridium,* and *Ruminococcus* are among the almost 200 genera of Firmicutes. Despite the health benefits of *Lactobacillus. Prevotella* and *Bacteroides* make up the majority of the Bacteroides genus. The less common phylum Actinobacteria is dominated by the health-promoting *Bifidobacterium.* Proteobacteria include *Shigella, Helicobacter, Salmonella,* and *Enterobacter*, among other infections. From birth until weaning, regular feeding, and old age, the gut flora evolves. During the first 2 weeks, facultative anaerobes, the initial gut bacteria, create anaerobic conditions for mandatory anaerobes such as *Bacteroides* and *Bifidobacterium.* Compared to formula-fed infants, breast-fed newborns showed a more stable microbiome with higher levels of *Bifidobacterium.* Following a healthy diet, gut microbiomes alter and Firmicutes populations increase. In the gut microbiome of adults aged 65 and up, *Enterobacteriaceae* proliferate while *Bifidobacteria* decline. The exterior mucin layer of healthy individuals contains *Akkermansia muciniphila*, whereas the inner mucin and epithelial surface show minimal microbial colonization. *Ruminococcus* species inhabit the liquid phase of the gut lumen, according to dietary fibers. The ileum and jejunum contained dominating bacterial groups such as *Proteobacteria* and *Bacteroides* as well as facultative anaerobes like *Streptococcus, Lactobacillus*, and *Enterococcus.* In addition to improving energy generation and metabolic efficiency, they are crucial for digestion (Aggarwal et al. 2022). Increased permeability allows virulence factors and microbial metabolites to enter the bloodstream, changing the microbiome and triggering immunological responses that can lead to autoimmune diseases, allergies, and inflammation. Good bacteria can control gut pathogens.

Innocent bacteria can become deadly if this balance is upset. *Clostridium* difficile that lives in the gut doesn't cause problems unless medications alter the microbiota. Diarrhea, colitis, sepsis, and mortality are caused by the breakdown and inflammation of the gut barrier caused by *C. difficile.* Another gut microbiota disease is inflammatory bowel disease, which encompasses Crohn's and ulcerative colitis (Yang et al. 2021). There is a decline in the production of butyrate, a vital source of energy for colon cells, by Firmicutes bacteria, such as *Faecalibacterium prausnitzii* and *Roseburia* (Oren, Aharon and George 2021). In contrast to mucosa, stomach juice is primarily composed of Actinobacteria, Fusobacteria, Firmicutes, and Bacteroidetes (Liu et al. 2018). The majority of *Helicobacter pylori* strains, the most common stomach bacterium, have the ability to change the gastrointestinal environment, which can have an impact on other microorganisms. Changes in the flora of the stomach may increase the risk of gastric cancer. Removing *H. pylori* might improve the diversity of stomach microbes.

1.4.3 Respiratory Tract Microbiome

The nasal cavity serves as a critical passageway among the human body along with the extrinsic environment. Breathing allows microorganisms, allergens, and pollutants to enter the nasal cavity. The location of nose bacteria varies in the respiratory system. Compared to the middle meatus and sphenoethmoidal recesses, the anterior nares, or front of the nose, exhibit a lower diversity of microorganisms. The anterior nares microbiome of healthy individuals is dominated by Actinobacteria, Firmicutes, and Proteobacteria (Jensen et al. 2021). The main bacterial genera found in this area include *Staphylococcus, Propionibacterium, Corynebacterium*, and *Moraxella.* However, a sinus and nasal infection called chronic rhinosinusitis (CRS) can be brought on by *Staphylococcus aureus.* Nasal polyps or the severity of chronic rhinosinusitis may be associated with *S. aureus* in the sinuses and nasal cavity. *S. aureus* levels were higher in persons with nasal polyps and chronic rhinosinusitis than in those without polyps. According to these early findings, nasal microbiota, specifically *S. aureus,* may play a role in CRS. Further investigation is required. It was believed to be microbial-free because germs found in the respiratory tract are difficult to thrive. However, environmental bacteria can enter the upper respiratory system (pharynx, larynx, etc.) and then travel through the mouth or nose to the lower respiratory system. According to research, infants who have *Moraxella catarrhalis, Haemophilus influenzae*, or *Streptococcus pneumoniae* in their throats are at an increased risk of developing asthma and recurrent wheezing. These microorganisms also exacerbate signs of asthma and chronic obstructive pulmonary disease (COPD). Despite these findings, additional research is necessary to gain a comprehensive understanding of the importance and function of the respiratory tract microbiome. Although breathing exposes the lungs to numerous environmental microbes, it was formerly thought that the lungs were totally sterile. Because it was difficult to cultivate bacteria from respiratory samples, it was thought that healthy lungs were germ-free. Research on the lung microbiota was slowed by the invasive methods required to collect clinical samples. High-throughput 16S rRNA gene sequencing is frequently used to study lung bacterial populations. *Prevotella, Veillonella, Streptococcus, Haemophilus, Neisseria,* and *Corynebacteria* are all present in healthy individuals (Drago et al. 2016). Certain microorganisms can flourish because mucus produces higher temperatures, lower oxygen levels, and nutrients. Immune responses brought on by these alterations may exacerbate inflammation and disturb the lung microbiota. The microbial population changes from Bacteroidetes to Proteobacteria and other pathogenic bacteria when lung ailments worsen.

1.4.4 Skin Microbiome

The epidermis microbiota has been categorized into categories: sticky, moist, and dry, based on its composition, which is relatively consistent across individuals and physiologies. The epidermis contains various beneficial microbes, bacteria, viruses, archaea, and parasites. The main body and arm sites, which have been colonized by *Aspergillus* spp., *Cryptococcus* spp., *Rhodotorula* spp., *Epicoccum* spp., and various other organisms, are more likely to contain fungal cultures than bacteria. The stratum corneum, sebum, and sweat make up the skin, which is relatively less nutrient-rich than other environments. *Propionibacterium acnes,* which flourishes in anoxic sebaceous glands, grows as a result. Because they feed on sebum lipid and *Staphylococcus* spp., stick to the skin for nourishment, other auxotrophic species including *Malassezia* and *Corynebacterium* are widely distributed in the skin mycobiome. The local flora on the skin will engage with these pathogenic germs to stop their colonization. *S. aureus*, which causes atopic dermatitis, will become much more prevalent if the resident microbiota declines. According to Drago et al. (2016), there are multiple indications that the skin microbiota plays a significant part in the etiology of disease.

1.4.5 Urinary System and Its Microbiome

Commensal urobiomes, which are shown to be comparable in both males and females belonging to the phylum Firmicutes, may now be easily detected thanks to improvements in sample and DNA sequencing procedures. Males were found to have *Escherichia coli, Enterococcus, Prevotella, Streptococcus, Citrobacter,* and *Pseudomonas,* whereas females were shown to have high levels of *Corynebacterium, Lactobacillus*, and *Streptococcus*. In fact, some studies have found microbiota overlap between urine and vaginal samples, suggesting that female urinary tract microbes are largely found in the vagina. Several bacterial species, including *Tepidimonas* and *Flavobacterium,* are found in the urobiome but not in the vaginal microbiome. Aggarwal et al. (2022) suggest other potential sources of urine microorganisms. This is caused by the anatomical proximity of the urinary tract and reproductive system openings, which also extends the traditional understanding of Urinary tract infection (UTI) pathophysiology and the co-infection of commensal gut microbiome members in females. Additionally, sporadic exposure to anaerobic *Gardnerella vaginalis* spp. triggers intracellular *E. coli* activation in the bladder, increasing the risk of urinary tract infections or vaginal-associated bacterial diseases. So far, only one archaea species *Methanobrevibacter smithii* has been linked to urinary tract infections. The bladder and kidney stones of women with urgency urinary incontinence (UUI) have been found to have bacteriophages, such as those that infect *Pseudomonas aeruginosa* and *Escherichia coli,* which are found in a urine virome. Interestingly, there appears to be a correlation between the higher prevalence of these bacteria in the stomach and a higher risk of developing a urinary tract infection (McKinnon et al. 2019).

1.4.6 Reproductive System and Its Microbiome

Lactobacillus, a single genus, dominates the human vaginal microbiome, which inhibits the growth of other bacteria and lowers pH compared to other body parts. Numerous pathogens from facultative and obligatory anaerobes, including *Gardnerella, Prevotella, Atopobium*, and *Sneathia*, are linked to acquired immunodeficiency syndrome (AIDS) and Sexually transmitted infection (STIs). This implies that a population that is not dominated by *Lactobacillus* may be less resistant to experiencing negative health consequences (Agarwal and Lewis 2021). Hormonal changes also have an impact on vaginal physiology because the number of mucins and glycogen in the tissues of the epithelium acts as a protective layer and a source of nutrients for the microbiome. However, their levels vary greatly during the menstrual cycle and menopause, which significantly changes the vaginal microbiome's microenvironment (Vagios and Mitchell 2021; Lacroix et al. 2020).

1.5 MICROECOLOGICAL INTERVENTION: PROBIOTICS AND PREBIOTICS

Probiotic and prebiotic therapies have shown efficacy in the management of metabolic, allergy, and autoimmune juvenile disorders (Sanders et al. 2019). Probiotics comprise living microbes that offer health benefits when ingested in sufficient quantities, whereas prebiotics are selectively used substrates that promote the growth of host microorganisms (Nicolucci et al. 2017). Insulin supplementation has been shown to diminish body and trunk fat in obese adolescents, while enhancing *Bifidobacterium* and reducing *Bacteroides vulgatus* (Bosco et al. 2023).

Moreover, distinct changes were seen based on the feeding style, characterized by a majority of *Bifidobacteria* in breastfed babies and a more diverse microbiota in formula-fed infants, with a predominance of *Bacteroides, Staphylococci, Clostridia, Enterobacteria*, and *Enterococci* (Wang et al. 2020). Probiotics have also demonstrated beneficial effects in pediatric conditions. A combination of species of *Bifidobacterium* and *Lactobacillus* significantly reduced autism symptoms and *GI* issues (Huang, Chie, and Wang 2018). *Lactobacillus acidophilus* supplementation in stunted children led to improved physical growth by weight, height, and reduced morbidity related to diarrhea, fever, cough, and cold. Furthermore, *Lactobacillus paracasei* and *Lactobacillus fermentum* were found to alleviate asthma symptoms, with their combination proving more effective than individual use (Savilahti et al. 2018). To ensure safe probiotic interventions, further research is needed to understand their mechanisms of action. The use of organoids engineered with microbiota niches and related animal models can provide deeper insights into probiotic functions and safety.

1.6 ENABLING TECHNOLOGIES FOR MICROBIOME RESEARCH AND ENGINEERING

Advancements in technologies have revolutionized microbiome research, offering profound insights into its development and stabilization during several life stages. The evolution of the microbiome from infancy to maturity has been elucidated by high-throughput sequencing methods, including 16S rRNA sequencing, metagenomics, meta-transcriptomics, and single-cell sequencing. These methodologies enable researchers to delineate microbial colonization patterns, identify pivotal species contributing to homeostasis, and monitor temporal fluctuations in microbial diversity. Microbiome engineering to restore and sustain microbial equilibrium during disturbances in the gut microbiota. Probiotics and prebiotics facilitate the proliferation of advantageous microbes,

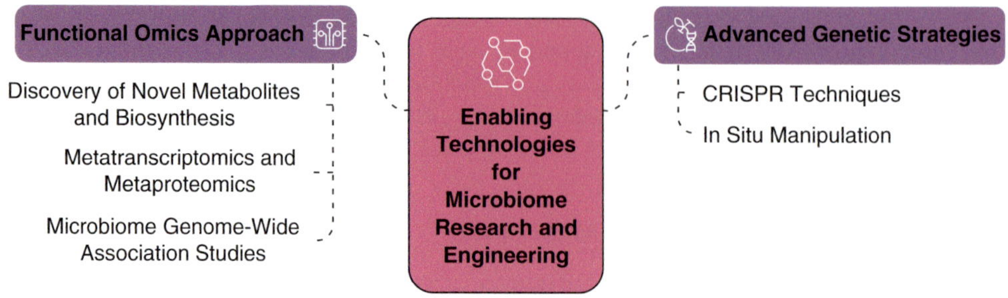

FIGURE 1.2 Next-generation technologies for microbiome engineering.

but fecal microbiota transplantation is seeing an increase in prominence as an effective method to restore a compromised microbiome with a robust microbial community. Clustered Regularly Interspaced Short Palindromic Repeats (CRISPR) based microbiome editing enables exact alterations of microbial genes, possibly allowing for the elimination of detrimental bacteria while maintaining advantageous ones (Figure 1.2). Synthetic biology methods concentrate on designing advantageous bacteria capable of colonizing the gut, synthesizing vital metabolites, and improving immune function, presenting a possible avenue for tailored and targeted microbiome therapeutics (Aggarwal et al. 2022).

1.6.1 Discovery of Novel Metabolites through Coculture

A sophisticated microbial cultivation strategy that replicates their natural ecological interactions is known as coculture (Yao et al. 2016). This method enables interspecies communication by facilitating metabolic exchange, competitive or cooperative interactions, and signaling molecules. One of the most significant benefits of coculture is its capacity to activate silent biosynthetic gene clusters (BGCs) that are not expressed under conventional monoculture conditions. Secondary metabolite production can be modulated, stress responses can be triggered, and quorum-sensing mechanisms can be induced during interactions between microbial species. This phenomenon is especially advantageous in the identification of novel bioactive compounds. Actinomycetes is an example of novel metabolites that generate bioactive secondary metabolites, novel antibiotics, antifungals, anticancer drugs, and siderophores, which have been discovered as a result of the activation of quiet bioactive compounds generated (BGC) by coculture techniques combining actinomycetes with other microbial species. Coculture interactions activate metabolic pathways that are inactive in monocultures, leading to the production of molecules with distinctive structures and important pharmacological properties.

Microbial cocultures are essential for studying ecological interactions, enhancing enzymatic production, and discovering novel metabolites. Bacteria–bacteria interactions help explore competition, symbiosis, and amensalism (Figure 1.3). Bacteria–fungi cocultures enhance enzyme production and promote metabolite discovery (Jaiswal et al. 2023). Bacteria–plant cell interactions contribute to plant growth and symbiotic relationships. Fungi–fungi cocultures aid in biomass degradation by producing synergistic enzymatic cocktails, making them valuable for industrial and environmental applications (Long et al. 2020).

Metatranscriptomics and metaproteomics are functional omics approaches that analyze gene expression and protein synthesis within microbial communities to provide critical insights into the activity of the human microbiome. Complementing this, Microbiome Genome-Wide Association Studies (mGWAS) investigate the genetic factors that influence the composition and function of the microbiome by revealing how host genetic variations affect microbial diversity and interactions. mGWAS offers valuable insights into microbiome-associated diseases and prospective therapeutic strategies by connecting genetic predisposition with microbiome diversity. When combined, these

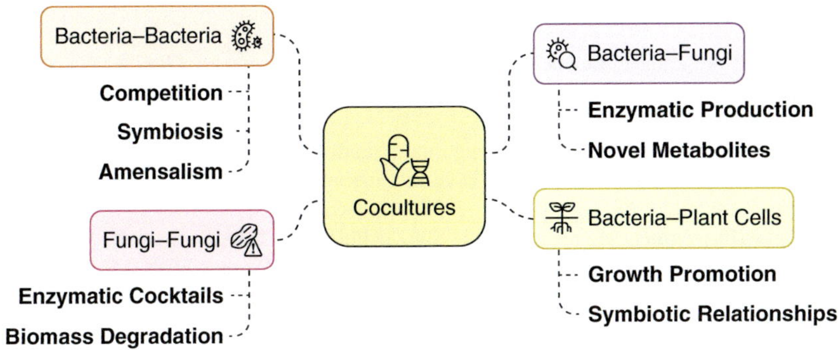

FIGURE 1.3 Microbial interactions with cellular cocultures.

methods facilitate the stabilization of the microbiome and the development of personalized medicine by fostering a more profound comprehension of the interactions between hosts and microbes (Zhang et al. 2020; Mills et al. 2022; Voorhees et al. 2020).

1.6.2 Advanced Genetic Strategies: CRISPR and In Situ Manipulation for Microbiome Development and Stability

CRISPR technology enables accurate genetic alterations to augment advantageous bacteria, eradicate detrimental ones, and modulate microbial gene expression. This facilitates the creation of advanced probiotics designed to enhance digestion, bolster immunological function, and improve disease resistance by altering microbial communities directly in their native habitat, maintaining microbial diversity and assuring microbiome stability. In situ techniques provide real-time modifications of the microbiome, reinstating lost microbial capabilities and adaptively reacting to environmental fluctuations (Voorhees et al. 2020). Tailored approaches to individual microbiome compositions offer personalized therapeutic strategies for gut health, chronic disease management, and overall well-being. Researchers can meticulously manipulate microbial genomes while preserving ecosystem stability, facilitating microbiome-based therapies for metabolic disorders, inflammatory illnesses, and neurological problems (Zheng et al. 2022). The breakthroughs in microbial engineering provide significant promise for personalized medicine, probiotic creation, and microbiome-based therapeutics, providing unique methods to enhance long-term microbiome stability and human health.

1.7 CONCLUSION

Future developments in microbiome research will be propelled by multi-omics technologies, artificial intelligence (AI)-driven analytics, and synthetic biology, which will enable the accurate manipulation of the microbiome for commercial and medical uses. CRISPR-based gene editing and in situ microbial engineering will enable targeted therapies for neurological, immunological, and metabolic problems, while coculture methods will uncover new bioactive compounds with potential medical and environmental applications. Microbiome-based therapies will be improved by AI-powered data analysis, leading to individualized treatments and sustainable bioremediation and agricultural interventions. These developments will address the ecological and health issues facing the world community by enhancing the microbiome's stability, resilience, and adaptability. As research progresses, the combination of omics, gene editing, and artificial intelligence will transform microbiome therapies and position them as a cornerstone of biotechnology and precision medicine. This will guarantee the environment's and health's long-term viability.

REFERENCES

Agarwal, Kavita, and Amanda L Lewis. "Vaginal Sialoglycan Foraging by Gardnerella Vaginalis: Mucus Barriers as a Meal for Unwelcome Guests?" *Glycobiology* 31, no. 6 (April 5, 2021): 667–80. https://doi.org/10.1093/glycob/cwab024.

Aggarwal, Nikhil, Shohei Kitano, Ginette Ru Ying Puah, Sandra Kittelmann, In Young Hwang, and Matthew Wook Chang. "Microbiome and Human Health: Current Understanding, Engineering, and Enabling Technologies." *Chemical Reviews* 123, no. 1 (2022): 31–72.

Baumann-Dudenhoeffer, Aimee M., Alaric W. D'Souza, Phillip I. Tarr, Barbara B. Warner, and Gautam Dantas. "Infant Diet and Maternal Gestational Weight Gain Predict Early Metabolic Maturation of Gut Microbiomes." *Nature Medicine* 24, no. 12 (October 25, 2018): 1822–29. https://doi.org/10.1038/s41591-018-0216-2.

Bäckhed, Fredrik, Josefine Roswall, Yangqing Peng, Qiang Feng, Huijue Jia, Petia Kovatcheva-Datchary, Yin Li et al. "Dynamics and Stabilization of the Human Gut Microbiome During the First Year of Life." *Cell Host & Microbe* 17, no. 5 (2015): 690–703.

Berg, Gabriele, Daria Rybakova, Doreen Fischer, Tomislav Cernava, Marie-Christine Champomier Vergès, Trevor Charles, Xiaoyulong Chen et al. "Microbiome Definition Re-visited: Old Concepts and New Challenges." *Microbiome* 8, no. 1 (June 30, 2020). https://doi.org/10.1186/s40168-020-00875-0.

Black, Mairead, Siladitya Bhattacharya, Sam Philip, Jane E. Norman, and David J. McLernon. "Planned Cesarean Delivery at Term and Adverse Outcomes in Childhood Health." *JAMA* 314, no. 21 (December 1, 2015): 2271. https://doi.org/10.1001/jama.2015.16176.

Bokulich, Nicholas A., Jennifer Chung, Thomas Battaglia, Nora Henderson, Melanie Jay, Huilin Li, Arnon D. Lieber et al. "Antibiotics, Birth Mode, and Diet Shape Microbiome Maturation During Early Life." *Science Translational Medicine* 8, no. 343 (June 15, 2016). https://doi.org/10.1126/scitranslmed.aad7121.

Bosco, Alice, Michele Loi, Giulia Pinna, Roberta Pintus, Vassilios Fanos, and Angelica Dessì. "Childhood Obesity and the Cryptic Language of the Microbiota: Metabolomics' Upgrading." *Metabolites* 13, no. 3 (March 11, 2023): 414. https://doi.org/10.3390/metabo13030414.

Chelimo, Carol, Carlos A. Camargo, Susan M. B. Morton, and Cameron C. Grant. "Association of Repeated Antibiotic Exposure up to Age 4 Years with Body Mass at Age 4.5 Years." *JAMA Network Open* 3, no. 1 (January 22, 2020): e1917577. https://doi.org/10.1001/jamanetworkopen.2019.17577.

Craig, Sarah J. C., Daniel Blankenberg, Alice Carla Luisa Parodi, Ian M. Paul, Leann L. Birch, Jennifer S. Savage, Michele E. Marini et al. "Child Weight Gain Trajectories Linked to Oral Microbiota Composition." *Scientific Reports* 8, no. 1 (September 13, 2018). https://doi.org/10.1038/s41598-018-31866-9.

Dekaboruah, Elakshi, Mangesh Vasant Suryavanshi, Dixita Chettri, and Anil Kumar Verma. "Human Microbiome: An Academic Update on Human Body Site Specific Surveillance and Its Possible Role." *Archives of Microbiology* 202, no. 8 (June 10, 2020): 2147–67. https://doi.org/10.1007/s00203-020-01931-x.

Dominguez-Bello, Maria Gloria, Filipa Godoy-Vitorino, Rob Knight, and Martin J Blaser. "Role of the Microbiome in Human Development." *Gut* 68, no. 6 (January 22, 2019): 1108–14. https://doi.org/10.1136/gutjnl-2018-317503.

Drago, Lorenzo, Roberta De Grandi, Gianfranco Altomare, Paolo Pigatto, Oliviero Rossi, and Marco Toscano. "Skin Microbiota of First Cousins Affected by Psoriasis and Atopic Dermatitis." *Clinical and Molecular Allergy* 14, no. 1 (January 21, 2016). https://doi.org/10.1186/s12948-016-0038-z.

Herath, Madushani, Suzanne Hosie, Joel C. Bornstein, Ashley E. Franks, and Elisa L. Hill-Yardin. "The Role of the Gastrointestinal Mucus System in Intestinal Homeostasis: Implications for Neurological Disorders." *Frontiers in Cellular and Infection Microbiology* 10 (May 28, 2020). https://doi.org/10.3389/fcimb.2020.00248.

Hill, Jennifer H., and June L. Round. "SnapShot: Microbiota Effects on Host Physiology." *Cell* 184, no. 10 (May 1, 2021): 2796–2796.e1. https://doi.org/10.1016/j.cell.2021.04.026.

Huang, Chian-Feng, Wei-Chu Chie, and I- Jen Wang. "Efficacy of Lactobacillus Administration in School-Age Children with Asthma: A Randomized, Placebo-Controlled Trial." *Nutrients* 10, no. 11 (November 5, 2018): 1678. https://doi.org/10.3390/nu10111678.

Jaiswal, Durgesh Kumar, Jay Prakash Verma, Tarun Belwal, Arthur Prudêncio De Araujo Pereira, and Avinash Bapurao Ade. "Editorial: Microbial Co-Cultures: A New Era of Synthetic Biology and Metabolic Engineering." *Frontiers in Microbiology* 14 (June 23, 2023). https://doi.org/10.3389/fmicb.2023.1235565.

Jensen, Emilija D., Caitlin A. Selway, Gabrielle Allen, Jana Bednarz, Laura S. Weyrich, Sam Gue, Alexia S. Peña, and Jennifer Couper. "Early Markers of Periodontal Disease and Altered Oral Microbiota Are

Associated With Glycemic Control in Children With Type 1 Diabetes." *Pediatric Diabetes* 22, no. 3 (December 8, 2020): 474–81. https://doi.org/10.1111/pedi.13170.

Kim, Sung Ae, Bo Ri Kim, Mi Young Chun, and Sang Woong Youn. "Relation between pH in the Trunk and Face: Truncal pH Can Be Easily Predicted from Facial pH." *Annals of Dermatology* 28, no. 2 (January 1, 2016): 216. https://doi.org/10.5021/ad.2016.28.2.216.

Korpela, Katri, Anne Salonen, Lauri J. Virta, Riina A. Kekkonen, Kristoffer Forslund, Peer Bork, and Willem M. De Vos. "Intestinal Microbiome Is Related to Lifetime Antibiotic Use in Finnish Pre-school Children." *Nature Communications* 7, no. 1 (January 26, 2016). https://doi.org/10.1038/ncomms10410.

Kuhle, Stefan, Oskar S. Tong, and Christy G. Woolcott. "Association between Caesarean Section and Childhood Obesity: A Systematic Review and Meta-analysis." *Obesity Reviews* 16, no. 4 (March 5, 2015): 295–303. https://doi.org/10.1111/obr.12267.

Lacroix, Guillaume, Valérie Gouyer, Frédéric Gottrand, and Jean-Luc Desseyn. "The Cervicovaginal Mucus Barrier." *International Journal of Molecular Sciences* 21, no. 21 (November 4, 2020): 8266. https://doi.org/10.3390/ijms21218266.

Lee, Learn-Han, Sunny Hei Wong, Siok-Fong Chin, Vishal Singh, and Nurul-Syakima Ab Mutalib. "Editorial: Human Microbiome: Symbiosis to Pathogenesis." *Frontiers in Microbiology* 12 (February 17, 2021). https://doi.org/10.3389/fmicb.2021.605783.

Leong, Claudia, Jillian J. Haszard, Blair Lawley, Anna Otal, Rachael W. Taylor, Ewa A. Szymlek-Gay, Elizabeth A. Fleming et al. "Mediation Analysis as a Means of Identifying Dietary Components That Differentially Affect the Fecal Microbiota of Infants Weaned by Modified Baby-Led and Traditional Approaches." *Applied and Environmental Microbiology* 84, no. 18 (July 10, 2018). https://doi.org/10.1128/aem.00914-18.

Liu, Xiaosun, Li Shao, Xia Liu, Feng Ji, Ying Mei, Yiwen Cheng, Fengping Liu, Chongxian Yan, Lanjuan Li, and Zongxin Ling. "Alterations of Gastric Mucosal Microbiota Across Different Stomach Microhabitats in a Cohort of 276 Patients With Gastric Cancer." *EBioMedicine* 40 (December 21, 2018): 336–48. https://doi.org/10.1016/j.ebiom.2018.12.034.

Long, Shuping, Yi Yang, Chengpin Shen, Yiwen Wang, Anmei Deng, Qin Qin, and Liang Qiao. "Metaproteomics Characterizes Human Gut Microbiome Function in Colorectal Cancer." *NPJ Biofilms and Microbiomes* 6, no. 1 (March 24, 2020). https://doi.org/10.1038/s41522-020-0123-4.

Lv, Hanying, Lijiang Zhang, Yuqiu Han, Li Wu, and Baohong Wang. "The Development of Early Life Microbiota in Human Health and Disease." *Engineering* 12 (February 10, 2021): 101–14. https://doi.org/10.1016/j.eng.2020.12.014.

Maioli, Tatiani Uceli, Esther Borras-Nogues, Licia Torres, Sara Candida Barbosa, Vinicius Dantas Martins, Philippe Langella, Vasco Ariston Azevedo, and Jean-Marc Chatel. "Possible Benefits of Faecalibacterium Prausnitzii for Obesity-Associated Gut Disorders." *Frontiers in Pharmacology* 12 (December 2, 2021). https://doi.org/10.3389/fphar.2021.740636.

McKinnon, Lyle R., Sharon L. Achilles, Catriona S. Bradshaw, Adam Burgener, Tania Crucitti, David N. Fredricks, Heather B. Jaspan et al. "The Evolving Facets of Bacterial Vaginosis: Implications for HIV Transmission." *AIDS Research and Human Retroviruses* 35, no. 3 (January 14, 2019): 219–28. https://doi.org/10.1089/aid.2018.0304.

Mills, Robert H., Parambir S. Dulai, Yoshiki Vázquez-Baeza, Consuelo Sauceda, Noëmie Daniel, Romana R. Gerner, Lakshmi E. Batachari et al. "Multi-Omics Analyses of the Ulcerative Colitis Gut Microbiome Link Bacteroides Vulgatus Proteases with Disease Severity." *Nature Microbiology* 7, no. 2 (January 27, 2022): 262–76. https://doi.org/10.1038/s41564-021-01050-3.

Nicolucci, Alissa C., Megan P. Hume, Inés Martínez, Shyamchand Mayengbam, Jens Walter, and Raylene A. Reimer. "Prebiotics Reduce Body Fat and Alter Intestinal Microbiota in Children Who Are Overweight or with Obesity." *Gastroenterology* 153, no. 3 (June 5, 2017): 711–22. https://doi.org/10.1053/j.gastro.2017.05.055.

Oren, Aharon, and George M. Garrity. "Valid Publication of the Names of Forty-Two Phyla of Prokaryotes." *International Journal of Systematic and Evolutionary Microbiology* 71, no. 10 (2021): 005056.

Rinninella, Emanuele, Pauline Raoul, Marco Cintoni, Francesco Franceschi, Giacinto Abele Donato Miggiano, Antonio Gasbarrini, and Maria Cristina Mele. "What Is the Healthy Gut Microbiota Composition? A Changing Ecosystem across Age, Environment, Diet, and Diseases." *Microorganisms* 7, no. 1 (January 10, 2019): 14. https://doi.org/10.3390/microorganisms7010014.

Sakanaka, Akito, Masae Kuboniwa, Shuichi Shimma, Samar A. Alghamdi, Shota Mayumi, Richard J. Lamont, Eiichiro Fukusaki, and Atsuo Amano. "Fusobacterium Nucleatum Metabolically Integrates Commensals and Pathogens in Oral Biofilms." *mSystems* 7, no. 4 (July 19, 2022). https://doi.org/10.1128/msystems.00170-22.

Sanders, Mary Ellen, Daniel J. Merenstein, Gregor Reid, Glenn R. Gibson, and Robert A. Rastall. "Probiotics and Prebiotics in Intestinal Health and Disease: From Biology to the Clinic." *Nature Reviews Gastroenterology & Hepatology* 16, no. 10 (July 11, 2019): 605–16. https://doi.org/10.1038/s41575-019-0173-3.

Savage, Jessica H., Kathleen A. Lee-Sarwar, Joanne E. Sordillo, Nancy E. Lange, Yanjiao Zhou, George T. O'Connor, Megan Sandel et al. "Diet During Pregnancy and Infancy and the Infant Intestinal Microbiome." *The Journal of Pediatrics* 203 (August 30, 2018): 47–54.e4. https://doi.org/10.1016/j.jpeds.2018.07.066.

Savilahti, Erkki, Taina Härkönen, Emma M. Savilahti, Kaarina Kukkonen, Mikael Kuitunen, and Mikael Knip. "Probiotic Intervention in Infancy Is Not Associated With Development of Beta Cell Autoimmunity and Type 1 Diabetes." *Diabetologia* 61, no. 12 (September 20, 2018): 2668–70. https://doi.org/10.1007/s00125-018-4738-4.

Sender, Ron, Shai Fuchs, and Ron Milo. "Revised Estimates for the Number of Human and Bacteria Cells in the Body." *PLoS Biology* 14, no. 8 (August 19, 2016): e1002533. https://doi.org/10.1371/journal.pbio.1002533.

Shreiner, Andrew B., John Y. Kao, and Vincent B. Young. "The Gut Microbiome in Health and in Disease." *Current Opinion in Gastroenterology* 31, no. 1 (November 12, 2014): 69–75. https://doi.org/10.1097/mog.0000000000000139.

Simon, Jean-Christophe, Julian R. Marchesi, Christophe Mougel, and Marc-André Selosse. "Host-Microbiota Interactions: From Holobiont Theory to Analysis." *Microbiome* 7, no. 1 (2019): 5. https://doi.org/10.1186/s40168-019-0619-4

Theis, Kevin R., Nolwenn M. Dheilly, Jonathan L. Klassen, Robert M. Brucker, John F. Baines, Thomas C. G. Bosch, John F. Cryan et al. "Getting the Hologenome Concept Right: An Eco-Evolutionary Framework for Hosts and Their *Microbiomes*." *mSystems* 1, no. 2 (March 30, 2016). https://doi.org/10.1128/msystems.00028-16.

Thompson, Amanda L., Andrea Monteagudo-Mera, Maria B. Cadenas, Michelle L. Lampl, and M. Andrea Azcarate-Peril. "Milk-and Solid-feeding Practices and Daycare Attendance Are Associated with Differences in Bacterial Diversity, Predominant Communities, and Metabolic and Immune Function of the Infant Gut Microbiome." *Frontiers in Cellular and Infection Microbiology* 5 (2015): 3. https://doi.org/10.3389/fcimb.2015.00003

Vagios, Stylianos, and Caroline M. Mitchell. "Mutual Preservation: A Review of Interactions between Cervicovaginal Mucus and Microbiota." *Frontiers in Cellular and Infection Microbiology* 11 (July 13, 2021). https://doi.org/10.3389/fcimb.2021.676114.

Voorhees, Peter J., Carlos Cruz-Teran, Jasmine Edelstein, and Samuel K. Lai. "Challenges & Opportunities for Phage-based in Situ Microbiome Engineering in the Gut." *Journal of Controlled Release* 326 (June 19, 2020): 106–19. https://doi.org/10.1016/j.jconrel.2020.06.016.

Wang, Lidan, Wei Zhang, Xianglu Wu, Xiao Liang, Lijie Cao, Jincheng Zhai, Yiyang Yang et al. "MIAOME: Human Microbiome Affect the Host Epigenome." *Computational and Structural Biotechnology Journal* 20 (January 1, 2022): 2455–63. https://doi.org/10.1016/j.csbj.2022.05.024.

Wang, Ying, Ning Li, Jun-Jie Yang, Dong-Mei Zhao, Bin Chen, Guo-Qing Zhang, Shuo Chen et al. "Probiotics and Fructo-Oligosaccharide Intervention Modulate the Microbiota-gut Brain Axis to Improve Autism Spectrum Reducing also the Hyper-serotonergic State and the Dopamine Metabolism Disorder." *Pharmacological Research* 157 (April 17, 2020): 104784. https://doi.org/10.1016/j.phrs.2020.104784.

Willis, Jesse R., and Toni Gabaldón. "The Human Oral Microbiome in Health and Disease: From Sequences to Ecosystems." *Microorganisms* 8, no. 2 (February 23, 2020): 308. https://doi.org/10.3390/microorganisms8020308.

Yao, Lu, Li-Ping Zhu, Xiao-Yan Xu, Ling-Ling Tan, Martin Sadilek, Huan Fan, Bo Hu et al. "Discovery of Novel Xylosides in Co-culture of Basidiomycetes Trametes Versicolor and Ganoderma Applanatum by Integrated Metabolomics and Bioinformatics." *Scientific Reports* 6, no. 1 (September 12, 2016). https://doi.org/10.1038/srep33237.

Yang, Irene, Robert Adam Arthur, Liping Zhao, Jasmine Clark, Yijuan Hu, Elizabeth J. Corwin, and James Lah. "The Oral Microbiome and Inflammation in Mild Cognitive Impairment." *Experimental Gerontology* 147 (2021): 111273.

Yatsunenko, Tanya, Federico E. Rey, Mark J. Manary, Indi Trehan, Maria Gloria Dominguez-Bello, Monica Contreras, Magda Magris et al. "Human Gut Microbiome Viewed across Age and Geography." *Nature* 486, no. 7402 (May 8, 2012): 222–27. https://doi.org/10.1038/nature11053.

Zafar, Hassan, and Milton H. Saier. "Gut Bacteroides Species in Health and Disease." *Gut Microbes* 13, no. 1 (January 1, 2021). https://doi.org/10.1080/19490976.2020.1848158.

Zhang, Xu, Zhibin Ning, Janice Mayne, Yidai Yang, Shelley A. Deeke, Krystal Walker, Charles L. Farnsworth et al. "Widespread Protein Lysine Acetylation in Gut Microbiome and Its Alterations in Patients with Crohn's Disease." *Nature Communications* 11, no. 1 (August 17, 2020). https://doi.org/10.1038/s41467-020-17916-9.

Zheng, Linggang, Yang Tan, Yucan Hu, Juntao Shen, Zepeng Qu, Xianbo Chen, Chun Loong Ho, Elaine Lai-Han Leung, Wei Zhao, and Lei Dai. "CRISPR/Cas-Based Genome Editing for Human Gut Commensal Bacteroides Species." *ACS Synthetic Biology* 11, no. 1 (January 6, 2022): 464–72. https://doi.org/10.1021/acssynbio.1c00543.

2 Health and Disorders
Modulation of Microbiome

Mandeep Kaur, Nidhi Singla, and Preeti Behl

2.1 INTRODUCTION

It was in the early 1990s when the vast majority of microbes including bacteria, viruses, and fungi were discovered to be present on the human body at various sites such as the gut, skin, lungs, and oral cavity and were called "microbiota". It is also referred to as the hidden organ or the last organ as it contributes 150 times more genetic material than present in the entire human genome (Hou et al. 2022; Berg et al. 2020). "Microbiota" and "microbiome" are the two terms that have been often used synonymously; however, they differ in certain respects. While "microbiota" encompasses the various microorganisms present in specific habitats, such as the oral cavity and the gut, the "microbiome" refers to the total genetic content of microorganisms in a specific habitat. The microbiome also includes its structures, metabolites, and the environmental factors that affect them (Hou et al. 2022). In simple words, the microbiome is a house and the microbiota is the people living there (Berg et al. 2020).

The microbial colonization of various body sites starts since birth and is vertically transferred to the next generations. Although the microbial communities have been found to colonize all body sites, the highest numbers are found in the gut and have been studied extensively (de Vos et al. 2022).

2.2 DEVELOPMENT OF HUMAN MICROBIOTA

Starting from birth, a long-term symbiotic association is established between the human body and its resident microbial flora that plays a vital role in supporting good health and well-being (Ogunrinola et al. 2020).

2.2.1 Gut Microbiota

The human gut microbiota is predominantly made up of strict anaerobic microorganisms, which outnumber the facultative anaerobes and aerobic microbes by as much as 100-fold (Bull and Plummer 2014; Okolie et al. 2024; Sasso et al. 2023). The microbial populations, along with their genetic material and metabolic products, form what is known as the gut microbiome. A neonate is born sterile and the microbial colonization begins right after birth, during, and after delivery. The gut bacteria continue to evolve as the child grows and are modulated by different environmental factors such as microbes present in the immediate environment, the type of food eaten, and the feeding habits, in addition to maternal microbiota (Bull and Plummer 2014; Tanaka and Nakayama 2017). Despite the abundant factors that influence the composition of gut microbiota, the human gut remains fairly stable at the phylum level. Six primary bacterial phyla, namely, Bacteroidetes, Proteobacteria, Firmicutes, Fusobacteria, Actinobacteria, and Verrucomicrobia, are commonly found in the gut. The bacterial number and its composition vary along the length of the GIT ranging from 10 to 100 bacteria per gram of stomach to more than 1,000 bacteria per gram in the large intestine. Firmicutes and Bacteroidetes are the predominant bacterial species present in the small intestine and colon, respectively (Bull and Plummer 2014; Hou et al. 2022). Apart from bacterial species, the gut environment also encompasses a wide range of other microbial species such

 DOI: 10.1201/9781003610274-3

as fungi, viruses, bacteriophages, and archaeal species. Fungi such as *Candida*, *Saccharomyces*, *Malassezia*, and *Cladosporium* form the gut mycobiota while *Methanobrevibacter smithii* is an important archaeon species present in the human intestine (Hou et al. 2022).

The bidirectional communication of the gut with other body systems like nervous, endocrine, immune, and metabolic is essential for maintaining overall balance and good health. Hence, when the gut microbial balance gets disturbed in a condition known as dysbiosis, the body gets prone to various diseases including obesity, cardiovascular, hypertension, endocrine and neurological diseases, inflammatory bowel diseases, and some neoplasms (Afzaal et al. 2022).

2.2.2 Oral Microbiota

The oral cavity houses a diverse and unique community of microbes and contains the second-highest number of microbial flora in the body after the gut microbiome (Willis and Gabaldón 2020). There is a considerable difference in the microbial communities harbored at different sites in the oral cavity. Tongue, teeth, gums, buccal mucosa, hard and soft palate, gingival area, and saliva have a similar microbiome with small-scale differences. The environmental factors, pH changes, genetic mutations, and bacterial interactions may be the cause of the difference in the microbial populations at these sites (Hou et al. 2022). Largely, Fusobacteria, Chlamydiae, Spirochaetes, Bacteriodetes, Firmicutes, Proteobacteria, and Actinobacteria are the major bacterial species isolated from a healthy oral cavity while some saprophytic protozoans like *Entamoeba gingivalis* and *Trichomonas tenax* and fungi like *Candida* are other microbial species forming a healthy oral microbiome. Cumulatively, these diverse microbial communities are responsible for maintaining oral homeostasis as the dysbiosis in the oral microbiome has been linked to several diseases like gingivitis, periodontitis, and dental caries (Willis and Gabaldón 2020; Lê et al. 2023). Oral microbiome usually exists as biofilms present on different surfaces and is crucial in preserving the balance of oral microenvironment, protecting the tissues, and reducing the risk of many oral diseases (Jia et al. 2018; Deo and Deshmukh 2019).

2.2.3 Skin Microbiota

The skin provides habitat to diverse microbiota such as bacteria, viruses, fungi, and microeukaryotes. The skin microbiota is adapted to thrive in unique microenvironments, providing defense against pathogenic and opportunistic microorganisms through mutualistic and commensal interactions with mammalian cells. The native microbes compete for the nutrients and create an environment that offers resistance against colonization by harmful pathogens. Protective invaginations like pilosebaceous unit provide a microaerophilic environment for obligate anaerobes and facultative aerobes. *Cutibacterium acnes* present on skin can break down sebum lipids to generate nutrients and also lower skin pH protecting against colonization. Thus, distinct microbial communities exist on the skin depending upon differences in moisture, pH, and temperature across various regions (Flowers and Grice 2020). The major bacterial phyla colonizing skin include Actinobacteria, Firmicutes, Bacteroidetes, Proteobacteria, and Cyanobacteria. Skin microbiota, like gut microbiota, plays a role in the development of the innate immune system. *Staphylococcus epidermidis*, an important skin commensal, contributes to skin immunity by promoting the expression of antimicrobial peptides like human β defensin and by strengthening the function of skin lymphocytes (Hou et al. 2022).

2.2.4 Genital Tract Microbiota

The genital microbiota, particularly in the female reproductive tract, plays a crucial role in maintaining reproductive and overall health, disruption of which may lead to many reproductive tract infections including pelvic inflammatory diseases (PID), preterm birth, and infertility (Chen et al. 2021).

Females: The vaginal microbiome is a dynamic and complex microecosystem that constantly undergoes fluctuations during entire woman's life. Hormonal changes occurring during the menstrual cycle and pregnancy, the use of contraceptives, and nutrition shape the vaginal microbiome at various stages of life. The vaginal microbiome thrives on mutualistic interaction with host cells offering benefits to each other (Holdcroft et al. 2023). Lactobacillus genus, comprising species such as *Lactobacillus crispatus*, *L. iners*, *L. gasseri*, and *L. jensenii*, is the most predominant bacterial species in a healthy vaginal ecosystem (Pathak et al. 2023). These bacteria by producing lactic acid create an acidic environment in vagina, thereby limiting the growth of harmful pathogens. Therefore, disruptions of vaginal microbiome have been associated with many sexually transmitted infections including candidiasis, bacterial vaginosis, and PID (Chen et al. 2021; Alhamlan et al., 2024; Vinderola et al., 2024).

Males: Until now, genital microbiota has been the synonymous term used for female genital tract microbiota with little interest to study the importance of microbes in the male reproductive tract. But recently, this overlooked area of research has gained the attention of many scientists across the world (Zuber et al. 2023). Microbiota in males is generally present in the lower genital tract whereas the upper genital tract is sterile except in infections. Microbiome in the male genital tract is influenced by many factors such as sexual behavior, circumcision, and microbiota of the sexual partner. The bacterial communities commonly found within the male genital tract include *Staphylococcus*, *Corynebacterium*, *Lactobacillus*, *Prevotella*, *Peptoniphilus*, *Finegoldia*, and *Anaerococcus*. The emerging evidence suggests that the disruptions of microbiota in the male genital tract greatly affect male reproductive health and fertility and have implications in many infectious conditions like urethritis and prostatitis (Gonçalves et al. 2022).

2.2.5 Microbiota at Other Body Sites

The dominant bacterial species present in the urinary tract of healthy women are the same as in genital microbiota, i.e., *Lactobacillus* species followed by *Gardnerella vaginalis*. Some strains of *G. vaginalis* are harmless while others may have pathogenic potential, causing urinary tract infections mostly in women but sometimes in men too (Wu et al. 2017).

The microbial communities present in the nasal tract are diverse with bacteria belonging to phyla Actinobacteria, Firmicutes, and Proteobacteria with small populations of Fusobacteria and Bacteroidetes form the predominant resident flora that help in maintaining local respiratory homeostasis. But this microbial composition fluctuates over time under the influence of many factors like age, smoking habits, underlying health diseases, medication use, and due to interactions with other microbial ecosystems (Konovalovas et al. 2024). Even the lung that is believed to be sterile harbors some complex but dynamic resident microbiota belonging to Actinobacteria, Bacteroidetes, Firmicutes, and Proteobacteria. The pulmonary microbiome provides a state of immune tolerance and protects against inflammatory responses (Hou et al. 2022).

2.3 MICROBIOME IN HEALTH

A healthy balance of intestinal microorganisms is essential for overall well-being and is strongly connected to both the prevention and progression of many human diseases. The production of many bioactive compounds, viz. vitamins, amino acids, and lipids, significantly relies on the activity of the gut microbiota. The human microbiota offers protection to the host from external agents not only by producing mucus, antimicrobial peptides, and secretory IgA but also by maintaining the integrity of mucosal barriers (Hou et al. 2022; Sobti et al. 2023). Early exposure to microbiota helps "educate" and shape the immune system, especially in infants. Gut microbiota maintains a state of homeostasis between proinflammatory and anti-inflammatory responses and also prevents autoimmune diseases by promoting the T-regulatory cells, which prevent unnecessary immune responses (Table 2.1).

TABLE 2.1
Role of Microbiota in Health and Diseases

Body Site	Normal Microbiota Role	Disruption (Dysbiosis)	Associated Health Ailments
Gut	• Aids digestion, produces vitamins • Regulates immunity	• Antibiotics • Poor diet • Infections	• IBS • IBD (Crohn's, ulcerative colitis) • Obesity • Diabetes • Colorectal cancer
Skin	• Prevents pathogen colonization • Maintains pH balance	• Over-washing • Application of harsh chemicals	• Acne • Eczema • Psoriasis • Dermatitis
Oral cavity	• Maintains oral health • Prevents overgrowth of pathogens	• Poor hygiene • Sugar-rich diet	• Dental caries • Gingivitis • Periodontitis
Respiratory tract	• Modulates immune response to airborne pathogens	• Smoking • Pollution • Infections	• Asthma • Chronic bronchitis • Respiratory allergies
Vaginal tract	• Maintains acidic pH to prevent infections	• Antibiotics • Hormonal imbalance	• Bacterial vaginosis • Yeast infections • Preterm birth
Urinary tract	• Prevents pathogen adhesion and colonization	• Catheter use • Antibiotic use	• Urinary tract infections (UTIs)
Brain–gut axis	• Microbiota influence neurotransmitter production • Mood regulation	• Chronic stress, diet	• Depression • Anxiety • Neurodevelopmental disorders (e.g., autism)

Sekirov et al. (2010) and Cryan and Dinan (2012).
IBS, irritable bowel syndrome; IBD, inflammatory bowel disease.

2.4 DYSBIOSIS AND DISEASES

Irreversible alterations in human gut microbiota have been associated with serious detrimental effects on health of the host and has been linked to gut barrier dysfunction and gastrointestinal, renal, liver, metabolic, and behavioral disorders such as inflammatory bowel disease, malnutrition, diabetes, and liver cirrhosis including behavioral and mood disorders (Safarchi et al. 2025) (Figure 2.1).

2.4.1 Inflammatory Bowel Disease (IBD)

The microbial community within the gut is critically involved in triggering, advancing, and sustaining IBD such as ulcerative colitis (UC) and Crohn's disease (CD). The disease process involves dysregulated immune responses, genetic mutations, and environmental factors (Hou et al. 2022; Biagioli et al. 2024).

The gut microbiota plays an important role in the development of IBD, where imbalances can trigger abnormal immune responses, inflammation, and impair the integrity of the intestinal barrier (Hou et al. 2022). The tight junctions present around epithelial cells keep the barrier intact, which is vital for protection against invasive pathogens. Dysbiosis can result in damaged mucus layer covering the epithelium, reduction in antimicrobial peptides, and decreased tight junction proteins making the intestinal barrier susceptible to bacterial invasion. The presence of pathogenic

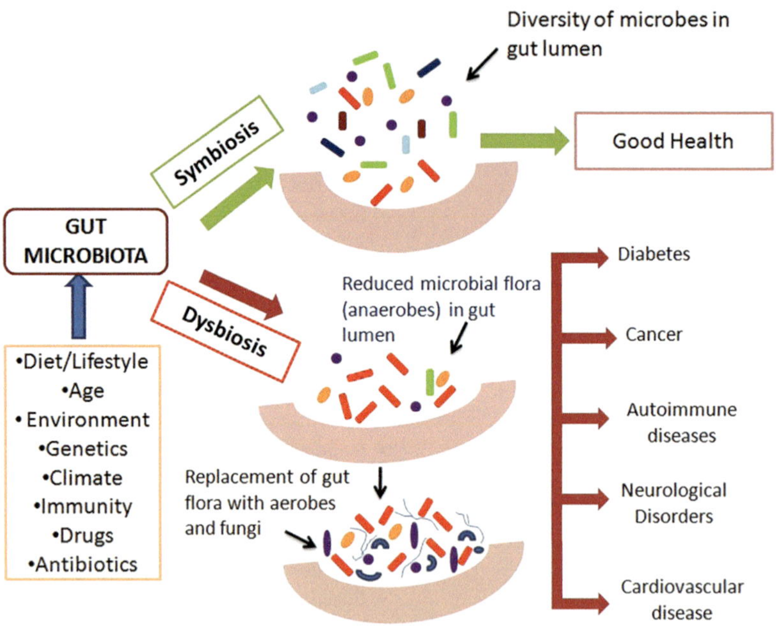

FIGURE 2.1 Influence of gut microbiota balance on health and disease.

bacteria often compromises the intestinal permeability as is suggested in a study wherein the abundance of adherent-invasive *Escherichia coli* present in CD patients, further altered the microbial composition and also induced the gut inflammation (Ahmed et al. 2016; Hou et al. 2022). In IBD patients, a reduction in Firmicutes and an increase in Proteobacteria have been observed that results in high levels of proinflammatory cytokines (e.g., IL-12 and interferon gamma (IFN-γ)) and a reduction in anti-inflammatory cytokines (e.g., IL-10) (Kleessen et al. 2002). *Bacteroides*, *Eubacterium*, and *Lactobacillus* are protective gut bacteria that show significant reduction in both CD and UC patients (Granados-Martinez et al. 2024), further validating their role in the prevention of IBD. Some helminthic infections such as *Trichuris suis* or *Necator americanus* have been used in clinical trials to achieve remission in IBD due to their active modulation of the immune system by shifting it to anti-inflammatory responses (T-regulatory and Th2 immune responses), altering gut microbiota, increasing mucus production and gut motility (Ogunirola et al. 2020). So far, no causal link between the development of IBD and dysbiosis has been studied rather, some studies suggest that the altered microbiome could be the response to the intestinal inflammation. This warrants further exploration and investigation to study the causal association of IBD with gut dysbiosis.

2.4.2 Cancer

The role of microbiota in oncogenesis has been widely studied. More than 20 cancers have one or more infectious agents associated with their development. Carcinogenesis is marked by abnormal and uncontrolled proliferation of cells with a reduction of apoptotic activity in the cells (Hou et al. 2022).

Gut microbiota can significantly influence cellular proliferation and regeneration under various conditions by producing metabolites that can either promote or inhibit carcinogenesis, especially in colorectal cancer (CRC). Gut bacteria ferment dietary fiber to produce metabolites like short-chain fatty acids (SCFAs) particularly butyrate that exerts tumor suppressive properties by inducing anti-inflammatory T-reg. immune response, maintaining intestinal epithelial barrier thus preventing inflammation, and regulating the expression of genes involved in cell proliferation and cell death. Commensal bacteria in the gut may also produce certain tumor-promoting agents such as secondary bile acids derived from fatty foods that can produce reactive oxygen radicals that trigger inflammation and DNA damage. Similarly, N-nitroso compounds, by-products of the fermentation of amino acids following high protein intake, can induce alkylation and mutation in the host (O'Keefe 2016; Bhatt et al. 2017; Khan et al. 2022). The carcinogenic effect of gut microbiota can go beyond intestines affecting other body systems such as hepatocellular carcinoma, which may occur due to dissemination of these microbes to liver (Khan et al. 2022). It has been seen that the same commensal bacteria such as *Clostridium* species produce tumor-promoting secondary bile acids with fatty meal and butyrate, an antitumor metabolite with a fiber-rich diet. Hence, labeling commensal gut bacteria as good or bad may not be right as it mostly depends on one's dietary intake (Bhatt et al. 2017).

Many microorganisms have been directly associated with cancer development like *Helicobacter pylori* with gastric adenocarcinoma and mucosa-associated lymphoid tissue (MALT) lymphoma. Relatively high bacterial load of *Fusobacterium* and *Clostridiodes* in gastric cancer, *Staphylococcus* in breast cancer, *Chlamydia trachomatis* in cervical cancer, *Escherichia coli* in CRC, *Salmonella enterica* in gall bladder, and *Bacteroides massiliensis* in prostate cancer have been observed among healthy population (Ogunirola et al. 2020; Khan et al. 2022; Hou et al. 2022).

2.4.3 Cardiovascular Diseases

Worldwide, cardiovascular disease (CVD) still remains a major cause of mortality and morbidity. Efforts are being made to identify all possible causes and devise preventive strategies to avoid the onset and progression of CVD. The role of gut microbiota in the pathogenesis of CVDs is largely unexplored (Nesci et al. 2023). The gut microbes produce various metabolites through different signaling pathways such as trimethylamine (TMA)/trimethylamine-N oxide (TMAO), SCFAs, and primary and secondary bile acid production that play a role in the pathogenesis of CVDs. These bacteria break down nutrients like L-carnitine, choline, and phosphatidylcholine into TMA, which gets converted to TMAO inside the liver in turn causing impaired lipid metabolism, increased arterial stiffness, and foam cell formation citing its pathogenic role in atherosclerosis (Tang et al. 2017; Nesci et al. 2023).

The presence of bacterial DNA in atherosclerotic plaque and the presence of similar bacterial species in the patient's gut or oral microbiota suggest its role in the pathophysiology of CVD. An altered gut microbiome especially reduced levels of genus Roseburian has been associated with increased production of proinflammatory cytokines and reduced production of anti-inflammatory carotenes. Elevated TMA/TMAO levels tend to have a positive correlation with the size of atherosclerotic plaque and are associated with major adverse cardiovascular events (Tang et al. 2017). The disrupted intestinal barrier due to the release of proinflammatory cytokines allows lipopolysaccharides secreted by Gram-negative bacteria in the gut to enter the bloodstream and initiate inflammatory responses, further helping in the formation of atherosclerotic plaques (Tang et al. 2017; Nesci et al. 2023).

Hypertension is the most important modifiable risk factor for CVD. From the recent studies, a strong association has been observed between gut dysbiosis and hypertension (Tang et al. 2017). SCFAs generated by gut microbiota have been shown to regulate blood pressure (BP). It increases BP by stimulating renin release from kidneys but also through separate receptors may cause vasodilation, thus creating a balanced effect.

Hypertensive individuals tend to harbor increased numbers of *Clostridiales* and *Bacteroidales* and a reduction in *Robinsoniella* in fecal samples, highlighting its protective role in CVD. A hypertensive mouse model study showed an increased Firmicutes-to-Bacteroidetes ratio in hypertensive mice and strongly suggests that the gut dysbiosis may be a key factor in hypertension (Nesci et al. 2023).

The bidirectional role of gut microbiota is clearly explained by gut theory of heart failure. A failing heart results in peripheral edema and systemic congestion causing translocation of bacteria across the swollen intestinal epithelial lining, which allows bacterial endotoxins to enter the blood stream causing widespread inflammation. Therefore, individuals with chronic heart failure exhibit higher levels of harmful bacteria such as *Campylobacter*, *Salmonella*, *Shigella*, *Yersinia enterocolitica*, and *Candida*. On the other hand, altered microbiota produces metabolites like TMAO that causes progression of heart failure creating a vicious cycle (Tang et al. 2017; Chen et al. 2023).

2.4.4 Neuropsychiatric and Neurodegenerative Disorders

In recent years, gut microbiota has been recognized to play a crucial role in overall health. A two-way communication between the gut microbiome and the brain has been involved in many neurodevelopmental and neurological diseases (Ullah et al. 2023). The mechanisms by which gut microbiota influence these central nervous system disorders still remain unexplored. For many years, the gut microbiota has been linked to the hypothalamus–pituitary axis and there have been many studies to document its role in mood disorders such as bipolar and depressive disorders. The role of gut microbiota in schizophrenia is also an emerging area of interest (Bonnechère et al. 2022).

The gut microbiota is also thought to influence brain function, behavior, and mental health. The most widely accepted theory for such influence is through the generation of various metabolites by intestinal bacteria that makes the gut barrier permeable allowing endotoxins and other bacterial products to enter bloodstream, triggering systemic inflammation and immune response. A similar mechanism can help harmful substances to gain entry into the brain through a permeable blood brain barrier disrupting neurotransmitter functions. The metabolic products like SCFAs, gamma aminobutyric acid or serotonin, the neurotransmitters, and various modulators like quinolinic acid help in the modulation of immune system and also control the release of hormones like cortisol (Grau-Del Valle et al. 2023). Chronic stress influencing the composition and diversity of gut microbiota also plays an indirect role in these disorders (Madison and Kiecolt-Glaser 2019).

Many researchers have observed an altered gut microbiome in patients suffering from major depressive disorders such as an abundance of Bacteroidetes, Actinobacteria, and Proteobacteria while lower counts of *Lachnospiraceae* and *Ruminococcaceae* families within Firmicutes were observed in such patients. In a state of health, these bacterial families maintain the integrity of the intestinal barrier and release anti-inflammatory metabolites (Jiang et al. 2015). Autism spectrum disorder is another central nervous system disorder where the patient shows restricted or repetitive behaviors with difficulty in communication and interaction with other people. The research studies have shown that 90% of such Autistic individuals experience gut dysbiosis with a significant increase in inflammatory bacteria like *Clostridiodes*, *Bacteroidetes*, *Actinobacterium*, and *Proteobacterium* and a reduction in protective bacterial families like Firmicutes (Kang et al. 2017). The same has been further confirmed in a pilot study where the replacement of protective bacteria in the gut of these patients through fecal microbiota transplantation (FMT) has resulted in an improvement in gastrointestinal function and a reduction in behavioral symptoms in the study participants (Kang et al. 2017).

Alzheimer's disease (AD) is a gradually progressive neurodegenerative disorder resulting in dementia in patients. In the brain, β amyloid fibrils get deposited extracellulary around neurons leading to the formation of intracellular neurofibrillary tangles causing disruption of synaptic signaling (Jovel et al. 2018; Seo and Holtzman 2024). The microbiota alteration in AD patients has been noted in many studies (Li et al. 2017; Seo and Holtzman 2024). The abundance of proinflammatory such as *Escherichia/Shigella* against reduced anti-inflammatory bacterial communities like *Eubacterium rectale* in gut has been found to contribute to the pathogenesis of AD. Certain microbes can directly cause progression of AD like in Herpes virus infection of brain, amyloid production increases as a protective mechanism against the effect of viral infection while indirectly through the production of various metabolites such as SCFAs that may help in causing neuroinflammation that can activate glial cell activity and may propagate abnormal Aβ toxicity, amyloid accumulation, neuronal damage, and disease progression (Seo and Holtzman 2024). Certain other pathogens like

Toxoplasma gondii and *Chlamydophila pneumoniae* usually result in chronic neuroinflammation and contribute to the formation of neurofibrillary tangles and thus progression of AD (Contini et al. 2010; Zhao et al. 2024).

Similarly, in Parkisnson's disease patients, microbiota alterations have been attributed to the pathogenesis of this illness as seen in the results of the study conducted by Li et al. in China, where abundant numbers of *Prevotella* and *Akkermansia* and decreased numbers of Lactobacillus species were found (Li et al. 2017).

2.4.5 Other Systemic Illnesses Associated with Dysbiosis

Gut dysbiosis has been associated with many other systemic and metabolic disorders, especially diabetes mellitus and obesity (Bull and Plummer 2014). The human microbiome can affect metabolism in many ways. First, by producing SCFAs from otherwise indigestible fiber to provide energy, second, by disrupting the intestinal barrier through inflammatory cytokines; and lastly, by producing bioactive signal molecules to regulate human metabolism (Laudes et al. 2021).

In type 1 diabetes mellitus (T1DM) patients, the composition of oral and fecal microbiota is distinct from that of healthy controls. An abundance of *Christensenella* and *Bifidobacterium* is observed in fecal samples of diabetic patients, while a negative correlation was observed for fecal *Streptococcus* (de Groot et al. 2017). In type 2 diabetes mellitus too, a deranged Bacteroidetes to Firmicutes ratio has been observed, resulting in proinflammatory processes. Lipopolysaccharide from bacteria can bind to TLR4, triggering an inflammatory pathway causing inhibition of insulin secretion.

In various animal studies, altered microbial flora and mutation in the leptin gene have been found in obese mice as compared to lean mice. In experimental studies, an increased Firmicute/Bacteroidetes ratio has been noted in obese mice and a similar picture was observed in obese children where the proportion of Firmicutes was increased. The contrary findings were gathered in a research study where the Firmicutes/Bacteriodetes ratio was relatively lower in the obese population than in the lean group. Hence, there is a need to further explore gut microbiota association with obesity (Liu et al. 2021).

Microbiota dysbiosis has been linked to autoimmune conditions and chronic kidney and liver diseases. Disrupted intestinal flora may result in autoimmunity due to the disturbed balance between proinflammatory Th17 cells and T reg cells, while proinflammatory bacteria produce toxins like p-cresylsulfate and indoxyl sulfate that cause renal damage (Morrison and Preston 2016; Kanbay et al. 2018).

Nonalcoholic fatty liver disease is the result of inflammation and resultant fibrosis is caused by LPS derived from bacteria when microbial balance gets disturbed (Schwabe and Luedde 2018). Likewise, susceptibility to respiratory infections and other noninfective conditions like asthma and chronic obstructive pulmonary disease is greatly increased in gut dysbiosis primarily due to altered immune responses and increased inflammatory processes (Song et al. 2023).

2.5 MICROBIOME MODULATION AND ITS THERAPEUTIC ROLE

Microbiome modulation means to change the structure and/or function of microbes at a specific site in the body like human gut in order to restore the microbial balance and improve overall health (Hou et al. 2022).

Currently, many interventions and strategies are available that aim to modulate the gut microbial communities toward a healthier or more balanced state. Some of these strategies include FMT, the use of probiotics and prebiotics, antibiotic treatment, and engineered microbes that are specifically designed to bring about the change in the composition and function of gut microbiota (Schmidt et al. 2018; Hou et al. 2022; Table 2.2).

TABLE 2.2
Microbiome Modulation Strategies and Potential Health Benefits

Modulation Strategy	Mechanism of Action	Examples	Potential Health Benefits
Fecal microbiota transplantation (FMT)	Transfers healthy microbiota from donor to recipient	Used in recurrent *Clostridioides difficile* infections	Restores microbiome diversity, potential therapy for IBD, IBS
Probiotics	Introduce live beneficial microorganisms	*Lactobacillus*, *Bifidobacterium* spp.	Gut health, immune modulation, vaginal health, mental well-being
Prebiotics	Feed and stimulate the growth of beneficial microbes	Inulin, fructo-oligosaccharides (FOS), galacto-oligosaccharides (GOS)	Improved digestion, calcium absorption, reduced inflammation
Antibiotics	Eliminates or suppresses microbial populations (often nonselectively)	Broad- or narrow-spectrum antibiotics	Infection control (but can cause dysbiosis and side effects)
Phage therapy	Uses bacteriophages to target specific bacterial strains	Experimental use for drug-resistant infections	Selective pathogen control without affecting commensals
Dietary modification	Alters nutrient availability and pH, favoring beneficial species	High-fiber diets, polyphenol-rich foods	Weight management, reduced risk of metabolic diseases
Lifestyle interventions	Influences microbiota composition through stress, sleep, physical activity	Exercise, stress reduction	Supports microbiota diversity, improves mental and gut health
Engineered gut microbes	Create specifically designed microbiota through the delivery of genetically manipulated bacteria	*Lactococcus lactis* IL10, *E. coli* Nissl 1917, *Lactobacillus* strains, *E. coli* engineered with CRISPR	Modulate host immunity Restore balance or treat disease

Hou et al. (2022).
IBS, irritable bowel syndrome; IBD, inflammatory bowel disease.

2.5.1 Fecal Microbiota Transplantation

FMT is a type of nonspecific or untargeted intervention of microbiome modulation. It involves transplanting the fecal matter from a healthy donor to the recipient's gut with the aim of replacing the proinflammatory microbiota with healthy microbial flora and thereby restoring the microbial balance in the recipient. FMT has many advantages over other strategies. It increases the microbial diversity without disrupting the resident flora as observed with antibiotic use. The single-dose therapy and long-lasting effect are other important benefits of FMT unlike prebiotics and probiotics whose effects are often transient due to limited colonization (Baldi et al. 2021).

FMT has shown remarkable success in the treatment of recurrent CD infection, irritable bowel disease, *Clostridiodes difficile* infection, etc.; however, its efficacy in other illnesses like UC and metabolic syndrome is limited and inconsistent. Its potential to be used as a therapy for the management of various other diseases is being explored in more than 200 clinical trials (Schmidt et al. 2018; Choi and Cho 2016).

Due to the rapid introduction of FMT into clinical practice and the lack of large-scale prospective studies, sufficient evidence about its safety is limited. There have been reports of transmission of some gastrointestinal infections like norovirus gastroenteritis, cytomegalovirus infection, and

Escherichia coli bacteremia following FMT. But there is no conclusive evidence drawn in these studies to derive a causal association of FMT with these infections (Choi and Cho 2016). Hence, FMT is generally considered a safe intervention with mild side effects being reported such as constipation, diarrhea, belching, and abdominal discomfort that resolve over a few days. There is a need to fully investigate the potential of FMT in therapeutics and as a preventive medicine in an era of emerging diseases. More patient personalized and tailored approaches are needed to gain the maximum benefit of this intervention (Choi and Cho 2016).

2.5.2 Probiotics and Prebiotics

Probiotics are specific living microorganisms like bacteria or yeast that provides health benefits to the host when taken in adequate amounts (Sasso et al. 2023). They restore the microbiota by replacing the disrupted flora in the gut via active colonization and pitch competition to pathogenic microorganisms against the establishment of infection (Schmidt et al. 2018).

Probiotics containing specific strains interact with mucosal immunity strengthening barrier, influence metabolic processes affecting absorption of certain nutrients, and finally regulate the gut neuromuscular function. However, safety is a concern with many documented cases of bacteremia, fungemia, endocarditis, liver abscesses, and pneumonia being reported in the literature (Schmidt et al. 2018).

In addition to probiotics, prebiotics are indigestible compounds that function as substrates which serve as prerequisites for the multiplication of favorable bacteria in the intestine. The growth of such bacteria results in the production of metabolites like SCFAs, which act as important agents to maintain gut health. Certain prebiotics that have been extensively studied include galacto-oligosaccharides (GOS), fructo-oligosaccharides (FOS), oligofructose, chicory fiber, and inulin. Some prebiotics such as human milk oligosaccharides present naturally in breast milk support the growth of beneficial bacteria like *Bifidobacteria* in infants leading to improved health (Hou et al. 2022). Prebiotics can inhibit the colonization by pathogens by mimicking glycoconjugates present on intestinal epithelium in turn modulating gut immune response. But its clinical application is still debated and it should be used with caution, based on the evidence from a study which suggested the potential role of prebiotics in tumor progression among genetically susceptible population (Belcheva et al. 2014; Hou et al. 2022).

2.5.3 Antibiotic Therapy

Antibiotics are intended to kill harmful bacteria, but may often affect the "good" bacteria as well (DeGruttola et al. 2016). This is especially important in newborns, when early antibiotic exposure to them can lead to problems such as early-onset sepsis (Lathakumari et al. 2024). One clear sign of this disruption is the loss of certain healthy gut bacteria, especially the obligate anaerobes that play a key role in digestion and immunity (DeGruttola et al. 2016).

It is noteworthy that the therapeutic benefit of antibiotics is not limited to treating infections. In preclinical studies of melanoma and lung cancer, vancomycin has been shown to improve the effectiveness of radiotherapy, apparently through pushing the increased levels of CD8+ T cells and interferon-gamma to the tumor site. Moreover, there has been some evidence that antibiotics can slow down the progression of cancer like in a recent study where metronidazole eliminated *Fusobacterium* in the intestines and therefore delayed the development of CRC (Hou et al. 2022).

Antibiotics don't just shift the balance of bacteria in your gut—they also change the gut's metabolome, i.e., the collection of small molecules produced by the gut bacteria that play a big role in how your body functions. Phage therapy is another promising way to modify the microbiome. Phage therapy is a precision-targeted approach to kill harmful bacteria only sparing the beneficial bacteria unlike antibiotics' indiscriminate kill (Ugwu et al. 2024).

2.5.4 Engineered Gut Microbes

Engineered gut microbes are synthetic bacteria that can be introduced into the body for a desired effect. These "designed" bacteria have shown promise in the management of colitis, diabetes, obesity, and certain infectious conditions. One such bacterium is *Lactobacillus jannaschii* found in the female genital tract that secretes IV-resistant cyanovirin-N that have shown a reduction in IV infection by 63 in rhesus monkeys (Hou et al. 2022). These engineered bacteria have a unique advantage over conventional chemotherapies in cancer management due to their ability to target tumor cells selectively without harming normal tissue cells. These bacteria can be used as vectors to transfer genetic material in tumor cells. Newer technologies like gene-editing techniques and CRISPER, i.e.,, clustered regularly interspaced short palindromic repeats, are being used to create antimicrobial therapies meant for specific target killing (Hwang et al. 2014). The potential of bacterial Type 3 secretion systems to introduce therapeutic proteins inside the cells is also an emerging area of interest (Bai et al. 2018).

2.6 CONCLUSION

The understanding of human body interactions with the microbiome is an interesting area of research. The newer insights from this field throw a potential to develop new therapeutic strategies against various systemic illnesses. The microbial balance gets disturbed due to the frequent use of antibiotics resulting in deleterious effects on the body and further elevating the risk of infectious as well as noninfectious pathologies. The interventions such as probiotics and prebiotics targeted to support and re-establish the microbiome should be promoted as therapeutic agents in the management of various infectious diseases. The role of microbiome on neurological health needs more attention along with exploring the contributions of viromes and fungal communities in sustaining a healthy microbiome.

ACKNOWLEDGMENTS

None.

FUNDING SOURCE

None.

CONFLICT OF INTEREST

None.

REFERENCES

Afzaal, M., F. Saeed, Y. A. Shah, M. Hussain, R. Rabail, C. T. Socol, A. Hassoun, M. Pateiro, J. M. Lorenzo, A. V. Rusu, and R. M. Aadil. 2022. "Human Gut Microbiota in Health and Disease: Unveiling the Relationship." *Frontiers in Microbiology* 13: 999001. https://doi.org/10.3389/fmicb.2022.999001.

Ahmed, I., B. C. Roy, S. A. Khan, S. Septer, and S. Umar. 2016. "Microbiome, Metabolome and Inflammatory Bowel Disease." *Microorganisms* 4 (2): 20. https://doi.org/10.3390/microorganisms4020020.

Alhamlan, F. S., I. A. Albadawi, A. A. Al-Qahtani, K. A. Awartani, D. A. Obeid, and A. M. Tulbah. 2024. "Cervicovaginal and Gastrointestinal Microbiomes in Gynecological Cancers and Their Roles in Therapeutic Intervention." *Frontiers in Microbiology* 15: 1489942. https://doi.org/10.3389/fmicb.2024.1489942.

Bai, F., Z. Li, A. Umezawa, N. Terada, and S. Jin. 2018. "Bacterial Type III Secretion System as a Protein Delivery Tool for a Broad Range of Biomedical Applications." *Biotechnology Advances* 36 (2): 482–493. https://doi.org/10.1016/j.biotechadv.2018.01.016.

Baldi, S., T. Mundula, G. Nannini, and A. Amedei. 2021. "Microbiota Shaping: The Effects of Probiotics, Prebiotics, and Fecal Microbiota Transplant on Cognitive Functions: A Systematic Review." *World Journal of Gastroenterology* 27 (39): 6715–6732. https://doi.org/10.3748/wjg.v27.i39.6715.

Belcheva, A., T. Irrazabal, S. J. Robertson, C. Streutker, H. Maughan, S. Rubino, E. H. Moriyama, J. K. Copeland, A. Surendra, S. Kumar, B. Green, K. Geddes, R. C. Pezo, W. W. Navarre, M. Milosevic, B. C. Wilson, S. E. Girardin, T. M. S. Wolever, W. Edelmann, D. S. Guttman, D. J. Philpott, and A. Martin. 2014. "Gut Microbial Metabolism Drives Transformation of MSH2-Deficient Colon Epithelial Cells." *Cell* 158 (2): 288–299. https://doi.org/10.1016/j.cell.2014.04.051.

Berg, G., D. Rybakova, D. Fischer, T. Cernava, M. C. Vergès, T. Charles, X. Chen, L. Cocolin, K. Eversole, G. H. Corral, M. Kazou, L. Kinkel, L. Lange, N. Lima, A. Loy, J. A. Macklin, E. Maguin, T. Mauchline, R. McClure, B. Mitter, M. Ryan, I. Sarand, H. Smidt, B. Schelkle, H. Roume, G. S. Kiran, J. Selvin, R. S. C. Souza, L. van Overbeek, B. K. Singh, M. Wagner, A. Walsh, A. Sessitsch, and M. Schloter. 2020. "Microbiome Definition Re-Visited: Old Concepts and New Challenges." *Microbiome* 8 (1): 103. https://doi.org/10.1186/s40168-020-00875-0.

Bhatt, A. P., M. R. Redinbo, and S. J. Bultman. 2017. "The Role of the Microbiome in Cancer Development and Therapy." *CA: A Cancer Journal for Clinicians* 67 (4): 326–344. https://doi.org/10.3322/caac.21398.

Biagioli, M., C. Di Giorgio, C. Massa, S. Marchianò, R. Bellini, M. Bordoni, G. Urbani, R. Roselli, G. Lachi, E. Morretta, F. Dal Piaz, B. Charlier, B. Fiorillo, B. Catalanotti, L. Cari, G. Nocentini, P. Ricci, E. Distrutti, C. Festa, V. Sepe, A. Zampella, M. Chiara Monti, and S. Fiorucci. 2024. "Microbial-Derived Bile Acid Reverses Inflammation in IBD via GPBAR1 Agonism and RORγt Inverse Agonism." *Biomedicine & Pharmacotherapy* 181: 117731. https://doi.org/10.1016/j.biopha.2024.117731.

Bonnechère, B., N. Amin, and C. van Duijn. 2022. "The Role of Gut Microbiota in Neuropsychiatric Diseases: Creation of an Atlas Based on Quantified Evidence." *Frontiers in Cellular and Infection Microbiology* 12: 831666. https://doi.org/10.3389/fcimb.2022.831666.

Bull, M. J., and N. T. Plummer. 2014. "Part 1: The Human Gut Microbiome in Health and Disease." *Integrative Medicine* 13 (6): 17–22.

Chen, A. T., J. Zhang, and Y. Zhang. 2023. "Gut Microbiota in Heart Failure and Related Interventions." *Imeta* 2 (3): e125. https://doi.org/10.1002/imt2.125.

Chen, X., Y. Lu, T. Chen, and R. Li. 2021. "The Female Vaginal Microbiome in Health and Bacterial Vaginosis." *Frontiers in Cellular and Infection Microbiology* 11: 631972. https://doi.org/10.3389/fcimb.2021.631972.

Choi, H. H., and Y. S. Cho. 2016. "Fecal Microbiota Transplantation: Current Applications, Effectiveness, and Future Perspectives." *Clinical Endoscopy* 49 (3): 257–265. https://doi.org/10.5946/ce.2015.117.

Contini, C., S. Seraceni, R. Cultrera, M. Castellazzi, E. Granieri, and E. Fainardi. 2010. "Chlamydophila pneumoniae Infection and Its Role in Neurological Disorders." *Interdisciplinary Perspectives on Infectious Diseases* 2010: 273573. https://doi.org/10.1155/2010/273573.

Cryan, J. F., and T. G. Dinan. 2012. "Mind-Altering Microorganisms: The Impact of the Gut Microbiota on Brain and Behaviour." *Nature Reviews Neuroscience* 13 (10): 701–712. https://doi.org/10.1038/nrn3346.

de Groot, P. F., C. Belzer, Ö. Aydin, E. Levin, J. H. Levels, S. Aalvink, F. Boot, F. Holleman, D. H. van Raalte, T. P. Scheithauer, S. Simsek, F. G. Schaap, S. W. M. Olde Damink, B. O. Roep, J. B. Hoekstra, W. M. de Vos, and M. Nieuwdorp. 2017. "Distinct Fecal and Oral Microbiota Composition in Human Type 1 Diabetes, an Observational Study." *PLoS One* 12 (12): e0188475. https://doi.org/10.1371/journal.pone.0188475.

de Vos, W. M., H. Tilg, M. Van Hul, and P. D. Cani. 2022. "Gut Microbiome and Health: Mechanistic Insights." *Gut* 71 (5): 1020–1032. https://doi.org/10.1136/gutjnl-2021-326789.

DeGruttola, A. K., D. Low, A. Mizoguchi, and E. Mizoguchi. 2016. "Current Understanding of Dysbiosis in Disease in Human and Animal Models." *Inflammatory Bowel Diseases* 22 (5): 1137–1150. https://doi.org/10.1097/MIB.0000000000000750.

Deo, P. N., and R. Deshmukh. 2019. "Oral Microbiome: Unveiling the Fundamentals." *Journal of Oral and Maxillofacial Pathology* 23 (1): 122–128. https://doi.org/10.4103/jomfp.JOMFP_304_18.

Flowers, L., and E. A. Grice. 2020. "The Skin Microbiota: Balancing Risk and Reward." *Cell Host & Microbe* 28 (2): 190–200. https://doi.org/10.1016/j.chom.2020.06.017.

Gonçalves, M. F. M., Â. R. Fernandes, A. G. Rodrigues, and C. Lisboa. 2022. "Microbiome in Male Genital Mucosa (Prepuce, Glans, and Coronal Sulcus): A Systematic Review." *Microorganisms* 10 (12): 2312. https://doi.org/10.3390/microorganisms10122312.

Granados-Martinez, C., N. Alfageme-Lopez, M. Navarro-Oviedo, C. Nieto-Vaquero, M. I. Cuartero, B. Diaz-Benito, M. A. Moro, I. Lizasoain, M. Hernandez-Jimenez, and J. M. Pradillo. 2024. "Gut Microbiota, Bacterial Translocation, and Stroke: Current Knowledge and Future Directions." *Biomedicines* 12 (12): 2781. https://doi.org/10.3390/biomedicines12122781.

Grau-Del Valle, C., J. Fernández, E. Solá, I. Montoya-Castilla, C. Morillas, and C. Bañuls. 2023. "Association between Gut Microbiota and Psychiatric Disorders: A Systematic Review." *Frontiers in Psychology* 14: 1215674. https://doi.org/10.3389/fpsyg.2023.1215674.

Holdcroft, A. M., D. J. Ireland, and M. S. Payne. 2023. "The Vaginal Microbiome in Health and Disease—What Role Do Common Intimate Hygiene Practices Play?" *Microorganisms* 11 (2): 298. https://doi.org/10.3390/microorganisms11020298.

Hou, K., Z. X. Wu, X. Y. Chen, J. Q. Wang, D. Zhang, C. Xiao, D. Zhu, J. B. Koya, L. Wei, J. Li, and Z. S. Chen. 2022. "Microbiota in Health and Diseases." *Signal Transduction and Targeted Therapy* 7 (1): 135. https://doi.org/10.1038/s41392-022-00974-4.

Hwang, I. Y., M. H. Tan, E. Koh, C. L. Ho, C. L. Poh, and M. W. Chang. 2014. "Reprogramming Microbes to Be Pathogen-Seeking Killers." *ACS Synthetic Biology* 3 (4): 228–237. https://doi.org/10.1021/sb400077j.

Jia, G., A. Zhi, P. F. H. Lai, G. Wang, Y. Xia, Z. Xiong, H. Zhang, N. Che, and L. Ai. 2018. "The Oral Microbiota: A Mechanistic Role for Systemic Diseases." *British Dental Journal* 224 (6): 447–455. https://doi.org/10.1038/sj.bdj.2018.217.

Jiang, H., Z. Ling, Y. Zhang, H. Mao, Z. Ma, Y. Yin, W. Wang, W. Tang, Z. Tan, J. Shi, L. Li, and B. Ruan. 2015. "Altered Fecal Microbiota Composition in Patients with Major Depressive Disorder." *Brain, Behavior, and Immunity* 48: 186–194. https://doi.org/10.1016/j.bbi.2015.03.016.

Jovel, J., L. A. Dieleman, D. Kao, A. L. Mason, and E. Wine. 2018. "The Human Gut Microbiome in Health and Disease." In *Metagenomics*, edited by Muniyandi Nagarajan, 197–213. London: Academic Press. https://doi.org/10.1016/B978-0-08-102268-9.00010-0.

Kanbay, M., E. M. Onal, B. Afsar, T. Dagel, A. Yerlikaya, A. Covic, and N. D. Vaziri. 2018. "The Crosstalk of Gut Microbiota and Chronic Kidney Disease: Role of Inflammation, Proteinuria, Hypertension, and Diabetes Mellitus." *International Urology and Nephrology* 50 (8): 1453–1466. https://doi.org/10.1007/s11255-018-1873-2.

Kang, D. W., J. B. Adams, A. C. Gregory, T. Borody, L. Chittick, A. Fasano, A. Khoruts, E. Geis, J. Maldonado, S. McDonough-Means, E. L. Pollard, S. Roux, M. J. Sadowsky, K. S. Lipson, M. B. Sullivan, J. G. Caporaso, and R. Krajmalnik-Brown. 2017. "Microbiota Transfer Therapy Alters Gut Ecosystem and Improves Gastrointestinal and Autism Symptoms: An Open-Label Study." *Microbiome* 5 (1): 10. https://doi.org/10.1186/s40168-016-0225-7.

Khan, A. A., A. T. Sirsat, H. Singh, and P. Cash. 2022. "Microbiota and Cancer: Current Understanding and Mechanistic Implications." *Clinical and Translational Oncology* 24 (2): 193–202. https://doi.org/10.1007/s12094-021-02690-x.

Kleessen, B., A. J. Kroesen, H. J. Buhr, and M. Blaut. 2002. "Mucosal and Invading Bacteria in Patients with Inflammatory Bowel Disease Compared with Controls." *Scandinavian Journal of Gastroenterology* 37 (9): 1034–1041. https://doi.org/10.1080/003655202320378220.

Konovalovas, A., J. Armalytė, L. Klimkaitė, T. Liveikis, B. Jonaitytė, E. Danila, D. Bironaitė, D. Mieliauskaitė, E. Bagdonas, and R. Aldonytė. 2024. "Human Nasal Microbiota Shifts in Healthy and Chronic Respiratory Disease Conditions." *BMC Microbiology* 24 (1): 150. https://doi.org/10.1186/s12866-024-03294-5.

Lathakumari, R. H., L. KakithakaraVajravelu, A. Satheesan, S. Ravi, and J. Thulukanam. 2024. "Antibiotics and the Gut Microbiome: Understanding the Impact on Human Health." *Medicine in Microecology* 20: 100106. https://doi.org/10.1016/j.medmic.2024.100106.

Laudes, M., C. Geisler, N. Rohmann, J. Bouwman, T. Pischon, and K. Schlicht. 2021. "Microbiota in Health and Disease—Potential Clinical Applications." *Nutrients* 13 (11): 3866. https://doi.org/10.3390/nu13113866.

Lê, S., C. Cecchin-Albertoni, C. Thomas, P. Kemoun, M. Minty, and V. Blasco-Baque. 2023. "The Role of Dysbiotic Oral Microbiota in Cardiometabolic Diseases: A Narrative Review." *Diagnostics* 13 (20): 3184. https://doi.org/10.3390/diagnostics13203184.

Li, W., X. Wu, X. Hu, T. Wang, S. Liang, Y. Duan, F. Jin, and B. Qin. 2017. "Structural Changes of Gut Microbiota in Parkinson's Disease and Its Correlation with Clinical Features." *Science China Life Sciences* 60 (11): 1223–1233. https://doi.org/10.1007/s11427-016-9001-4.

Liu, B. N., X. T. Liu, Z. H. Liang, and J. H. Wang. 2021. "Gut Microbiota in Obesity." *World Journal of Gastroenterology* 27 (25): 3837–3850. https://doi.org/10.3748/wjg.v27.i25.3837.

Madison, A., and J. K. Kiecolt-Glaser. 2019. "Stress, Depression, Diet, and the Gut Microbiota: Human-Bacteria Interactions at the Core of Psychoneuroimmunology and Nutrition." *Current Opinion in Behavioral Sciences* 28: 105–110. https://doi.org/10.1016/j.cobeha.2019.01.011.

Morrison, D. J., and T. Preston. 2016. "Formation of Short Chain Fatty Acids by the Gut Microbiota and Their Impact on Human Metabolism." *Gut Microbes* 7 (3): 189–200. https://doi.org/10.1080/19490976.2015.1134082.

Nesci, A., C. Carnuccio, V. Ruggieri, A. D'Alessandro, A. Di Giorgio, L. Santoro, A. Gasbarrini, A. Santoliquido, and F. R. Ponziani. 2023. "Gut Microbiota and Cardiovascular Disease: Evidence on the Metabolic and

Inflammatory Background of a Complex Relationship." *International Journal of Molecular Sciences* 24 (10): 9087. https://doi.org/10.3390/ijms24109087.

O'Keefe, S. J. D. 2016. "Diet, Microorganisms and Their Metabolites, and Colon Cancer." *Nature Reviews Gastroenterology & Hepatology* 13 (12): 691–706. https://doi.org/10.1038/nrgastro.2016.165.

Ogunrinola, G. A., J. O. Oyewale, O. O. Oshamika, and G. I. Olasehinde. 2020. "The Human Microbiome and Its Impacts on Health." *International Journal of Microbiology* 2020: 8045646. https://doi.org/10.1155/2020/8045646.

Okolie, M., G. Edo, I. Ainyanbhor, A. Jikah, P. Akpoghelie, E. Yousif, K. Zainulabdeen, E. Isoje, U. Igbuku, J. Othuke Orogu, J. Owheruo, A. Essaghah, and U. Huzaifa. 2024. "Gut Microbiota and Immunity in Health and Diseases: A Review." *Proceedings of the Indian National Science Academy* 91. https://doi.org/10.1007/s43538-024-00355-1.

Pathak, Y. V., S. Pathak, and C. Stough, eds. 2023. *Anxiety, Gut Microbiome, and Nutraceuticals: Recent Trends and Clinical Evidence*, 1st ed. Boca Raton, Florida: CRC Press. https://doi.org/10.1201/9781003333821.

Safarchi, A., G. Al-Qadami, C. Tran, and M. Conlon. 2025. "Understanding Dysbiosis and Resilience in the Human Gut Microbiome: Biomarkers, Interventions, and Challenges." *Frontiers in Microbiology* 16. https://doi.org/10.3389/fmicb.2025.1559521.

Sasso, J. M., R. M. Ammar, R. Tenchov, S. Lemmel, O. Kelber, M. Grieswelle, and Q. Angela Zhou. 2023. "Gut Microbiome–Brain Alliance: A Landscape View into Mental and Gastrointestinal Health and Disorders." *ACS Chemical Neuroscience* 14 (3): 1717–1763. https://doi.org/10.1021/acschemneuro.3c00127.

Schmidt, T. S. B., J. Raes, and P. Bork. 2018. "The Human Gut Microbiome: From Association to Modulation." *Cell* 172 (6): 1198–1215. https://doi.org/10.1016/j.cell.2018.02.044.

Schwabe, R. F., and T. Luedde. 2018. "Apoptosis and Necroptosis in the Liver: A Matter of Life and Death." *Nature Reviews Gastroenterology & Hepatology* 15 (12): 738–752. https://doi.org/10.1038/s41575-018-0065-y.

Sekirov, I., S. L. Russell, L. C. Antunes, and B. Brett Finlay. 2010. "Gut Microbiota in Health and Disease." *Physiological Reviews* 90 (3): 859–904. https://doi.org/10.1152/physrev.00045.2009.

Seo, Do, and D. M. Holtzman. 2024. "Current Understanding of the Alzheimer's Disease-Associated Microbiome and Therapeutic Strategies." *Experimental & Molecular Medicine* 56: 86–94. https://doi.org/10.1038/s12276-023-01146-2.

Sobti, R. C., R. Chander Kuhad, R. Lal, and P. Rishi. 2023. *Role of Microbes in Sustainable Development: Human Health and Diseases.* 1st ed. Singapore: Springer Nature Singapore. https://doi.org/10.1007/978-981-99-3126-2.

Song, W., Y. Yue, and Q. Zhang. 2023. "Imbalance of Gut Microbiota Is Involved in the Development of Chronic Obstructive Pulmonary Disease: A Review." *Biomedicine & Pharmacotherapy* 165: 115150. https://doi.org/10.1016/j.biopha.2023.115150.

Tanaka, M., and J. Nakayama. 2017. "Development of the Gut Microbiota in Infancy and Its Impact on Health in Later Life." *Allergology International* 66 (4): 515–522. https://doi.org/10.1016/j.alit.2017.07.010.

Tang, W. H. W., T. Kitai, and S. L. Hazen. 2017. "Gut Microbiota in Cardiovascular Health and Disease." *Circulation Research* 120 (7): 1183–1196. https://doi.org/10.1161/CIRCRESAHA.117.309715.

Ugwu, O. P., E. U. Alum, M. B. Okon, and E. I. Obeagu. 2024. "Mechanisms of Microbiota Modulation: Implications for Health, Disease, and Therapeutic Interventions." *Medicine (Baltimore)* 103 (19): e38088. https://doi.org/10.1097/MD.0000000000038088.

Ullah, H., S. Arbab, Y. Tian, C. Q. Liu, Y. Chen, L. Qijie, M. I. U. Khan, I. U. Hassan, and K. Li. 2023. "The Gut Microbiota–Brain Axis in Neurological Disorder." *Frontiers in Neuroscience* 17: 1225875. https://doi.org/10.3389/fnins.2023.1225875.

Vinderola, G., A. Ouwehand, S. Salminen, and A. von Wright, eds. 2024. *Lactic Acid Bacteria: Microbiological and Functional Aspects,* 6th ed. Boca Raton, FL: CRC Press. https://doi.org/10.1201/9781003352075.

Willis, J. R., and T. Gabaldón. 2020. "The Human Oral Microbiome in Health and Disease: From Sequences to Ecosystems." *Microorganisms* 8 (2): 308. https://doi.org/10.3390/microorganisms8020308.

Wu, P., Y. Chen, J. Zhao, G. Zhang, J. Chen, J. Wang, and H. Zhang. 2017. "Urinary Microbiome and Psychological Factors in Women with Overactive Bladder." *Frontiers in Cellular and Infection Microbiology* 7: 488. https://doi.org/10.3389/fcimb.2017.00488.

Zhao, M., Y. Wang, Y. Shen, C. Wei, G. Zhang, and L. Sun. 2024. "A Review of the Roles of Pathogens in Alzheimer's Disease." *Frontiers in Neuroscience* 18: 1439055. https://doi.org/10.3389/fnins.2024.1439055.

Zuber, A., A. Peric, N. Pluchino, D. Baud, and M. Stojanov. 2023. "Human Male Genital Tract Microbiota." *International Journal of Molecular Sciences* 24 (8): 6939. https://doi.org/10.3390/ijms24086939.

3 Microbiome Metabolites
Direct and Indirect Communication and Relation with Host

Varshika Singh, Kashish Julka, Sonali, and Wafa Aziz

3.1 INTRODUCTION

The human gut microbiome comprises trillions of microorganisms including bacteria, archaea, viruses and fungi and functions as a dynamic metabolic organ integral to host health (Thursby and Juge 2017). This microbial ecosystem orchestrates a vast array of biochemical transformations, producing a diverse repertoire of metabolites that influence host physiology, immune regulation and neurobehavioral processes (Ouyang et al. 2025). The microbiota-derived metabolites (short-chain fatty acids (SCFAs), bile acids (BAs) and tryptophan derivatives) are released as end products of microbial enzymatic activity on dietary components and host-secreted substrates (Feng et al. 2022).

These metabolites engage in sophisticated communication with host systems through direct and indirect signaling pathways (Jyoti and Dey 2025). Direct signaling involves localized interactions with intestinal epithelial cells (IECs), modulating barrier integrity, mucosal immunity and hormone secretion (Legan et al. 2022). Indirect signaling, however, enables systemic effects by influencing distant organs via circulation, impacting metabolic homeostasis, immune responses and even brain function (Legan et al. 2022). Understanding these dual modes of communication is pivotal for unraveling the mechanisms of host–microbe symbiosis and identifying therapeutic targets for microbial dysbiosis and chronic inflammation conditions.

3.2 THE MICROBIOME METABOLOME

Microbiome metabolites are structurally and functionally diverse and are responsible for communication with the host. The most known are the SCFAs, which comprise acetate, propionate and butyrate. These are primarily produced from the fermentation of dietary fibers (Fujisaka et al. 2023). Other critical classes include secondary BAs, indoles and their derivatives. BAs are produced because of microbial deconjugation and dehydroxylation of host primary BAs; whereas indole and its derivatives are produced from tryptophan metabolism (Shang et al. 2025). The specific composition and relative abundance of the gut metabolome is highly dependent on individual diet, microbial community structure and transit time (Graf et al. 2015). Hence, the host–microbe communication is intrinsically a highly personalized process with profound diagnostic and therapeutic implications.

3.2.1 SCFAs

SCFAs are known as the potent systemic signaling molecules (Figure 3.1) and the primary mechanisms through which they operate are

- **GPCR binding**: SCFAs bind to host G-protein-coupled receptors (GPCRs), specifically GPR41 and GPR43, present at enteroendocrine, adipose and immune cell (Masse and Lu 2023).

 DOI: 10.1201/9781003610274-4

- **HDAC inhibition and immune modulation**: Butyrate and propionate function as histone deacetylase (HDAC) inhibitors, governing host gene expression and promoting the differentiation of anti-inflammatory regulatory T cells (Tregs) in the gut lamina propria (Lange et al. 2023).

3.2.2 Secondary Bile Acids (BAs)

Secondary BAs result from the microbial deconjugation of host primary BAs by 7α-dehydroxylating bacteria (Kim et al. 2022). These BAs are key transducers of the gut–liver axis (Figure 3.1), signaling through nuclear receptors like the Farnesoid X Receptor (FXR) and the membrane-bound TGR5 to regulate host lipid, glucose and energy metabolism (Masse and Lu 2023).

3.2.3 Tryptophan Derivatives

Derivatives of tryptophan, such as indole-3-propionic acid (IPA) and indole-3-aldehyde (IAld), are crucial modulators of intestinal homeostasis. Their primary host target is the Aryl Hydrocarbon Receptor (AhR), activation of which reinforces the epithelial barrier and modulates innate and adaptive immunity (Wu et al. 2025).

3.2.4 The Choline-Derived Trimethylamine N-Oxide (TMAO)

TMAO is synthesized by microbial enzymes and oxidized in the liver to yield a product strongly linked to cardiovascular risk, atherosclerosis and has been linked to insulin resistance and BCAA

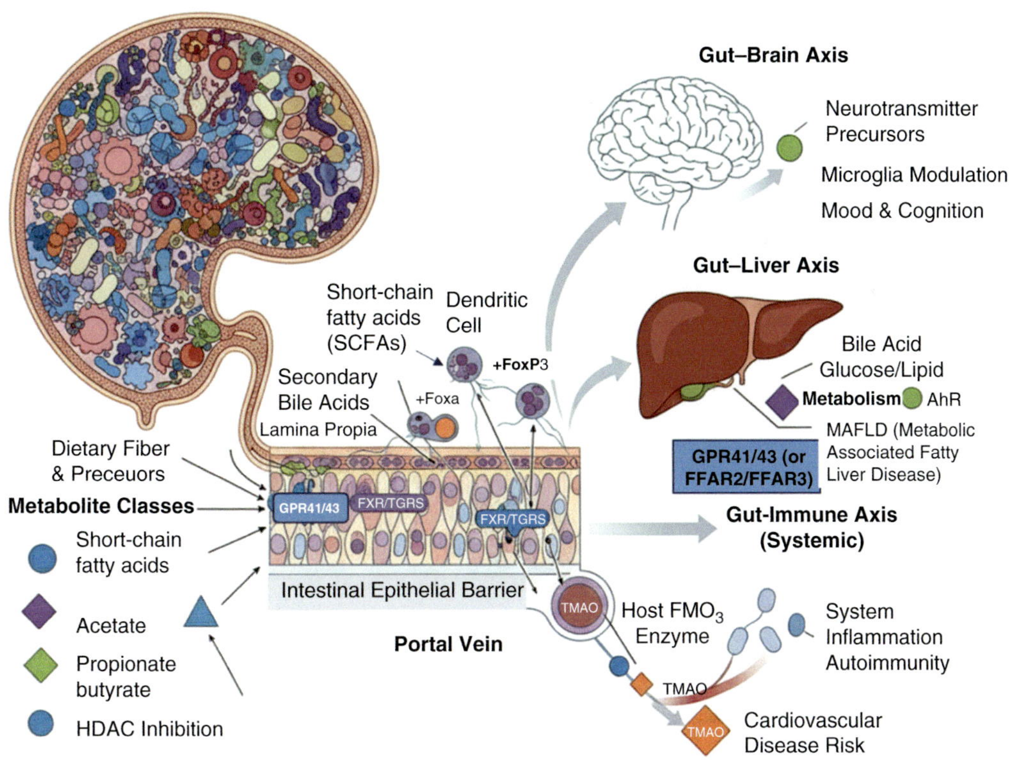

FIGURE 3.1 The systemic communication of gut microbiome metabolites with host organs and pathways.

TABLE 3.1
Key Microbiome Metabolites: Classes, Production and Mechanisms

Metabolite Class	Example(s)	Microbial Precursor/ Production	Primary Host Receptor/ Mechanism	Major Physiological Role(s)	Reference(s)
Short-chain fatty acids (SCFAs)	Acetate, propionate, butyrate	Dietary fibers (fermentation)	GPR41 (FFAR3), GPR43 (FFAR2); HDAC inhibition	Colonocyte energy, treg promotion, epigenetic regulation	Masse and Lu (2023)
Secondary bile acids (BAs)	Deoxycholate, lithocholate	Primary BAs (deconjugation/ dehydroxylation)	Farnesoid X receptor (FXR), TGR5 (GPBAR1)	Regulation of glucose/lipid metabolism, gut–liver axis signaling	Masse and Lu (2023)
Indole derivatives	Indole-3-propionic acid (IPA), indole-3-aldehyde (IAld)	Tryptophan (catabolism)	Aryl hydrocarbon receptor (AhR)	Intestinal barrier integrity, immunoregulation	Whitfield-Cargile et al. (2016), Wu et al. (2025)
Co-metabolites	Trimethylamine N-oxide (TMAO)	Choline, L-carnitine (microbial TMA to host FMO3 oxidation)	Endothelial function, inflammation	Cardiovascular disease risk, metabolic dysfunction	Liu et al. (2022)

(branched-chain amino acids) catabolites (Liu et al. 2022). TMAO mediates sophisticated cross-talk across multiple organ systems: Gut–brain, gut–liver (conversion of the bacterially produced TMA into the circulating TMAO) and gut–immune (influence systemic inflammation and immune responses) axes. The four major classes of metabolites discussed so far are summarized in Table 3.1 with their origin, primary host targets and physiological effects.

3.3 DIRECT COMMUNICATION: LOCALIZED EPITHELIAL SIGNALING

Direct communication in the gut involves metabolites acting locally within the intestinal lumen or adjacent IECs and underlying immune cells (Yoo et al. 2020). This localized signaling is crucial for maintaining the integrity of the intestinal barrier and regulating the immediate immune environment. The direct communication types are further discussed below:

3.3.1 G-Protein Coupled Receptors (GPCRs) and SCFAs

The GPCRs are expressed on various host cell types, including colonocytes, adipocytes and immune cells (van der Hee and Wells 2021). SCFAs act as ligands for GPCRs family in the direct mechanism of SCFA signaling (Figure 3.2). The roles of various GPCRs and SCFAs in direct signaling are as follows:

- **GPR43 (FFAR2)** and **GPR41 (FFAR3)** receptors are activated by all three major SCFAs (acetate, propionate and butyrate). GPR43 shows the highest affinity for acetate and GPR41 for propionate. Activation of GPR43 on L cells in the gut stimulates the secretion of glucagon-like peptide-1 (GLP-1) and peptide YY (PYY). This leads to an enhanced feeling of fullness and improves glucose homeostasis. This indicates the crucial direct link

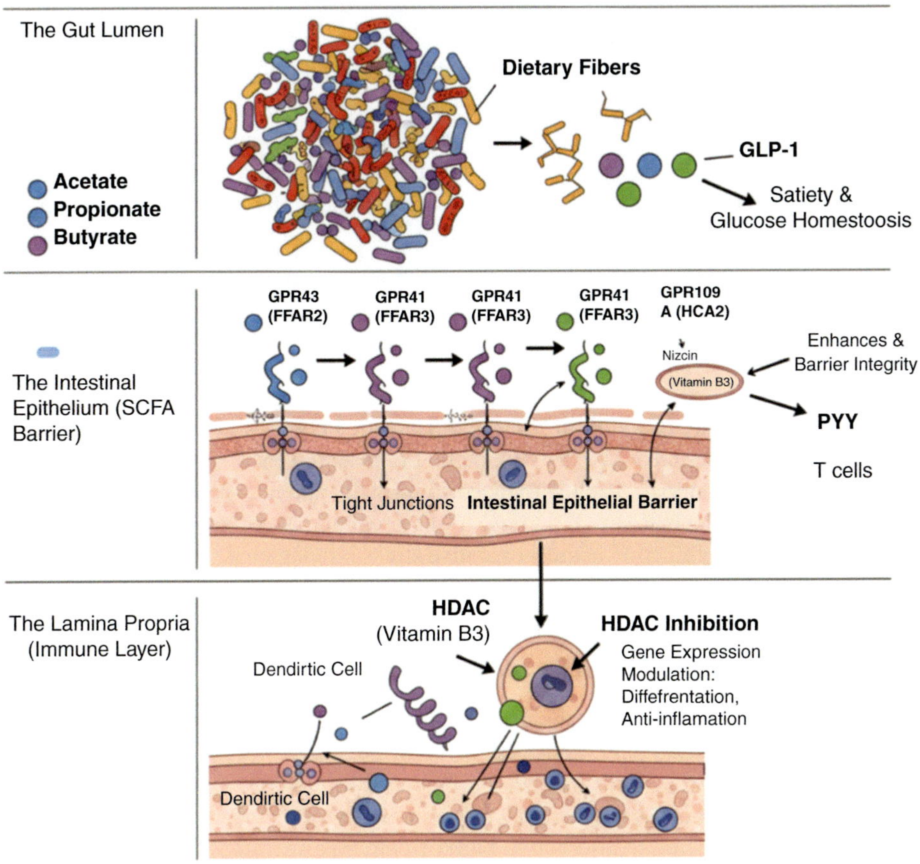

FIGURE 3.2 Direct communication and localized epithelial signaling.

between the microbiome and the enteroendocrine system (Chao et al. 2025). However, the binding of SCFAs to GPR43 on the surface of IECs has been shown to enhance tight junction assembly, reducing intestinal permeability and reinforcing barrier function (Venkatesh et al. 2014).

- **GPR109A (Hydroxycarboxylic acid receptor 2)** is selectively activated by butyrate and niacin (Vitamin B3) and is prominently expressed on colonocytes and immune cells.
- GPR109A activation leads to anti-inflammatory effects and protection against colitis-associated cancer (Gong et al. 2021). Butyrate acts as a HDAC inhibitor within IECs and offers defense against carcinogenesis. It alters gene expression to promote differentiation, suppress inflammation and regulate cell cycle progression (Fernandes and Vinolo 2024).

3.3.2 Nuclear Receptors and Bile Acids (BAs)

Secondary BAs are powerful direct signaling molecules. They signal host nuclear receptors to modulate gene expression to regulate metabolism, immunity and barrier function.

- **Farnesoid X receptor (FXR)** is highly expressed in the ileum, liver and kidney and is activated by various BAs. In the ileum, FXR activation stimulates the release of fibroblast growth factor 15/19 (FGF15/19) in a key feedback loop (Katafuchi and Makishima 2022). FGF19 then travels to the liver, where it represses the host enzyme cholesterol -hydroxylase

(CYP7A1), which acts as a rate-limiting enzyme in BA synthesis, thus leading to a host microbe–host regulatory circuit situation.

- **Takeda G-protein-coupled receptor 5 (TGR5)** is a membrane-bound GPCR and is expressed in immune cells (e.g., macrophages), L cells and brown adipose tissue. It is activated by secondary BAs, such as deoxycholic acid (DCA) and lithocholic acid (LCA), on L cells and further stimulates GLP-1 release (Zangerolamo et al. 2025). Whereas in macrophages, TGR5 activation has been shown to dampen the inflammatory response.

3.3.3 Aryl Hydrocarbon Receptor (AhR) and Tryptophan Metabolites

The AhR is a transcription factor activated by ligand molecules from associated metabolic pathways. Microbial tryptophan derivatives like IAld, IAA and IPA are potent AhR ligands (Ma et al. 2020).

- **AhR acts in barrier function** in IECs and dendritic cells by inducing the production of Interleukin-22 (IL-22). IL-22 is a cytokine that is essential for antimicrobial peptide synthesis and epithelial repair, leading to mucosal integrity and host defense (Mar et al. 2023).
- Microbial **indole derivatives** (IPA) bind to the pregnane X receptor on IECs, tightening cell junctions and protecting against oxidative stress (Huang et al. 2022).

3.4 INDIRECT COMMUNICATION: SYSTEMIC MODULATION OF DISTANT ORGANS

During indirect communication, microbiome metabolites get absorbed into the systemic circulation and reach distant host tissues and organs (Legan et al. 2022). Indirect communication thus influences broader physiological systems like gut–brain, gut–liver and gut–immune axes.

Secondary BAs are critical systemic regulators of indirect communication. After microbial modification in the colon, they are reabsorbed in bloodstream and interact with host receptors— FXR and the Takeda G-protein-coupled receptor 5 (TGR5) (Brandl et al. 2017). These receptors are mainly found in the liver, pancreas and adipose tissues.

Activation of FXR and TGR5 modulates lipid, glucose and energy homeostasis. In the liver, secondary BAs regulate host BA synthesis through a negative feedback loop and influence gluconeogenesis (Chiang and Ferrell 2020). Various influences of indirect communication are discussed below:

3.4.1 SCFAs Systemic Effect and Histone Deacetylase (HDAC) Inhibition

- **Systemic effect**: SCFAs can also circulate systemically and travel to bone marrow and peripheral lymphoid tissues. They bind to GPCRs on immune cells and drive the differentiation of anti-inflammatory regulatory T cells (Tregs) in the colon and spleen. This creates a potent, system-wide anti-inflammatory effect that protects against autoimmune conditions (Liu et al. 2023).
- **HDAC inhibition**: SCFAs; particularly butyrate, are potent inhibitors of HDACs. This increases the acetylation of histones, leading to a more open chromatin structure and enhanced transcription of genes promoting the differentiation of Tregs (Steliou et al. 2012; Mukhopadhya and Louis 2025). This is a profound, indirect epigenetic mechanism by which the microbiota controls host immune fate (Figure 3.3).

3.4.2 Tryptophan Metabolites and T-Cell Fate

- **AhR activation by indole** derivatives influences the balance of proinflammatory cells and anti-inflammatory Tregs. IPA generally bolsters the intestinal barrier and has antioxidant properties (Li et al. 2021).

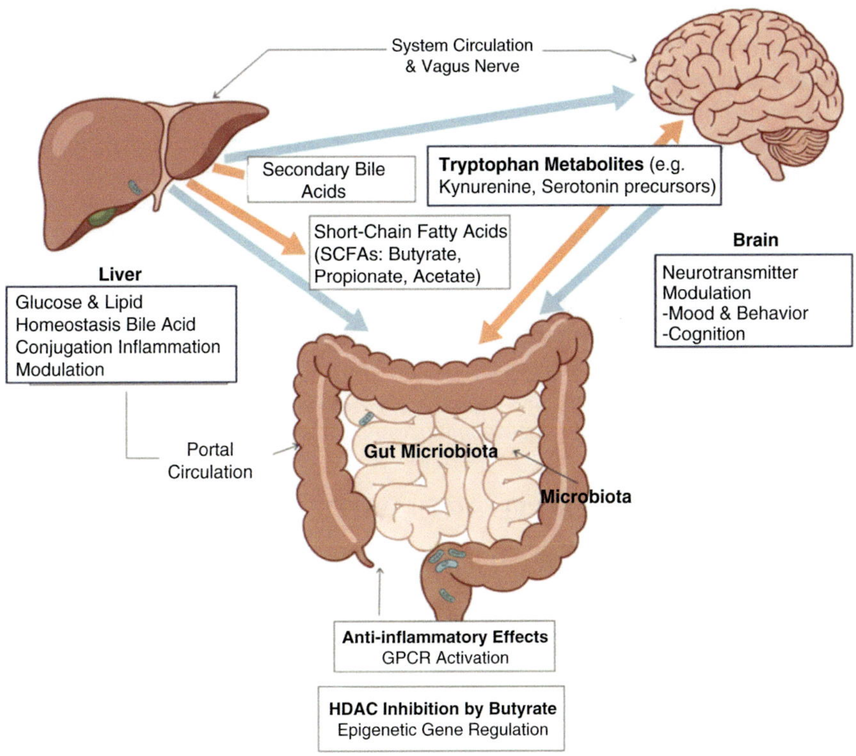

FIGURE 3.3 Indirect communication: system modulation of distant organs.

3.4.3 The Gut–Brain Axis (GBA)

Microbial metabolites also connect the gut and the central nervous system. Metabolites derived from tryptophan, such as kynurenine or microbial-derived neuroactive substances, can cross the blood–brain barrier (BBB) and modulate vagal nerve signaling (Shaw et al. 2023). This indirect communication is fundamental to the GBA, influencing mood, cognitive function and the pathophysiology of neurodegenerative disorders.

- **The GBA** serves as a bidirectional communication network where microbial metabolites such as SCFAs, TMAO and lipopolysaccharides (LPS) directly influence DNA methylation, histone modifications and noncoding RNA (ncRNA) activity in brain cells (Zheng et al. 2023).
- **Gut dysbiosis** is characterized by an overgrowth of proinflammatory bacteria and reduced beneficial microbes. This leads to decreased production of butyrate, which is a key SCFA that inhibits HDACs. HDAC overexpression further results in reduced acetylation of histone of genes like BDNF, ultimately impairing synaptic plasticity and memory formation (Anand et al. 2022).
- **Gut-derived LPS** induces DNA hypomethylation at the BACE1 gene promoter, increasing amyloid-beta production and accelerating plaque deposition, which is a hallmark of AD pathology (Jin et al. 2023).

3.4.4 Noncoding RNAs

Dysregulated ncRNAs, particularly microRNAs (miRNAs), act as molecular bridges between gut inflammation and neurodegeneration.

- **Chronic gut inflammation** upregulates miR-34a, which suppresses SIRT1 gene vital for neuronal survival and tau protein regulation (Alramadneh et al. 2025). This suppression promotes tau hyperphosphorylation and neurofibrillary tangle formation.
- **Epigenetic disruptions** lead to neuroinflammation, synaptic dysfunction and cognitive decline. The long noncoding RNA—BACE1-AS is stabilized by gut-derived inflammatory signals and enhances BACE1 mRNA stability, further driving amyloid-beta accumulation. This is a key component of the plaques found in the brains of individuals with Alzheimer's disease (AD) (Sayad et al. 2022).

The gut microbiota's role in modulating these epigenetic changes highlights potential therapeutic avenues, such as probiotic supplementation, dietary interventions (e.g., high-fiber diets to boost SCFAs) or HDAC inhibitors, which may help restore epigenetic balance and slow AD progression (Li et al. 2022). Understanding these mechanisms underscores the importance of maintaining gut health as a preventive strategy against neurodegenerative diseases.

3.5 GUT MICROBIOME AND RELATION WITH HOST

3.5.1 Gut Microbiome Key Players in Neurological Health

Bacteria, fungi, viruses, and archaea are among the many microorganisms that live in the gut and all play a part in maintaining general health (Chavoya-Guardado et al. 2022). These microorganisms have long been linked to immunity and digestion, but new studies have shown how important they are for brain function (Table 3.2). Some of the millions of microbial species that live in the gut are particularly notable for having a direct impact on mental health and neurological function:

- **Bacteroides**: Consider these microorganisms to be metabolic engineers. They generate butyrate, one of the SCFAs that have been connected to brain protection (Facchin et al. 2024). Butyrate contributes to neuroprotection, lowers inflammation, and strengthens the BBB, all of which may prevent the onset of neurodegenerative illnesses.
- ***Lactobacillus* and *Bifidobacterium***: These powerful probiotics actively support the synthesis of neurotransmitters in addition to helping with digestion (Kezer et al. 2025). Gamma-aminobutyric acid (GABA), a crucial neurotransmitter with calming effects on the brain, may be synthesized by both. These microorganisms aid in lowering anxiety, encouraging relaxation and even improving the quality of sleep by raising GABA levels.
- ***Firmicutes* and *Actinobacteria***: These bacterial groups promote gut integrity by preserving the mucosal lining, which prevents hazardous compounds from entering the circulation (Chavoya-Guardado et al. 2022). They also create anti-inflammatory chemicals, which may help combat the chronic inflammation that is commonly associated with neurological illnesses such as Alzheimer's and Parkinson's.

3.5.2 Microbial Metabolites and Their Impact on the Brain

Microbes in the gut are not simply passive residents; they actively produce a range of compounds that interact with the brain (Parker et al. 2020). Among the most influential are SCFAs, neurotransmitters and other bioactive molecules that influence cognitive and emotional wellness (Chen et al. 2021). These molecules have the unique capacity to pass the BBB and directly influence neuronal processes (Table 3.2).

- **Fighting inflammation**: One of the most important ways microbial metabolites benefit brain function is through their anti-inflammatory effects. SCFAs such as butyrate, acetate and propionate decrease neuroinflammation by regulating immune responses (Morris et al. 2017).

TABLE 3.2
Gut Dysbiosis and Neurodegenerative/Mood Disorders

Neuro Condition	Gut Microbiota Changes	Key factors (Genes/Proteins/Metabolite/Pathways)	Mechanistic Insight	References
Parkinson's disease (PD)	↓ Prevotella, ↑ enterobacteriaceae	α-synuclein, SCFAs, LPS, TLR4	Microbial metabolites influence α-syn aggregation via inflammation; vagus nerve implicated	Sampson et al. (2016)
Alzheimer's disease (AD)	↓ Bifidobacterium, ↑ proteobacteria	Aβ, IL-6, LPS, TREM2, GSK-3β	Gut inflammation elevates peripheral cytokines, increases BBB permeability and Aβ deposition	Harach et al. (2017)
Major depressive disorder	↓ Faecalibacterium, ↑ alistipes	BDNF, IL-1β, IL-6, cortisol	Dysbiosis leads to HPA axis dysregulation and reduced neurogenesis via inflammatory cytokines	Jiang et al. (2015)
Autism spectrum disorders	↑ Clostridia, ↓ bacteroides	GABA, serotonin, SCFAs, zonulin	Gut barrier dysfunction alters neurodevelopmental signaling through metabolite changes	Hsiao et al. (2013)
Multiple sclerosis (MS)	↓ SCFA-producing bacteria	IL-17, Tregs, akkermansia, prevotella	Gut microbiota modulate T-cell differentiation; reduced SCFAs impair immune tolerance	Cekanaviciute et al. (2017)
Bipolar disorder	Altered firmicutes/bacteroidetes ratio	TNF-α, S100B, tryptophan metabolism	Microbial impact on tryptophan–kynurenine pathway modulates mood and neuroinflammation	Painold et al. (2019)

Because chronic inflammation is a frequent feature of many brain disorders, keeping a healthy gut microbiome may help prevent or treat conditions including multiple sclerosis, depression and cognitive loss.

- **Regulating mood and cognition**: Neurotransmitters influence emotion, behavior and cognitive performance. Gut microorganisms can produce neurotransmitters like serotonin (the "happy" chemical), dopamine (which is linked to pleasure and motivation) and GABA (which has relaxing properties) (Loh et al. 2024). The gut microbiome influences neurotransmitter availability, which has a significant impact on mental health and may reduce symptoms of anxiety, depression and stress-related diseases.
- **Shaping early brain development**: The connection between gut microbiota and brain health begins early in life. During infancy and childhood, exposure to specific microbial metabolites can shape brain development in ways that influence cognitive function, emotional regulation and vulnerability to neurological disorders later in life (Sarkar et al. 2021). For example,

SCFAs play a role in the formation of neural circuits, while gut-derived immune signals help shape a child's stress response and resilience.

Understanding the intricate relationship between the gut microbiome and the brain opens new avenues for therapeutic strategies. By modulating gut microbial composition through diet, probiotics or lifestyle changes, it may be possible to improve mental well-being and reduce the risk of neurological disorders (Thangaleela et al. 2022). As research continues, the potential for harnessing gut microbes as allies in brain health looks increasingly promising.

3.5.3 Epigenetic Modulation of Mood, Anxiety, and Neurodevelopmental Disorders Via the Gut–Brain Axis

Emerging research highlights the GBA as a critical mediator of epigenetic changes in mood, anxiety and neurodevelopmental disorders like autism spectrum disorder (ASD) (Wang et al. 2023). The gut microbiota influences DNA methylation, histone modifications and ncRNA expression, thereby regulating stress responses, neurotransmitter signaling and neuroinflammation. In mood and anxiety disorders, microbial metabolites (e.g., SCFAs) modulate DNA methylation of genes like *NR3C1* (glucocorticoid receptor) and *BDNF*, impacting HPA axis function and synaptic plasticity (Watkeys et al. 2018). Similarly, dysbiosis-driven histone deacetylation in *BDNF* and serotonin-related genes contributes to depressive and anxiety-like behaviors. In ASD, gut microbiome alterations correlate with aberrant methylation of *MECP2* and *UBE3A*, disrupting neurodevelopment and excitatory-inhibitory balance (Sabit et al. 2021). These findings suggest that gut microbiota–epigenome interactions play a pivotal role in psychiatric and neurodevelopmental conditions, offering potential for microbiome-targeted therapies (e.g., probiotics and prebiotics) to restore epigenetic balance.

The GBA also links neuroinflammation and epigenetic dysregulation in these disorders. Gut-derived inflammatory signals (e.g., LPS) induce DNA hypermethylation of IL-6 and *TNF-α*, exacerbating neuroinflammation in depression and ASD (Nohesara et al. 2023). Conversely, microbiota-dependent histone modifications regulate synaptic genes (*CREB*, *DLGAP2*), while microbial miRNAs (e.g., miR-146a) cross the gut barrier to modulate neuronal gene expression. Therapeutic strategies like HDAC inhibitors and fecal microbiota transplantation (FMT) aim to reverse these epigenetic disruptions. By targeting the GBA, precision interventions could alleviate symptoms of mood disorders, anxiety and ASD through epigenetic reprogramming of stress, immune and neuroplasticity pathways (Komada and Nishimura 2022).

3.5.4 Epigenetic Dysregulation in ASD through the Gut–Brain Axis: Histone Modifications and Noncoding RNAs

The gut microbiome influences histone modifications in ASD by producing metabolites like SCFAs, which modulate HDAC activity (Sabit et al. 2021). Disrupted gut microbiota can lead to abnormal histone acetylation and methylation patterns in key synaptic genes (e.g., *SHANK3*), impairing neuronal connectivity and contributing to ASD-related social deficits (Zhou et al. 2025). Similarly, microbial imbalances alter repressive histone marks (*H3K27me3*) in neurodevelopmental genes (CHD8, ADNP), affecting cortical development. Noncoding RNAs (ncRNAs), regulated by gut-derived signals, further contribute to ASD pathogenesis: dysregulated miRNAs (miR-132, miR-134) disrupt synaptic plasticity, while elevated miR-146a promotes neuroinflammation. The GBA also modulates lncRNAs (*BDNF-AS*), which suppress neurotrophic factors like BDNF, exacerbating cognitive impairments (Maqsood et al. 2016). Targeting microbial-epigenetic crosstalk (e.g., with probiotics or HDAC inhibitors) may restore synaptic and immune homeostasis in ASD.

Microbiome-directed interventions could correct histone modifications and ncRNA dysregulation in ASD. For instance, butyrate (an SCFA) enhances histone acetylation at synaptic genes, while FMT may normalize miRNA profiles linked to neuroinflammation (miR-146a) and synaptic function (miR-132). Precision therapies targeting the GBA, including epigenetic drugs and microbiome engineering, hold promise for alleviating ASD symptoms by rewiring neurodevelopmental pathways (Wang et al. 2023).

3.5.5 The Gut Microbiome and Cancer

The gut microbiome has a crucial role in cancer development and therapy through its interactions with the immune system and tumor microenvironment (Nobels et al. 2025). Various animal experiments have confirmed that microbes facilitate the initiation and progression of various types of cancer including gastric cancer (GC) (Table 3.3) (Kwon et al. 2022; Palrasu et al. 2020), colorectal cancer (Kumar et al. 2017; Yang et al. 2022), hepatocellular carcinoma (Dapito et al. 2012), breast cancer (Parida et al. 2021; Fu et al. 2022) and pancreatic ductal adenocarcinoma (Aykut et al. 2019; Yu et al. 2022). Furthermore, poor response after receiving cancer treatment, including chemotherapy (Geller et al. 2017;), radiotherapy (Dong et al. 2021; Li et al. 2022), surgery and immunotherapy (Routy et al. 2018; Derosa et al. 2020), can be partially ascribed to some microbes, which was also confirmed in mice.

The gut microbiome acts as a critical determinant of systemic inflammation, metabolic health and immune homeostasis, influencing distal malignancies through defined molecular axes. A systematic comprehension of the underlying mechanisms, from structural barrier failure to microbial metabolite signaling, is paramount for advancing precision oncology. Microbial dysbiosis, characterized by a disruption of the balanced composition and function of the microbiota, is intimately linked to carcinogenesis (Zeng et al. 2024). This imbalance often involves a loss of beneficial, commensal species and an overgrowth or translocation of pathogenic or opportunistic taxa. In patients diagnosed with GC, studies consistently demonstrate a gradual reduction in microbial diversity and richness within the stomach, as well as in associated oral and fecal samples, when compared to healthy control individuals.

3.6 CONCLUSION

Microbiome metabolites are the principal vocabulary of the ongoing dialog between the human host and its resident microbial community. This communication is characterized by both direct, localized signaling that maintains the integrity of the intestinal barrier and mucosal immunity, and indirect, systemic signaling that modulates distant metabolic, immune and neurological functions. The duality of this metabolic signaling underscores the symbiotic relationship, where microbial metabolism is inextricably linked to host health. Disruptions in metabolite profiles, often termed dysbiosis, are increasingly implicated in a spectrum of diseases, from inflammatory bowel disease and metabolic syndrome to neurological disorders. Future research focusing on the precision manipulation of the microbial metabolome holds immense promise for developing targeted therapeutic strategies.

DECLARATION OF GENERATIVE AI AND AI-ASSISTED TECHNOLOGIES IN THE WRITING PROCESS

During the preparation of this work, the author(s) used AI assistance to improve the readability and language of the manuscript. Graphical abstract has been generated using the AI tool. The authors reviewed and edited the final content and took full responsibility for the content of the published article.

TABLE 3.3
Key Gut Microbes Influencing Cancer Pathogenesis and Therapeutic Response

Microbe (Taxa/Genus/Species)	Associated Cancer Type	Role/Outcome	Detailed Molecular Mechanism/Target	Relevant Context	References
Helicobacter pylori	Gastric cancer (GC), PDAC (risk)	Carcinogenesis & progression (pathogen)	CagA targets the YAP pathway (downregulating E-cadherin, driving EMT); OipA activates β-catenin; induces chronic inflammation (IL-17A, TNF-α).	Virulence factors, chronic inflammation	Duan et al. (2025)
Fusobacterium nucleatum	Colorectal cancer (CRC), breast cancer (BC)	Progression, immune evasion, metastasis (pathogen)	FadA adhesin activates Wnt/β-catenin signaling; Upregulates host PD-L1 expression (CRC); Fap2 lectin promotes metastasis by binding Gal-GalNAc (BC).	Oncogenic signaling, immunotherapy resistance	Wang et al. (2025)
Gammaproteobacteria (e.g., *Aggregatibacter actinomycetemcomitans*)	Pancreatic ductal adenocarcinoma (PDAC)	Chemotherapy resistance (pathogen)	Intratumoral expression of cytidine deaminase (CDDL/CDDS) enzyme, which enzymatically metabolizes and inactivates Gemcitabine	Enzymatic drug inactivation	Geller et al. (2017), Bjånes et al. (2020)
Clostridium scindens/specific *Clostridium* spp.	Hepatocellular carcinoma (HCC)	Progression/immune suppression (pathogen)	Drives overproduction of secondary BAs (DCA); DCA promotes M2-like macrophage polarization and regulatory T-cell (Treg) proliferation	Gut–liver axis, immunosuppressive TME	Lin et al. (2023)
beta-Glucuronidase-producing bacteria (e.g., *Bacteroides*, *Escherichia*, *Clostridium*)	Breast cancer (BC), endometrial	Increased carcinogenic risk (pathogen effect)	Production of β-glucuronidase (GUS); leads to deconjugation of estrogen metabolites, enhancing enterohepatic recirculation and systemic active estrogen load.	Hormone-driven carcinogenesis (estrobolome)	Simpson et al. (2024)
Bifidobacterium spp. (e.g., *B. bifidum*)	Various cancers	Enhanced immunotherapy efficacy (immunomodulator)	Activates STING signaling pathway; promotes DC maturation and Th1 differentiation; enhances CD8+ T-cell cytotoxicity (IFN-γ production).	Immune priming, systemic antitumor response	Sharma et al. (2023)
Akkermansia muciniphila	Various cancers (e.g., PDAC, BC)	Enhanced immunotherapy/chemotherapy efficacy (immunomodulator)	Improves gut barrier integrity; modulates T/B-cell adaptive immune responses; associated with enhanced efficacy of cyclophosphamide and ICI treatment.	Barrier function, immune modulation	Fang et al. (2024)
Faecalibacterium prausnitzii	Various cancers	Enhanced immunotherapy efficacy (immunomodulator)	Produces butyrate; direct activation of host immune cells (DC/CD4+ T cell); restores anti-PD-L1 response in dysbiotic models.	SCFA production, T-cell activation	Ding et al. (2025)
Butyrate-producing bacteria (SCFA producers)	Various cancers	Enhanced radiotherapy efficacy (immunomodulator)	SCFA (butyrate) acts as a histone deacetylase inhibitor (HDACi); correlates with better radiosensitivity by promoting apoptosis and modulating the cGAS-STING pathway.		Sun et al. (2024)

REFERENCES

Alramadneh TN, Abdulsahib WK, Ismael SH, Roopashree R, Nayak PP, Nandhini SU, Arora V, Chauhan AS, Ruziyev O. 2025. Crossroads of cell fate: miR-34-mediated regulation of apoptosis and autophagy in glioblastoma. *Discov Oncol.* 16(1):2177. doi:10.1007/s12672-025-04021-w.

Anand N, Gorantla VR, Chidambaram SB. 2022. The role of gut dysbiosis in the pathophysiology of neuropsychiatric disorders. *Cells.* 12(1):54. doi:10.3390/cells12010054.

Aykut B, Pushalkar S, Chen R, Li Q, Abengozar R, Kim JI, Shadaloey SA, Wu D, Preiss P, Verma N, Guo Y, Saxena A, Vardhan M, Diskin B, Wang W, Leinwand J, Kurz E, Kochen Rossi JA, Hundeyin M, Zambrinis C, Li X, Saxena D, Miller G. 2019. The fungal mycobiome promotes pancreatic oncogenesis via activation of MBL. *Nature.* 574:264–267.

Bjånes TK, Jordheim LP, Schjøtt J, Kamceva T, Cros-Perrial E, Langer A, Ruiz de Garibay G, Kotopoulis S, McCormack E, Riedel B. 2020. Intracellular cytidine deaminase regulates gemcitabine metabolism in pancreatic cancer cell lines. *Drug Metab Dispos.* 48(3):153–158. doi:10.1124/dmd.119.089334.

Brandl K, Kumar V, Eckmann L. 2017. Gut-liver axis at the frontier of host-microbial interactions. *Am J Physiol Gastrointest Liver Physiol.* 312(5):G413–G419. doi: 10.1152/ajpgi.00361.2016.

Cekanaviciute E, Yoo BB, Runia TF, Debelius JW, Singh S, Nelson CA, Kanner R, Bencosme Y, Lee YK, Hauser SL, Crabtree-Hartman E, Sand IK, Gacias M, Zhu Y, Casaccia P, Cree BAC, Knight R, Mazmanian SK, Baranzini SE. 2017. Gut bacteria from multiple sclerosis patients modulate human T cells and exacerbate symptoms in mouse models. *Proc Natl Acad Sci USA.* 114(40):10713–10718. doi:10.1073/pnas.1711235114.

Chao J, Coleman RA, Keating DJ, Martin AM. 2025. Gut microbiome regulation of gut hormone secretion. *Endocrinology.* 166(4):bqaf004. doi:10.1210/endocr/bqaf004.

Chavoya-Guardado Alejandra M, Vasquez-Garibay EM, Ruiz-Quezada SL, Ramírez-Cordero MI, Larrosa-Haro A, Castro-Albarran J. 2022. Firmicutes, bacteroidetes and actinobacteria in human milk and maternal adiposity. *Nutrients.* 14:2887. doi:10.3390/nu14142887.

Chen Y, Jinying X, Yu C. 2021. Regulation of neurotransmitters by the gut microbiota and effects on cognition in neurological disorders. *Nutrients.* 13(6):2099. doi:10.3390/nu13062099.

Chiang JYL, Ferrell JM. 2020. Bile acid receptors FXR and TGR5 signaling in fatty liver diseases and therapy. *Am J Physiol Gastrointest Liver Physiol.* 318(3):G554–G573. doi:10.1152/ajpgi.00223.2019.

Dapito DH, Mencin A, Gwak GY, Pradere JP, Jang MK, Mederacke I, Caviglia JM, Khiabanian H, Adeyemi A, Bataller R, Lefkowitch JH, Bower M, Friedman R, Sartor RB, Rabadan R, Schwabe RF. 2012. Promotion of hepatocellular carcinoma by the intestinal microbiota and TLR4. *Cancer Cell* 17;21(4):504–516. doi: 10.1016/j.ccr.2012.02.007.

Derosa L, Mondragón L, Iribarren K, Bonvalet M, Terrisse S, Rauber C, Goubet AG, Daillère R, Lemaitre F, Reni A, Casu B, Alou MT, Alves Costa Silva C, Raoult D, Fizazi K, Escudier B, Kroemer G, Albiges L, Zitvogel L. 2020. Gut bacteria composition drives primary resistance to cancer immunotherapy in renal cell carcinoma patients. *Eur Urol.* 78:195–206.

Ding K, Huo Y, Fu K, Chen Y, Xia L, Zhan J, Liu J, Liu J, Liu Y, Zhang M, Wu X, Choe H, Zhao D, Ma J, Dai C, Yu Z, Peng Y, Ma X, Zhang L. 2025. *Faecalibacterium prausnitzii* suppresses ovarian cancer by inducing ferroptosis via phenylalanine metabolism activation. *Microbiol Res.* 304:128342. doi:10.1016/j.micres.2025.128342.

Dong J, Li Y, Xiao H, Zhang S, Wang B, Wang H, Li Y, Fan S, Cui M. 2021. Oral microbiota affects the efficacy and prognosis of radiotherapy for colorectal cancer in mouse models. *Cell Rep.* 37:109886.

Duan Y, Xu Y, Dou Y, Xu D. 2025. Helicobacter pylori and gastric cancer: mechanisms and new perspectives. *J Hematol Oncol.* 18(1):10. doi:10.1186/s13045-024-01654-2.

Facchin, S, Bertin L, Bonazzi E, Lorenzon G, De Barba C, Barberio B, Zingone F, Maniero D, Scarpa M, Ruffolo C, Angriman I, Savarino EV. 2024. Short-chain fatty acids and human health: from metabolic pathways to current therapeutic implications. *Life.* 5:559. doi:10.3390/life14050559.

Fang J, Zhang H, Zhang X, Lu X, Liu J, Li H, Huang J. 2024. *Akkermansia muciniphila* improves gastric cancer treatment by modulating the immune microenvironment. *Future Microbiol.* 19(6):481–494. doi:10.2217/fmb-2023-0210. Epub 2024 Apr 17.

Feng W, Liu J, Cheng H, Zhang D, Tan Y, Peng C. 2022. Dietary compounds in modulation of gut microbiota-derived metabolites. *Front Nutr.* 9:939571. doi:10.3389/fnut.2022.939571.

Fernandes MF, Vinolo MAR. 2024. Histone acylations as a mechanism for regulation of intestinal epithelial cells. *Dig Med Res.* 7:4. doi:10.21037/dmr-23-3.

Fu X, Tan W, Song Q, Pei H, Li J. 2022. BRCA1 and breast cancer: molecular mechanisms and therapeutic strategies. *Front Cell Dev Biol.* 10:813457. doi:10.3389/fcell.2022.813457.

Fujisaka S, Watanabe Y, Tobe K. 2023. The gut microbiome: a core regulator of metabolism. *J Endocrinol.* 256(3):e220111. doi:10.1530/JOE-22-0111.

Geller LT, Barzily-Rokni M, Danino T, Jonas OH, Shental N, Nejman D, Gavert N, Zwang Y, Cooper ZA, Shee K, Thaiss CA, Reuben A, Livny J, Avraham R, Frederick DT, Ligorio M, Chatman K, Johnston SE, Mosher CM, Brandis A, Fuks G, Gurbatri C, Gopalakrishnan V, Kim M, Hurd MW, Katz M, Fleming J, Maitra A, Smith DA, Skalak M, Bu J, Michaud M, Trauger SA, Barshack I, Golan T, Sandbank J, Flaherty KT, Mandinova A, Garrett WS, Thayer SP, Ferrone CR, Huttenhower C, Bhatia SN, Gevers D, Wargo JA, Golub TR, Straussman R. 2017. Potential role of intratumor bacteria in mediating tumor resistance to the chemotherapeutic drug gemcitabine. *Science*. 357(6356):1156–1160. doi:10.1126/science.aah5043.

Gong Y, Jin X, Yuan B, Lv Y, Yan G, Liu M, Xie C, Liu J, Tang Y, Gao H, Zhu Y, Huang Y, Wang WG. 2021. Protein-coupled receptor 109A maintains the intestinal integrity and protects against ETEC mucosal infection by promoting IgA secretion. *Front Immunol*. 11:583652. doi:10.3389/fimmu.2020.583652.

Graf D, Di C R, Fåk F, Flint HJ, Nyman M, Saarela M, Watzl B. 2015. Contribution of diet to the composition of the human gut microbiota. *Microb Ecol Health Dis*. 26:26164. doi:10.3402/mehd.v26.26164.

Harach T, Marungruang N, Duthilleul N, Cheatham V, Mc Coy KD, Frisoni G, Neher JJ, Fåk F, Jucker M, Lasser T, Bolmont T. 2017. Reduction of Abeta amyloid pathology in APPPS1 transgenic mice in the absence of gut microbiota. *Sci Rep*. 7:46856. doi:10.1038/srep46856.

Hsiao EY, McBride SW, Hsien S, Sharon G, Hyde ER, McCue T, Codelli JA, Chow J, Reisman SE, Petrosino JF, Patterson PH, Mazmanian SK. 2013. Microbiota modulate behavioral and physiological abnormalities associated with neurodevelopmental disorders. *Cell*. 155(7):1451–1463. doi:10.1016/j.cell.2013.11.024.

Huang ZB, Hu Z, Lu CX, Luo SD, Chen Y, Zhou ZP, Hu JJ, Zhang FL, Deng F, Liu KX. 2022. Gut microbiota-derived indole 3-propionic acid partially activates aryl hydrocarbon receptor to promote macrophage phagocytosis and attenuate septic injury. *Front Cell Infect Microbiol*. 12:1015386. doi:10.3389/fcimb.2022.1015386.

Jiang H, Ling Z, Zhang Y, Mao H, Ma Z, Yin Y, Wang W, Tang W, Tan Z, Shi J, Li L, Ruan B. 2015. Altered fecal microbiota composition in patients with major depressive disorder. *Brain Behav Immun*. 48:186–194. doi:10.1016/j.bbi.2015.03.016.

Jin J, Xu Z, Zhang L, Zhang C, Zhao X, Mao Y, Zhang H, Liang X, Wu J, Yang Y, Zhang J. 2023. Gut-derived β-amyloid: likely a centerpiece of the gut-brain axis contributing to Alzheimer's pathogenesis. *Gut Microbes*. 15(1):2167172. doi:10.1080/19490976.2023.2167172.

Jyoti S, Dey P. 2025. Mechanisms and implications of the gut microbial modulation of intestinal metabolic processes. *NPJ Metab Health Dis*. 3:24. doi:10.1038/s44324-025-00066-1.

Katafuchi T, Makishima M. 2022. Molecular basis of bile acid-FXR-FGF15/19 signaling axis. *Int J Mol Sci*. 23(11):6046. doi:10.3390/ijms23116046.

Kezer G, Paramithiotis S, Khwaldia K, Harahap IA, Čagalj M, Šimat V, Smaoui S, Elfalleh W, Ozogul F, Esatbeyoglu T. 2025. A comprehensive overview of the effects of probiotics, prebiotics and synbiotics on the gut-brain axis. *Front Microbiol*. 16:1651965. doi:10.3389/fmicb.2025.1651965.

Kim KH, Park D, Jia B, Baek JH, Hahn Y, Jeon CO. 2022. Identification and characterization of major bile acid 7α-dehydroxylating bacteria in the human gut. *mSystems* 7:e00455–e00422. doi:10.1128/msystems.00455-22.

Komada M, Nishimura Y. 2022. Epigenetics and neuroinflammation associated with neurodevelopmental disorders: a microglial perspective. *Front Cell Dev Biol*. 10:852752. doi:10.3389/fcell.2022.852752.

Kumar R, Herold JL, Schady D, Davis J, Kopetz S, Martinez-Moczygemba M, Murray BE, Han F, Li Y, Callaway E, Chapkin RS, Dashwood WM, Dashwood RH, Berry T, Mackenzie C, Xu Y. 2017. *Streptococcus gallolyticus subsp. gallolyticus* promotes colorectal tumor development. *PLoS Pathog*. 13:e1006440. doi:10.1371/journal.ppat.1006440.

Kwon S-K., Park JC, Kim KH, Yoon J, Cho Y, Lee B, Lee JJ, Jeong H, Oh Y, Kim SH, Lee SD, Hwang BR, Chung Y, Kim JF, Nam KT, Lee YC. 2022. Human gastric microbiota transplantation recapitulates premalignant lesions in germ-free mice. *Gut*. 71:1266–1276. doi:10.1136/gutjnl-2021-324489.

Lange O, Proczko-Stepaniak M, Mika A. 2023. Short-chain fatty acids-a product of the microbiome and its participation in two-way communication on the microbiome-host mammal line. *Curr Obes Rep*. 12(2): 108–126. doi:10.1007/s13679-023-00503-6.

Legan TB, Lavoie B, Mawe GM. 2022. Direct and indirect mechanisms by which the gut microbiota influence host serotonin systems. *Neurogastroenterol Motil*. 34(10):e14346. doi:10.1111/nmo.14346.

Li D, Li Y, Yang S, Lu J, Jin X, Wu M. 2022. Diet-gut microbiota-epigenetics in metabolic diseases: from mechanisms to therapeutics. *Biomed Pharmacother*. 153:113290. doi:10.1016/j.biopha.2022.113290.

Li J, Zhang L, Wu T, Li Y, Zhou X, Ruan Z. 2021. Indole-3-propionic acid improved the intestinal barrier by enhancing epithelial barrier and mucus barrier. *J Agric Food Chem*. 69(5):1487–1495. doi:10.1021/acs.jafc.0c05205.

Li Z, Zhang Y, Hong W, Wang B, Chen Y, Yang P, Zhou J, Fan J, Zeng Z, Du S. 2022. Gut microbiota modulate radiotherapy-associated antitumor immune responses against hepatocellular carcinoma Via STING signaling. *Gut Microbes.* 14:2119055.

Lin S, Wang S, Wang P, Tang C, Wang Z, Chen L, Luo G, Chen H, Liu Y, Feng B, Wu D, Burrin DG, Fang Z. 2023. Bile acids and their receptors in regulation of gut health and diseases. *Prog Lipid Res.* 89:101210. doi:10.1016/j.plipres.2022.101210.

Liu J, Tan Y, Cheng H, Zhang D, Feng W, Peng C. 2022. Functions of gut microbiota metabolites, current status and future perspectives. *Aging Dis.* 13(4):1106–1126. doi:10.14336/AD.2022.0104.

Liu XF, Shao JH, Liao YT, Wang LN, Jia Y, Dong PJ, Liu ZZ, He DD, Li C, Zhang X. 2023. Regulation of short-chain fatty acids in the immune system. *Front Immunol.* 14:1186892. doi:10.3389/fimmu.2023.1186892.

Loh JS, Mak WQ, Tan LKS, Ng CX, Chan HH, Yeow SH, Foo JB, Ong YS, How CW, Khaw KY. 2024. Microbiota–gut–brain axis and its therapeutic applications in neurodegenerative diseases. *Sig Transduct Target Ther* 9:37. doi:10.1038/s41392-024-01743-1.

Ma N, He T, Johnston LJ, Ma X. 2020. Host–microbiome interactions: the aryl hydrocarbon receptor as a critical node in tryptophan metabolites to brain signaling. *Gut Microbes.* 11(5):1203–1219. doi:10.1080/19490976.2020.1758008.

Maqsood R, Stone TW. 2016. The gut-brain axis, BDNF, NMDA and CNS disorders. *Neurochem Res.* 41(11):2819–2835. doi:10.1007/s11064-016-2039-1.

Mar JS, Ota, N, Pokorzynski, ND, Peng Y, Jaochico A, Sangaraju D, Skippington E, Lekkerkerker AN, Rothenberg ME, Tan MW, Yi T, Keir ME. 2023. IL-22 alters gut microbiota composition and function to increase aryl hydrocarbon receptor activity in mice and humans. *Microbiome.* 11:47 doi:10.1186/s40168-023-01486-1.

Masse KE, Lu VB 2023. Short-chain fatty acids, secondary bile acids and indoles: gut microbial metabolites with effects on enteroendocrine cell function and their potential as therapies for metabolic disease. *Front. Endocrinol* 14. doi:10.3389/fendo.2023.1169624.

Morris G, Berk M, Carvalho A, Caso JR, Sanz Y, Walder K, Maes M. 2017. The role of the microbial metabolites including tryptophan catabolites and short chain fatty acids in the pathophysiology of immune-inflammatory and neuroimmune disease. *Mol Neurobiol.* 54(6):4432–4451. doi:10.1007/s12035-016-0004-2.

Mukhopadhya I, Louis P. 2025. Gut microbiota-derived short-chain fatty acids and their role in human health and disease. *Nat Rev Microbiol.* 23(10):635–651. doi:10.1038/s41579-025-01183-w.

Nobels A, van Marcke C, Jordan BF, Van Hul M, Cani PD. 2025. The gut microbiome and cancer: from tumorigenesis to therapy. *Nat Metab.* 7(5):895–917. doi:10.1038/s42255-025-01287-w.

Nohesara S, Abdolmaleky HM, Zhou JR, Thiagalingam S. 2023. Microbiota-induced epigenetic alterations in depressive disorders are targets for nutritional and probiotic therapies. *Genes (Basel)* 14(12):2217. doi:10.3390/genes14122217.

Ouyang H, Yang Y, Zhang X, Cui Y, Zhang Y. 2025. Microbial orchestration of neuroimmune crosstalk: from homeostasis to disease. *Front Immunol.* 16:1679286. doi:10.3389/fimmu.2025.1679286.

Painold A, Mörkl S, Kashofer K, Halwachs B, Dalkner N, Bengesser S, Birner A, Fellendorf F, Platzer M, Queissner R, Schütze G, Schwarz MJ, Moll N, Holzer P, Holl AK, Kapfhammer HP, Gorkiewicz G, Reininghaus EZ. 2019. A step ahead: exploring the gut microbiota in inpatients with bipolar disorder during a depressive episode. *Bipolar Disord.* 21(1):40–49. doi:10.1111/bdi.12682.

Palrasu M, Zaika E, el-Rifai W, Garcia-Buitrago M, Piazuelo MB, Wilson KT, Peek RM Jr, Zaika AI. 2020. Bacterial CagA protein compromises tumor suppressor mechanisms in gastric epithelial cells. *J Clin Invest.* 130:2422–2434. doi:10.1172/JCI130015.

Parida S, Wu S, Siddharth S, Wang G, Muniraj N, Nagalingam A, Hum C, Mistriotis P, Hao H, Talbot CC Jr, Konstantopoulos K, Gabrielson KL, Sears CL, Sharma D. 2021. A procarcinogenic colon microbe promotes breast tumorigenesis and metastatic progression and concomitantly activates notch and β-catenin axes. *Cancer Discov.* 11:1138–1157.

Parker A, Fonseca S, Carding SR. 2020. Gut microbes and metabolites as modulators of blood-brain barrier integrity and brain health. *Gut Microbes.* 11(2):135–157. doi:10.1080/19490976.2019.1638722.

Routy B, Le Chatelier E, Derosa L, Duong CPM, Alou MT, Daillère R, Fluckiger A, Messaoudene M, Rauber C, Roberti MP, Fidelle M, Flament C, Poirier-Colame V, Opolon P, Klein C, Iribarren K, Mondragón L, Jacquelot N, Qu B, Ferrere G, Clémenson C, Mezquita L, Masip JR, Naltet C, Brosseau S, Kaderbhai C, Richard C, Rizvi H, Levenez F, Galleron N, Quinquis B, Pons N, Ryffel B, Minard-Colin V, Gonin P, Soria JC, Deutsch E, Loriot Y, Ghiringhelli F, Zalcman G, Goldwasser F, Escudier B, Hellmann MD, Eggermont A, Raoult D, Albiges L, Kroemer G, Zitvogel L. 2018. Gut microbiome influences efficacy of PD-1-based immunotherapy against epithelial tumors. *Science.* 359:91–97.

Sabit H, Tombuloglu H, Rehman S, Almandil NB, Cevik E, Abdel-Ghany S, Rashwan S, Abasiyanik MF, Yee Waye MM. 2021. Gut microbiota metabolites in autistic children: an epigenetic perspective. *Heliyon.* 7(1):e06105. doi:10.1016/j.heliyon.2021.e06105.

Sampson TR, Debelius JW, Thron T, Janssen S, Shastri GG, Ilhan ZE, Challis C, Schretter CE, Rocha S, Gradinaru V, Chesselet MF, Keshavarzian A, Shannon KM, Krajmalnik-Brown R, Wittung-Stafshede P, Knight R, Mazmanian SK. 2016. Gut microbiota regulate motor deficits and neuroinflammation in a model of Parkinson's disease. *Cell.* 167(6):1469–1480.e12. doi:10.1016/j.cell.2016.11.018.

Sarkar A, Yoo JY, Valeria Ozorio Dutra S, Morgan KH, Groer M. 2021. The association between early-life gut microbiota and long-term health and diseases. *J Clin Med.* 3:459. doi:10.3390/jcm10030459.

Sayad A, Najafi S, Hussen BM, Abdullah ST, Movahedpour A, Taheri M, Hajiesmaeili M. 2022. The Emerging roles of the β-secretase BACE1 and the long non-coding RNA BACE1-AS in human diseases: a focus on neurodegenerative diseases and cancer. *Front Aging Neurosci.* 14:853180. doi:10.3389/fnagi.2022.853180.

Shang KM, Liu R, Ni HB, Ma H, Su JW, Yu HL, Guo L, Chen BN, Zhang XX, Yang X. 2025. Bile acid pathways in Caprinae gut microbiota: adaptive shifts in microbial metabolism and community structure. *Front Microbiol.* 16:1648896. doi:10.3389/fmicb.2025.1648896.

Sharma D, Gajjar D, Seshadri S. 2023. Understanding the role of gut microfloral bifidobacterium in cancer and its potential therapeutic applications. *Microbiome Res Rep.* 3(1):3. doi:10.20517/mrr.2023.51.

Shaw C, Hess M, Weimer BC. 2023. Microbial-derived tryptophan metabolites and their role in neurological disease: anthranilic acid and anthranilic acid derivatives. *Microorganisms.* 11(7):1825. doi:10.3390/microorganisms11071825.

Simpson JB, Walker ME, Sekela JJ, Ivey SM, Jariwala PB, Storch CM, Kowalewski ME, Graboski AL, Lietzan AD, Walton WG, Davis KA, Cloer EW, Borlandelli V, Hsiao Y-C, Roberts LR, Perlman DH, Liang X, Overkleeft Bhatt AP, Lu K, Redinbo MR. 2024. Gut microbial β-glucuronidases influence endobiotic homeostasis and are modulated by diverse therapeutics. *Cell Host Microbe.* 32:6 doi:10.1016/j.chom.2024.04.018.

Steliou K, Boosalis MS, Perrine SP, Sangerman J, Faller DV. 2012. Butyrate histone deacetylase inhibitors. *Biores Open Access.* 1(4):192–8. doi:10.1089/biores.2012.0223.

Sun J, Chen S, Zang D, Sun H, Sun Y, Chen J. 2024. Butyrate as a promising therapeutic target in cancer: From pathogenesis to clinic (Review). *Int J Oncol.* 64(4):44. doi:10.3892/ijo.2024.5632.

Thangaleela S, Sivamaruthi BS, Kesika P, Chaiyasut C. 2022. Role of probiotics and diet in the management of neurological diseases and mood states: a review. *Microorg.* 10(11):2268. doi:10.3390/microorganisms10112268.

Thursby E, Juge N. 2017. Introduction to the human gut microbiota. *Biochem J.* 474(11):1823–1836. doi: 10.1042/BCJ20160510.

Van der Hee B, Wells JM. 2021. Microbial regulation of host physiology by short-chain fatty acids. *Trends Microbiol.* 29(8):700–712. doi:10.1016/j.tim.2021.02.001.

Venkatesh M, Mukherjee S, Wang H, Li H, Sun K, Benechet AP, Qiu Z, Maher L, Redinbo MR, Phillips RS, Fleet JC, Kortagere S, Mukherjee P, Fasano A, Le Ven J, Nicholson JK, Dumas ME, Khanna KM, Mani S. 2014. Symbiotic bacterial metabolites regulate gastrointestinal barrier function via the xenobiotic sensor PXR and Toll-like receptor 4. *Immunity.* 21;41(2):296–310. doi:10.1016/j.immuni.2014.06.014.

Wang L, Wang B, Wu C, Wang J, Sun M. 2013. Autism spectrum disorder: neurodevelopmental risk factors, biological mechanism, and precision therapy. *Int J Mol Sci.* 17;24(3):1819. doi:10.3390/ijms24031819.

Wang Q, Hu T, Zhang Q, Zhang Y, Dong X, Jin Y, Li J, Guo Y, Guo F, Chen Z, Zhong P, Yang Y, Ma Y. 2025. Fusobacterium nucleatum promotes colorectal cancer through neogenesis of tumor stem cells. *J Clin Invest.* 135(3):e181595. doi:10.1172/JCI181595.

Wang Q, Yang Q, Liu X. 2023. The microbiota-gut-brain axis and neurodevelopmental disorders. *Protein Cell.* 14(10):762–775. doi:10.1093/procel/pwad026.

Watkeys OJ, Kremerskothen K, Quidé Y, Fullerton JM, Green MJ. 2018. Glucocorticoid receptor gene (NR3C1) DNA methylation in association with trauma, psychopathology, transcript expression, or genotypic variation: a systematic review. *Neurosci Biobehav Rev.* 95:85–122. doi:10.1016/j.neubiorev.2018.08.017.

Whitfield-Cargile CM, Cohen ND, Chapkin RS, Weeks BR, Davidson LA, Goldsby JS, Hunt CL, Steinmeyer SH, Menon R, Suchodolski JS, Jayaraman A, Alaniz RC. 2016. The microbiota-derived metabolite indole decreases mucosal inflammation and injury in a murine model of NSAID enteropathy. *Gut Microbes.* 7(3):246–61. doi:10.1080/19490976.2016.1156827.

Wu G, Wang R, Wang Y, Sun S, Chen J, Zhang Q. 2025. Crosstalk among gut microbiota, microbial metabolites, and inflammatory cytokines: current understanding and future directions. *Foods.* 14:3836. doi:10.3390/foods14223836.

Yang J., Wei H, Zhou Y, Szeto CH, Li C, Lin Y, Coker OO, Lau HCH, Chan AWH, Sung JJY, Yu J. 2022. High-fat diet promotes colorectal tumorigenesis through modulating gut microbiota and metabolites. *Gastroenterology.* 162:135–149.e2. doi:10.1053/j.gastro.2021.08.041.

Yoo JY, Groer M, Dutra SVO, Sarkar A, McSkimming DI. 2020. Gut microbiota and immune system interactions. *Microorganisms.* 8(10):1587. doi:10.3390/microorganisms8101587.

Yu Q, Newsome RC, Beveridge M, Hernandez MC, Gharaibeh RZ, Jobin C, Thomas RM. 2022. Intestinal microbiota modulates pancreatic carcinogenesis through intratumoral natural killer cells. *Gut Microbes.* 14:2112881.

Yu T, Guo F, Yu Y, Sun T, Ma D, Han J, Qian Y, Kryczek I, Sun D, Nagarsheth N, Chen Y, Chen H, Hong J, Zou W, Fang JY. 2017. Fusobacterium nucleatum promotes chemoresistance to colorectal cancer by modulating autophagy. *Cell.* 170:548–563.e516.

Zangerolamo L, Carvalho M, Barbosa HCL. 2025. The critical role of the bile acid receptor TGR5 in energy homeostasis: insights into physiology and therapeutic potential. *Int J Mol Sci.* 26(14):6547. doi:10.3390/ijms26146547.

Zeng R, Gou H, Lau HCH, Yu J. 2024. Stomach microbiota in gastric cancer development and clinical implications. *Gut* 73:2062–2073.

Zheng Y, Bonfili L, Wei T, Eleuteri AM. 2023. Understanding the gut-brain axis and its therapeutic implications for neurodegenerative disorders. *Nutrients.* 15(21):4631. doi:10.3390/nu15214631.

Zhou M, Niu B, Ma J, Ge Y, Han Y, Wu W, Yue C. 2025. Intervention and research progress of gut microbiota-immune-nervous system in autism spectrum disorders among students. *Front Microbiol.* 16:1535455. doi:10.3389/fmicb.2025.1535455.

4 Current Status, Impact, and Significance of Gut-Brain Axis

Namrata Gupta, Preeti Solanki, Sachin Pahal, Sajpreet Kour, and Santosh Kumar

4.1 INTRODUCTION

Gut–brain axis (GBA) can be defined as a highly complex bidirectional network that allows communication between the central nervous system (CNS), the intestine, and the gut microbiota. GBA constitutes the autonomic and enteric nervous system (ENS), the endocrine system, the hypothalamic–pituitary–adrenal axis (HPA), the gut microbiota, and its associated metabolites. The metabolites produced by the gut microbiota range from essential vitamins, secondary bile acids, and amino acids to various short-chain fatty acids (SCFAs), which play a drastic role in influencing major immune pathways, resulting in alterations in an individual's behavior, memory, learning, locomotion, and brain development. Research has shown that any significant changes in the gut microbiota composition can highly alter the patient's response to a particular disease condition, further emphasizing the importance of its healthy maintenance (Rutsch et al., 2020).

Over 2000 years ago, Hippocrates famously stated, "All disease starts in the gut," highlighting the pivotal role of the gut in overall health. This statement marked the beginning of human curiosity about the gut and its associated microbiota. In the 1980s, technological advancements, particularly in brain imaging, enabled researchers to study the GBA in greater detail, providing insights into disorders like irritable bowel syndrome (IBS) and other conditions linked to dysregulated gut microbiota (Cryan et al., 2019). The gut and the brain are intricately connected through a complex network of neurotransmitters and signaling molecules, largely influenced by the 500 million neurons present in the intestine. This connection is primarily mediated by the nervous system, with the vagus nerve being the largest conduit for bidirectional communication between the gut and the brain. Research has demonstrated that disruptions in gut microbiota can significantly impact brain function and overall mental health (Pellissier et al., 2014).

Another study found that the production of gamma-aminobutyric acid (GABA)—a neurotransmitter synthesized by gut microbes can be enhanced through probiotic supplementation, potentially offering a therapeutic approach for managing anxiety and depression-like behaviors (Janik et al., 2016). Beyond neurotransmitter production, gut microbes contribute to metabolic processes that influence brain function. They help break down dietary fiber into SCFAs such as acetate, propionate, and butyrate, which play a key role in maintaining the blood–brain barrier (BBB) and regulating appetite suppression by inducing a sense of fullness. Additionally, these microbes assist in bile acid metabolism, producing compounds that impact brain activity (Figure 4.1).

Growing evidence has revealed the bidirectional communication between the gut microbiome and the CNS, known as the "microbiota–gut–brain axis" (MGBA) (Ahmed et al., 2022). Despite being anatomically distinct, the gut and the brain interact through multiple pathways. These include modulation of the immune system, vagus nerve, ENS, neuroendocrine signaling, and circulatory system, facilitated by the production of neuroactive compounds, metabolites, and hormones (Figure 4.2). Any major disruption in the GBA can lead to several disease conditions, such as several mood conditions, anxiety, depression, and autism spectrum disorder (ASD), along with major GI disorders, such as IBS, and neurodegenerative diseases like Alzheimer's disease and Parkinson's disease (PD) (Cryan et al., 2019). A deeper understanding of the GBA is essential for exploring the

DOI: 10.1201/9781003610274-5

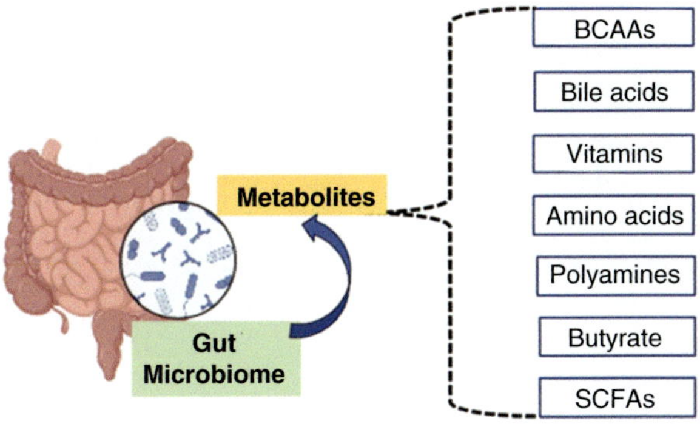

FIGURE 4.1 Gut microbiota and their metabolites. SCFAs, short-chain fatty acids; BCAAs, branched-chain amino acids.

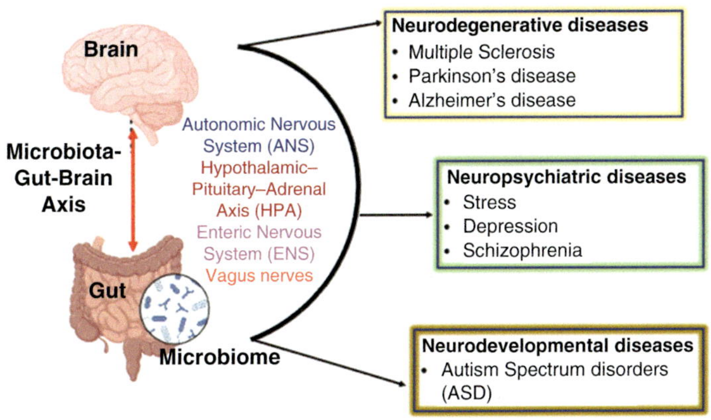

FIGURE 4.2 The gut microbiota–brain axis represents a bidirectional communication system between the gut microbiome and the brain, facilitated by the vagus nerve, ENS, HPA, and ANS.

physiological and neurocognitive effects of gut dysbiosis, and targeting gut microbiota may offer novel therapeutic strategies for improving mental and neurological health.

4.2 COMPONENTS OF THE GUT–BRAIN AXIS

The GBA is represented by a complex mechanism of dynamic communication between the ENS and the CNS, which is necessary for maintaining a constant state of homeostasis inside the body to allow proper control over body processes like digestion, immunity, and emotional responses (Obadia et al., 2024), helps in receiving, sending, and controlling major gastrointestinal (GI) system through the enteric, vagal, and spinal nerves interacting together through the sympathetic and parasympathetic nervous system (Figure 4.3).

4.2.1 Enteric Nervous System (ENS)

The ENS, often referred to as the "first brain" due to its evolutionary significance, is a highly intricate network of neurons and glial cells embedded within the walls of the GI tract. It also communicates regularly with both the sympathetic and parasympathetic nervous systems. The sympathetic system inhibits gut motility and secretion during stress, redirecting resources away from

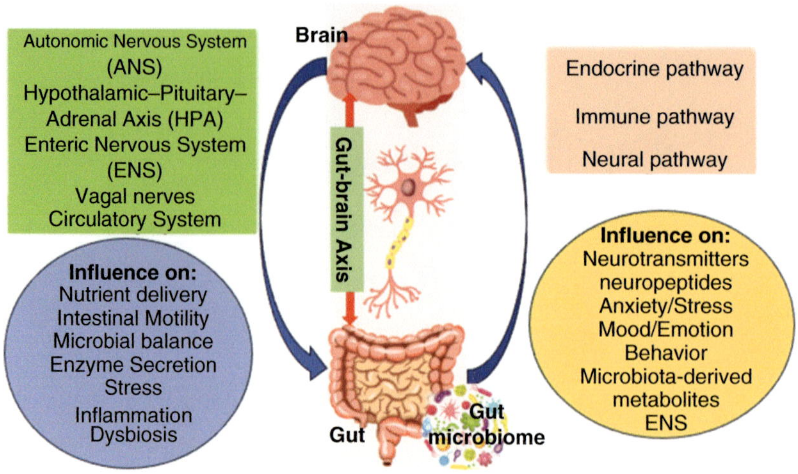

FIGURE 4.3 Schematic diagram illustrating the communication between the gut and the brain, incorporating the bidirectional pathways such as the autonomic nervous system (ANS), ENS, hypothalamic–pituitary–adrenal (HPA) axis, immune pathways, endocrine pathways, and neural pathways.

digestion when needed, whereas the parasympathetic system induces digestive functions. These autonomic inputs ensure that the gut adapts to changing physiological and environmental conditions (Figure 4.3) (Sternini, 1997; Million and Larauche, 2016).

4.2.2 Vagus Nerve

The main pathway for the transfer of information between the gut and the brain is the vagus nerve, which is necessary for the maintenance and regulation of key body functions such as digestion, the immune system, and food metabolism. It constitutes vagal sensory neurons present in various organs like the liver, pancreas, and kidney, where they help detect mechanical stimuli like gut distension or chemical signals like changes in nutritional levels and hormones (Browning et al., 2017; Berthoud and Neuhuber, 2000; Cao et al., 2024).

Largely vagal sensory neurons can be classified into different subtypes based on their response profiles and molecular markers present including: mechanosensitive—detect physical changes in the gut; chemosensitive—detect chemical signals in the gut; immunesensitive—detect immune signals and inflammatory response; neuroendocrine—interaction with neuroendocrine cells and enteroendocrine cells; polymodal neurons—detect multiple mechanical and chemical signals in response to the gut's internal environment (Wang et al., 2020) (Figure 4.3).

4.2.3 Hypothalamic–Pituitary–Adrenal (HPA)

The HPA acts as an essential component of stress responses by the body and also plays an essential role in maintaining the GBA through the gut microbiota. The activation of HPA is dependent on corticotropin-releasing hormone, secreted by the hypothalamus, which in turn triggers the pituitary gland to release adrenocorticotropic hormones responsible for the control of cortisol release by the adrenal gland. Increased HPA activity due to any chronic stress can lead to excessive cortisol secretion, which in turn affects the immune system as well as the gut microbiota, leading to an array of gut disorders due to its dysregulation (Rusch et al., 2023).

These studies highlight the crucial role of gut microbiota in modulating the HPA axis, especially during early developmental stages when both microbial colonization and stress-response systems are being established (Rosin et al., 2021). Additionally, probiotic strains from genera such as

Lactobacillus and Bifidobacterium have been demonstrated to attenuate HPA axis activity, leading to lower cortisol levels and mitigating stress-related behaviors (Figure 4.3) (Rusch et al., 2023).

4.2.4 Microbiome as a Mediator–Gut–Brain Communication and Health

The gut microbiome, composed of diverse microbial communities throughout the GI tract, plays a vital role in maintaining metabolic and physiological balance. While each individual has a unique microbial profile, healthy individuals exhibit a common pattern of microbial distribution. Notably, Firmicutes and Bacteroidetes dominate the gut microbiome, accounting for approximately 75% of the total microbial population (Xiong et al., 2023). These microorganisms are not passive inhabitants but actively regulate critical bodily functions, particularly through their influence on gut–brain communication (Kurdi et al., 2024).

Early evidence emerged from studies on hepatic encephalopathy patients, where treatment with oral antibiotics led to significant cognitive improvements, underscoring the gut–brain connection (Morgan, 1991). Since then, extensive research has demonstrated that gut microbiota imbalances, or dysbiosis, can impact mood, anxiety, depression, and neurodevelopmental disorders such as autism, with severity often correlating with microbial composition (Xiong et al., 2023). A deeper understanding of this intricate communication network offers promising avenues for therapeutic interventions targeting both neurological and GI conditions.

4.3 MECHANISMS OF COMMUNICATION

Neuromodulators and neurotransmitters are the main components involved in the maintenance of crosstalk between the CNS and the ENS, therefore acting as essential messengers impacting both the neurological and GI functions. These can be in the form of microbial metabolites, neurotransmitters, or hormones, which control and coordinate the brain and gut functions. The gut produces approximately 90% of the body's serotonin and acetylcholine, a key neuromodulator in the GBA, primarily regulating GI functions such as motility, secretion, pain perception, and gut smooth muscle contractions (Yano et al., 2015; Sam and Bordoni, 2023). Dopamine is also produced in the gut and is required for the proper regulation of gut movement and also for the protection of gut mucosal lining, highlighting its importance in the maintenance of gut immunity and the regulation of inflammatory responses (Hamamah et al., 2022; Moncrieff et al., 2023; Hanna-Jairala and Drossman, 2024).

GABA, a crucial inhibitory neurotransmitter in the CNS, is also synthesized in the gut, primarily by microbial species such as *Lactobacillus* and *Bifidobacterium*. Gut-derived GABA has been linked to the modulation of stress responses, as well as the regulation of depression and anxiety. In the ENS, the GABAergic system influences gut motility and secretions, while in the CNS, fluctuations in GABA levels are associated with various neuropsychiatric conditions (Liu et al., 2023).

Thus, by ensuring an intricate balance between neurotransmitters and neuromodulators, gut function can be maintained to ensure overall homeostasis through the GI tract, therefore proving the significance of gut–brain signaling pathways in overall health maintenance.

4.4 GUT MICROBIOTA AND BRAIN HEALTH

4.4.1 Composition of the Gut Microbiota

Approximately 2000 species of bacteria, fungi, archaea, and viruses can be found inside the human gut, forming a complex microbial community, which contains nearly 150 times more genetic material than the human genome. Their concentrations may range from 10^1 to 10^3 cells per gram in the upper intestine to 10^{11}–10^{12} per gram in the colon. The majority of bacteria found in the gut belong to phyla Firmicutes (species such as *Lactobacillus, Clostridium*, and *Enterococcus*) and Bacteroidetes (species such as *Bacteroides*) (Xiong et al., 2023; Kurdi et al., 2024). Bacteria from other phyla like Actinobacteria (Bifidobacteria), Proteobacteria (*Escherichia coli*), Fusobacteria, Verrucomicrobia, and Cyanobacteria are also present but comparatively in lower concentrations (Qin et al., 2010).

Research has shown that they help enhance the amino acids and vitamins production in infants helping to improve their immune system and disease prevention, with production starting as early as 4 months since birth, such as nutrient absorption, immunity development, maintenance of intestinal barrier, metabolic rates, production of antimicrobial compounds, metabolism of cholesterol, and drugs (Loh et al., 2024).

4.4.2 Dysbiosis and Its Consequences

It is essential to maintain proper and continuous crosstalk between the ENS and the CNS to allow proper integration of gut and brain functions to not only maintain the digestion process but also ensure proper mental health and immunity maintenance. Any disruptions in the communication can lead to dysbiosis, which is associated with a wide array of GI, nervous, and mental disorders. For instance, the majority of symptoms associated with GI disorders like IBS, inflammatory bowel disease, anxiety, and depression can be largely due to dysregulation of gut–brain signaling, leading to changes in gut motility, visceral hypersensitivity, and altered immune responses (Raskov et al., 2016; Chong et al., 2019).

Research has proved that proper microbial colonization is essential for the proper development of both ENS and CNS, which in turn affects neurotransmitter production ultimately regulating gut motor functions (Chen et al., 2021). Thus, by properly understanding this complex communication network, novel therapeutic approaches can be developed aimed at targeting both neurological and GI disorders with higher efficacy.

4.5 GBA AND DISEASE CONNECTIONS

Proper development and maintenance of gut microbiota at early stages of life is necessary to ensure positive long-lasting effects on various physiological aspects of an individual. In addition, various clinical, epidemiological, and immunological evidence have been established that prove that the gut microbiota can significantly affect the mental, emotional, and neurological functioning of an individual and is often associated with various neurological, GI, and mental disorders (Table 4.1).

4.5.1 Neurological Disorders

Parkinson's disease: PD is described as a progressive loss of motor and nonmotor features, ultimately causing the total loss of muscle control and mobility. The most commonly associated hallmark for PD diagnosis is a Lewy pathology body, which is revealed as an aggregate of misfolded α-synuclein (α-SYN) protein in the CNS (Dauer and Przedborski, 2003) and even in certain regions of the GI tract (Stokholm et al., 2016). Age is generally considered the single most important risk factor for PD. The median onset age observed in PD patients is 60 years, mainly due to accumulative oxidative stress. Largely, diagnosis of PD depends on differential and comprehensive analysis of the patient's history and physical examination using the UPDRS (Unified Parkinson's Disease Rating Scale) (Jankovic and Tan, 2020). In advanced PD patients, usually the entire GI tract is affected often leading to conditions like sialorrhea or dysphagia.

Recent medical studies have proved the multifaceted pathophysiological nature of PD as the majority of GI tract symptoms can be directly correlated with the disease conditions. These GI symptoms can also prevent proper diagnosis and interfere with efficient treatment of other symptoms like motor loss (Tan et al., 2023). Further, studies have found that alterations in microbiota from the gut can have an additional impact on the pathogenesis of PD in conditions where the relative increase in the numbers of Firmicutes, *Lactobacillus*, *Oscillospira*, *Bifidobacterium*, and Enteriobacteriaceae

TABLE 4.1
Correlation of Gut-Brain Axis and Disease Connections (Summarizing the Correlation Between the Gut–Brain Axis (GBA) and Various Diseases)

Disease	GBA Involvement	Mechanisms	References
Neurodegenerative diseases (Alzheimer's disease, Parkinson's disease, multiple sclerosis)	Gut dysbiosis and inflammation	Agglomeration of tau protein and amyloid-β plaques, α-synuclein aggregation/overexpression, immune activation, altered gut barrier integrity, vagus nerve signaling	Honda and Littman (2016), Malek (2019)
Autism spectrum disorders (ASDs)	Altered gut microbiota permeability	Microbial metabolites affecting neurotransmitters, immune dysregulation, increased intestinal permeability	Bundgaard et al. (2023)
Anxiety and depression	Gut microbial dysbiosis leading to altered neurochemistry	Changes in the concentration of serotonin, GABA production; inflammation; HPA axis dysregulation	Mohan et al. (2023)
Irritable bowel syndrome (IBS)	Disruption in gut microbiome and motility	Enteric nervous system dysfunction, disruption of neuronal sensitization, altered serotonin signaling, chronic low-grade inflammation	Honda and Littman (2016)
Chronic fatigue syndrome (CFS)	Gut microbial imbalance linked to energy metabolism	Microbiome-induced mitochondrial dysfunction, impaired energy production neuroinflammation, increased gut permeability	Martín et al. (2023)
Schizophrenia	Alteration in SCFAs, dysbiosis affecting neurotransmitter pathways	Brain abnormalities, gut-derived inflammation, changes in dopamine/glutamate signaling, oxidative stress, GABAergic dysfunction	Ju et al. (2023)
Stroke	Microbiota alterations and dysbiosis influencing recovery	Gut dysbiosis affecting neuroinflammation, BBB integrity, and metabolic responses	Olanow et al. (2014)
Obesity and metabolic syndrome	Gut microbiota affecting energy homeostasis	Changes in gut-derived hormones (GLP-1, PYY), systemic inflammation, insulin resistance, metabolism	Çakır et al. (2017)
Multiple sclerosis (MS)	Gut microbiota imbalance triggering immune response	CNS function, microbiota-induced autoimmunity, increased gut permeability, cytokine imbalance, systemic inflammation	Honda et al. (2016)

GLP-1, glucagon-like peptide 1; PYY, peptide YY.

was observed when compared to Bacteriods, *Clostridium*, and *Enterococcaceae* (Kuai et al., 2021). Therefore, proper and detailed research is needed in the field of PD to enable a proper understanding of the connection between microbiota and GI dysfunction in PD patients.

Alzheimer's disease (AD): Alzheimer is a type of dementia mainly associated with complex degradation of neurological tissues. The main causative reasons for Alzheimer's may vary from genetic makeup, education, age, and environment to a combination of two or more factors. It is mainly divided into two types, namely, Familial Alzheimer's Disease and Sporadic Alzheimer's Disease (Khodabakhsh et al., 2021). Apolipoprotein ε4 is the most common gene considered responsible for AD with the presence of around 40% of the total individuals suffering from AD. The currently used drugs for AD treatment approved by the FDA include acetylcholinesterase inhibitors and N-methyl-D-aspartate receptor antagonists (Samanta et al., 2006).

New studies have proved that modifications in microbiota composition or high load pathogenic infections can act as potential causes of AD (Rutsch et al., 2020). A significant reduction in levels of CFAs-producing species has been observed in the guts of AD patients along with decreased levels of serotonin and GABA in patients' urine and serum, thus proving that the role of gut microbiota in AD. This has led to the emergence of new therapeutic strategies focused on treating neurodegenerative diseases by restoring the gut microbiota to a hemostatic state using probiotics to provide protection against and prevent further neurodegeneration and cognitive impairment in AD patients (Loh et al., 2024).

Autism spectrum disorder (ASD): ASD comprises a group of complex neurodevelopmental disorders characterized by impaired social communication, repetitive behaviors, and restricted interests, with recent evidence suggests a strong connection between ASD and gut microbiota, reinforcing the bio-psycho-social approach in ASD treatment. GI dysfunction, including constipation and abdominal pain, is frequently reported in ASD patients, indicating gut dysbiosis and altered GI permeability.

Notably, gut microbiota transplantation studies have demonstrated that transferring microbiota from ASD individuals to mouse models induces ASD-like behaviors, reinforcing the potential causative role of gut microbiota in ASD pathology. Conversely, transferring healthy gut microbiota to affected individuals has been associated with symptomatic improvement, including reduced anxiety and irritability. Although a strong correlation has been established between gut microbiota, GI permeability, and ASD symptoms, the precise mechanisms of action and regulatory pathways remain unclear (Bundgaard et al., 2023).

Multiple sclerosis: Multiple sclerosis (MS) is a chronic autoimmune, inflammatory, and neurodegenerative disorder of the CNS, encompassing the brain and spinal cord, where the immune system mistakenly attacks the myelin sheath, the protective layer that insulates nerve fibers resulting in a broad spectrum of neurological symptoms. Recent studies have investigated various experimental models of MS and explored the efficacy of novel immunomodulatory therapies to better understand the involvement of the MGBA in the disease pathophysiology. Notably, the abnormal immune activity seen in MS can be influenced by bacterial metabolites produced in the gut. Emerging evidence indicates that dysbiosis and imbalance in gut microbial composition may contribute to the onset and progression of MS by promoting systemic inflammation and autoimmune responses (Suganya and Koo, 2020).

4.5.2 Psychiatric Disorders

Depression and anxiety: Based on Barlow's concept, anxiety can be described as a state of mind associated with preparation for any possible upcoming negative events leading to a state of false future-oriented mood. In contrast to this, depression is often associated with a sense of expression of ego-helplessness and ego-powerlessness (Schwasinger and Macaluso, 2019).

For instance, obsessive-compulsive disorder (OCD) is a debilitating neuropsychiatric disorder characterized by recurrent, intrusive obsessions that lead to compulsive, repetitive behaviors or avoidance strategies (Chen et al., 2023). Research indicates that OCD patients exhibit altered gut microbiota, with a reduced abundance of butyrate-producing bacteria, which may contribute to elevated GABA and serotonin synthesis or impaired hydroxybutyric acid metabolism. Additionally, dietary restrictions and gut dysbiosis in OCD patients may lead to deficiencies in essential micronutrients such as vitamin B12, E, and C, further compromising immune function (Chen et al., 2023). This growing body of evidence highlights the potential role of gut microbiota in OCD pathophysiology, suggesting that microbiome-targeted interventions could serve as adjunctive therapeutic strategies for managing OCD.

A new study has shown that probiotics can possibly be used to treat anxiety or depression conditions as they can help modify and improve the gut microbiota population thereby helping treat associated psychiatric symptoms but there is still a lack of proper data to prove the validity of this hypothesis (Haug et al., 2002).

Schizophrenia: Schizophrenia (SZ) is generally characterized by delusions, psychosis, and cognitive dysfunction, leading to a complex, chronic, and debilitating condition resulting in apathy and withdrawal causing both physical and social morbidity. The major associated cause is genetic factors but, in some cases, it can be caused due to early life stage infections, but the main mechanism associated with this is still unclear (Severance et al., 2015).

Most commonly, SZ patients also have very poor attention span with increased chances of developing disorders like depression and anxiety, leading to strong suicidal ideation. Postmortem examinations of SZ patients have also revealed increased oxidative stress, high lipid peroxidation, and antioxidant activity alteration along with brain abnormalities (Sawamura et al., 2022).

Currently, antipsychotic drugs and tranquilizers are mostly used for their treatment but a number of side effects like weight gain, diabetes, atherosclerosis, and glucose metabolism dysregulation arise due to the prolonged use of antipsychotics. These can even lead to gut microbial dysbiosis largely due to the long-term use of antibiotics and changes in diet leading to modified levels of neurotransmitters causing altered brain function (Munawar et al., 2021). New preclinical trials have shown that probiotics use on a regular basis can help improve GI symptoms associated with SZ patients, but still further research is needed to ensure their effectiveness and mechanism of action (Ju et al., 2023).

4.5.3 Gastrointestinal Disorders and Their Neurological Impact

IBS: IBS is a functional GI disorder mainly associated with abdominal pain, changes in the stool's appearance, the presence of mucus, diarrhea, or constipation or a combination of both, leading to irregular bowel activity which affects the overall well-being and health of the patient. Studies have shown that IBS patients are at a higher risk of suffering from conditions such as anxiety and depression; therefore, it is often called a disorder of gut–brain interaction (Nathani et al., 2025).

Studies found that an increase in the number of *Enterobacteriaceae* taxa inside the gut can potentially contribute to disease progression in affected individuals when compared to that of controls (Soussou et al., 2023). In addition to this, an increased serine protease activity can also lead to intestinal barrier disruption, leading to loss of its functioning and neuronal sensitization as reported in IBS cases (Zhai et al., 2023). Further, the connection between the IBS and the GBA was observed by higher plasma levels of epinephrine, norepinephrin, and catecholamines observed in IBS patients (Pellissier et al., 2014).

New studies have provided evidence that the gut microbiome is a possible causative link between IBS development and gut dysbiosis, with the GBA acting as the primary physiological link. This new information has allowed for the creation of a new therapeutic approach involving the use of probiotics supplementation with some trials even indicating improved mood and relieved GI symptoms (Nathani et al., 2025).

4.5.4 Influence of Gut Microbiota on the Gut–Brain Axis

Over the years, significant efforts have been made to understand the intricate connection between the brain and the GI tract. The gut microbiome plays a crucial role in this communication, influencing the CNS through various regulatory pathways. The MGBA represents a bidirectional network, primarily mediated by microbial metabolites such as polyphenolic derivatives, amines, vitamins, SCFAs, neuromodulators, and serotonin precursors produced via modifications in the tryptophan pathway (Gwak and Chang, 2021). Psychobiotics, a class of live microorganisms, have emerged as potential therapeutic agents for mental health disorders, offering benefits by modulating stress-related responses and reducing cortisol levels. These beneficial microbes contribute to the production of key neurotransmitters and hormones, including GABA, oxytocin, and serotonin, which interact along the GBA to regulate mood and cognitive function (Oroojzadeh et al., 2022). Several studies have demonstrated the positive impact of probiotics in alleviating anxiety and depression (Ng et al., 2018). Furthermore, research by Goyal et al. (2021) highlights the role of gut bacteria in influencing neural circuits that may delay the onset of neurodegenerative diseases such as Alzheimer's. Given the growing evidence of gut microbiota's impact on neurological health, further in-depth clinical research is necessary to elucidate the precise mechanisms underlying these interactions and their therapeutic potential.

4.6 SIGNIFICANCE FOR HEALTH AND MEDICINE

In recent decades, efforts have been made to properly understand the interconnectivity between the brain and the GI tract by studying the GBA using various regulatory pathways such as metabolic product secretion, modulation of immune mechanisms, and modulation of HPA. The GBA enables the interaction of gut microbe-derived metabolites like polyphenolic derivatives, amines, vitamins, SCFAs, neuromodulators, and serotonin precursors to regulate and alter brain functions and help prevent and mitigate disease conditions (Gwak and Chang, 2021).

Gut microbiota has a potential role in maintaining brain function through the production of the neurotransmitter that makes a connection either by BBB (only for GABA) or ENS system. Different gut microbe is responsible for the production of a wide range of neurotransmitters, which are necessary for proper mental health such as acetylcholine produced by *Lactobacillus* (Fujita et al., 2024), GABA produced by *Lactobacillus* and Bifidobacterium (Dinan and Cryan, 2017), dopamine from *Serratia* and *Bacillus* (Bhombal et al., 2023), and serotonin by *Enterococcus*, *Escherichia*, *Candida,* and *Streptococcus* (Kelly et al., 2015). Therefore, it is necessary to better understand the association between changes in gut microbiota and their potential benefits for illnesses through deeper and more precise clinical research.

4.7 CURRENT STATUS OF THE GBA

Currently, the GBA is represented as a bidirectional system of communication allowing for complex link formation between the CNS and the GI tract (Cryan et al., 2019). At present, the ENS, immune pathways, endocrine signals along with gut microbiota composition is believed to alter and influence its functioning although there is still a lack of concrete evidence to indicate their specific working mechanisms. Till this point, the focus of research has been the diverse population of microorganisms found in the GI tract called the gut microbiota with main aim of identifying their role in affecting the GBA and association with a number of metabolic, neurological, and mental disorders, through a wide range of bioactive compounds like dopamine, serotonin, and most studied GABA (Morais et al., 2021). In addition to this, microbial fermentation within the gut has been associated with the production of SCFAs through the process of dietary fiber breakdown, providing an array of health benefits ranging from maintenance of BBB to regulation of inflammatory responses (Mhanna et al., 2024).

4.8 EMERGING TRENDS IN GBA RESEARCH

In the recent decade, significant advances have been observed in the field of GBA research, allowing for greater understanding regarding its functioning and regulation. Psychobiotics have emerged as a promising field in GBA studies with promising results in the field of mental health by allowing maintenance of healthy gut microbiota levels through the use of specific probiotics and prebiotics mix (Sarkar et al., 2020). An emerging trend has been observed in the study and production of neuroactive substances through bacterial strains of *Bifidobacterium* and *Lactobacillus* for possible applications in the field of anxiety and depression treatment. Another major trend has been found in the field of personalized nutrition with increased focus on an individual's microbiome profile as a possible treatment method to cure or possibly prevent neuropsychiatric disorders (Dinan and Cryan, 2021). Recent advances in omics technology have created a new field involving the integration of genomics, proteomics, and metabolomics, to study and analyze the role of gut microbes in the GBA and how they impact their host.

4.9 CHALLENGES IN GBA RESEARCH

Although there have been significant advancements in our understanding regarding the GBA, there are still a wide range of challenges that need to be addressed to allow proper utilization of GBA data in actual clinical applications. The main challenge associated is the lack of evidence proving a direct relationship between the gut microbiota alterations and changes in brain function, with current literature only indicating correlations between the two (Foster et al., 2021). Also, most of the currently available data is through observational studies, resulting in determining whether any changes in microbiota are the cause of the neurological condition or its associated consequence, making it difficult to apply it for possible therapeutic purposes. The major challenge observed is ethical and regulatory concerns associated with interfering with an individual's microbiota as it may have potential long-term implications on patient safety and well-being. Thus, all the challenges need to be properly addressed to allow for utilization of GBA research to its full potential to enable more safe and efficient clinical applications.

4.10 CONCLUSION

In recent decades, the GBA has emerged as a novel field of research through its ability to influence both the gut microbiota and brain function of an individual along with its role in several GI, neurological, and mental disorders. Although, current research mainly focuses on determining possible roles of gut microbiota in determining mental health and neurological development of an individual thorough analyzing production of neurotransmitter, associated immune responses, and metabolic signaling; a significant trend is being observed in the field of personalized nutrition, psychobiotics, and microbiome-based therapies as possible clinical applications of GBA function to treat and prevent a wide range of neurological disorders with more efficiency. However, more detailed studies along with better ethical and regulatory guidelines are required to ensure proper utilization of the current knowledge on the GBA. Thus, future research and development in the field of GBA can have far reaching implications in medical science.

REFERENCES

Ahmed, H., Leyrolle, Q., Koistinen, V., Kärkkäinen, O., Layé, S., Delzenne, N., & Hanhineva, K. (2022). Microbiota-derived metabolites as drivers of gut–brain communication. *Gut Microbes*, 14(1), 2102878. https://doi.org/10.1080/19490976.2022.2102878.

Berthoud, H.-R., & Neuhuber, W. L. (2000). Functional and chemical anatomy of the afferent vagal system. *Autonomic Neuroscience*, 85(1), 1–17. https://doi.org/10.1016/S1566-0702(00)00215-0.

Bhombal, Y. N., Patil, Y. P., Barvkar, V. T., Mhaske, M. M., Jadhav, J. P., & Pawar, K. D. (2023). Prodigiosin from *Serratia rubidaea* MJ 24 impedes *Helicoverpa armigera* development by the dysregulation of Juvenile hormone-dopamine system. *Microbiological Research*, 274, 127422. https://doi.org/10.1016/j.micres.2023.127422.

Browning, K. N., Verheijden, S., & Boeckxstaens, G. E. (2017). The vagus nerve in appetite regulation, mood, and intestinal inflammation. *Gastroenterology*, 152(4), 730–744. https://doi.org/10.1053/j.gastro.2016.10.046.

Bundgaard-Nielsen, C., Lauritsen, M. B., Knudsen, J. K., Rold, L. S., Larsen, M. H., Hindersson, P., Villadsen, A. B., Leutscher, P. D. C., Hagstrøm, S., Nyegaard, M., & Sørensen, S. (2023). Children and adolescents with attention deficit hyperactivity disorder and autism spectrum disorder share distinct microbiota compositions. *Gut Microbes*, 15(1), 2211923. https://doi.org/10.1080/19490976.2023.2211923.

Çakır, M., Aksel İşbilen, A., Eyüpoğlu, İ., Sağ, E., Örem, A., Mazlum Şen, T., Kaklıkkaya, N., & Kaya, G. (2017). Effects of long-term synbiotic supplementation in addition to lifestyle changes in children with obesity-related non-alcoholic fatty liver disease. *Turkish Journal of Gastroenterology*, 28(5), 377–383. https://doi.org/10.5152/tjg.2017.17084.

Cao, Y., Li, R., & Bai, L. (2024). Vagal sensory pathway for the gut-brain communication. *Seminars in Cell & Developmental Biology*, 156, 228–243. https://doi.org/10.1016/j.semcdb.2023.07.009.

Chen, L. L., Abbaspour, A., Aspvall, K., Rück, C., Bulik, C. M., & Pascal, D. (2023). Longitudinal study of gut microbiome in obsessive–compulsive disorder. *Brain and Behavior*, 13(8), e3115. https://doi.org/10.1002/brb3.3115.

Chen, Y., Xu, J., & Chen, Y. (2021). Regulation of neurotransmitters by the gut microbiota and effects on cognition in neurological disorders. *Nutrients*, 13(6), 2099. https://doi.org/10.3390/nu13062099.

Chong, P. P., Chin, V. K., Looi, C. Y., Wong, W. F., Madhavan, P., Yong, V. C., & Ambu, S. (2019). The microbiome and irritable bowel syndrome: A review on the pathophysiology, current research and future therapy. *Frontiers in Microbiology*, 10, 1136. https://doi.org/10.3389/fmicb.2019.01136.

Cryan, J. F., Cowan, C. S. M., Sandhu, K. V., Bastiaanssen, T. F. S., Boehme, M., Codagnone, M. G., .. & Dinan, T. G. (2019). The microbiota-gut-brain axis. *Physiological Reviews*, 99(4), 1877–2013. https://doi.org/10.1152/physrev.00018.2018.

Dauer, W., & Przedborski, S. (2003). Parkinson's disease: Mechanisms and models. *Neuron, 39*(6), 889–909. https://doi.org/10.1016/s0896-6273(03)00568-3.

Dinan, T. G., & Cryan, J. F. (2017). The microbiome–gut–brain axis in health and disease. *Gastroenterology Clinics of North America*, 46(1), 77–89. https://doi.org/10.1016/j.gtc.2016.09.007.

Dinan, T. G., & Cryan, J. F. (2021). Gut microbiota: A missing link in psychiatry. *Australian & New Zealand Journal of Psychiatry*, 55(1), 23–29. https://doi.org/10.1002/wps.20726 (https://pmc.ncbi.nlm.nih.gov/articles/PMC6953563/).

Foster, J. A., Rinaman, L., & Cryan, J. F. (2021). Stress & the gut-brain axis: Regulation by the microbiome. *Neurobiology of Stress*, 14, 100325. https://doi.org/10.1016/j.ynstr.2021.100325.

Fujita, Y., Kosakamoto, H., & Obata, F. (2024). Microbiota-derived acetylcholine can promote gut motility in *Drosophila melanogaster*. *Philosophical Transactions of the Royal Society B: Biological Sciences*, 379(1901), 20230075. https://doi.org/10.1098/rstb.2023.0075.

Goyal, D., Ali, S. A., & Singh, R. K. (2021). Emerging role of gut microbiota in modulation of neuroinflammation and neurodegeneration with emphasis on Alzheimer's disease. *Progress in Neuropsychopharmacology & Biological Psychiatry*, 106, 110112. https://doi.org/10.1016/j.pnpbp.2020.110112.

Gwak, M. G., & Chang, S. Y. (2021). Gut-brain connection: Microbiome, gut barrier, and environmental sensors. *Immune Network*, 21(3), e20. https://doi.org/10.4110/in.2021.21.e20.

Hamamah, S., Aghazarian, A., Nazaryan, A., Hajnal, A., & Covasa, M. (2022). Role of microbiota-gut-brain axis in regulating dopaminergic signaling. *Biomedicines*, 10(2), 436. https://doi.org/10.3390/biomedicines10020436.

Hanna-Jairala, I., & Drossman, D. A. (2024). Central neuromodulators in irritable bowel syndrome: Why, how, and when. *American Journal of Gastroenterology*, 119(7), 1272. https://doi.org/10.14309/ajg.0000000000002800.

Haug, T. T., Mykletun, A., & Dahl, A. A. (2002). Are anxiety and depression related to gastrointestinal symptoms in the general population? *Scandinavian Journal of Gastroenterology*, 37(3), 294–298. https://doi.org/10.1080/003655202317284192.

Honda, K., & Littman, D. R. (2016). The microbiota in adaptive immune homeostasis and disease. *Nature*, 535(7610), 75–84. https://doi.org/10.1038/nature18848.

Janik, R., Thomason, L. A. M., Stanisz, A. M., Forsythe, P., Bienenstock, J., & Stanisz, G. J. (2016). Magnetic resonance spectroscopy reveals oral Lactobacillus promotion of increases in brain GABA, N-acetyl aspartate and glutamate. *NeuroImage*, 125, 988–995. https://doi.org/10.1016/j.neuroimage.2015.11.018.

Jankovic, J., & Tan, E. K. (2020). Parkinson's disease: Etiopathogenesis and treatment. *Journal of Neurology, Neurosurgery & Psychiatry*, 91, 795–808. https://doi.org/10.1136/jnnp-2019-322338.

Ju, S., Shin, Y., Han, S., Kwon, J., Choi, T. G., Kang, I., & Kim, S. S. (2023). The gut–brain axis in schizophrenia: The implications of the gut microbiome and SCFA production. *Nutrients*, 15(20), 4391. https://doi.org/10.3390/nu15204391.

Kelly, J. R., Kennedy, P. J., Cryan, J. F., Dinan, T. G., Clarke, G., & Hyland, N. P. (2015). Breaking down the barriers: The gut microbiome, intestinal permeability and stress-related psychiatric disorders. *Frontiers in Cellular Neuroscience*, 9, 392. https://doi.org/10.3389/fncel.2015.00392.

Khodabakhsh, P., Bazrgar, M., Dargahi, L., Mohagheghi, F., Asgari Taei, A., Parvardeh, S., & Ahmadiani, A. (2021). Does Alzheimer's disease stem in the gastrointestinal system? *Life Sciences*, 287, 120088. https://doi.org/10.1016/j.lfs.2021.120088.

Kuai, X. Y., Yao, X. H., Xu, L. J. Zhou, Y. Q., Zhang, L. P., Liu, Y., & Wu, S. L. (2021). Evaluation of fecal microbiota transplantation in Parkinson's disease patients with constipation. *Microbial Cell Factories*, 20, 98. https://doi.org/10.1186/s12934-021-01589-0 (https://pubmed.ncbi.nlm.nih.gov/33985520/).

Kurdi, M., Bajwa, S. J. S., Sharma, R., & Choudhary, R. (2024). Gut microbiota and probiotics in perioperative management: A narrative review. *Cureus*, 16(9), e68404. https://doi.org/10.7759/cureus.68404.

Liu, J., Zhang, P., Zheng, Z., Gao, J., He, P., Zhang, S., Zhang, Y., Yang, S., Kang, N., Afridi, M. I., Gao, S., Chen, C., & Tu, H. (2023). GABAergic signaling between enteric neurons and intestinal smooth muscle promotes innate immunity and gut defense in Caenorhabditis elegans. *Immunity*, 56(7), 1515–1532.e9. https://doi.org/10.1016/j.immuni.2023.06.004.

Loh, J. S., Mak, W. Q., Tan, L. K., Ng, C. X., Chan, H. H., Yeow, S. H., Foo, J. B., Ong, Y. S., How, C. W., & Khaw, K. Y. (2024). Microbiota–gut–brain axis and its therapeutic applications in neurodegenerative diseases. *Signal Transduction and Targeted Therapy*, 9(1), 1–53. https://doi.org/10.1038/s41392-024-01743-1.

Malek, N. (2019). Deep brain stimulation in Parkinson's disease. *Neurology India*, 67(4), 968–978. https://doi.org/10.4103/0028-3886.266268.

Martín, F., Zambrano, P., Cáceres, O., Almirall, M., Lobo, B., Maria, A., Santos, J., Domingo, J. C., & Jurek, J. (2023). Increased gut permeability and bacterial translocation are associated with fibromyalgia and myalgic encephalomyelitis/chronic fatigue syndrome: Implications for disease-related biomarker discovery. *Frontiers in Immunology*, 14, 1253121. https://doi.org/10.3389/fimmu.2023.1253121.

Mhanna, A., Martini, N., Hmaydoosh, G. Safiah, M., Zeineh, S., Alsaid, B., & Zakaria, H. (2024). The correlation between gut microbiota and both neurotransmitters and mental disorders: A narrative review. *Medicine*, 103(5). https://doi.org/10.1097/md.0000000000037114 (https://pubmed.ncbi.nlm.nih.gov/38306525/).

Million, M., & Larauche, M. (2016). Stress, sex, and the enteric nervous system. *Neurogastroenterology and Motility*, 28(9). https://doi.org/10.1111/nmo.12937.

Mohan, A., Godugu, S., Joshi, S. S., Shah, K. B., Vanka, S. C., Shakil, H., P., D., Veliginti, S., Sure, P. S., & Goranti, J. (2023). Gut-brain axis: Altered microbiome and depression review. *Annals of Medicine and Surgery (London)*, 85(5), 1784–1789. https://doi.org/10.1097/ms9.0000000000000573 (https://pubmed.ncbi.nlm.nih.gov/37228982/).

Moncrieff, J., Cooper, R. E., Stockmann, T., Amendola, S., Hengartner, M. P., & Horowitz, M. (2023). The serotonin theory of depression: A systematic umbrella review of the evidence. *Molecular Psychiatry*, 28(8), 3243–3256. https://doi.org/10.1038/s41380-022-01661-0.

Morais, L. H., Schreiber, H. L., & Mazmanian, S. K. (2021). The gut microbiota-brain axis in behavior and brain disorders. *Nature Reviews Microbiology*, 19(4), 241–255. https://doi.org/10.1038/s41579-020-00460-0 (https://pubmed.ncbi.nlm.nih.gov/33093662/).

Morgan, M. Y. (1991). The treatment of chronic hepatic encephalopathy. *Hepato-Gastroenterology*, 38(5), 377–387.

Munawar, N., Ahmad, A., Anwar, M. A., & Muhammad, K. (2021). Modulation of gut microbial diversity through non-pharmaceutical approaches to treat schizophrenia. *International Journal of Molecular Sciences*, 23(5), 2625. https://doi.org/10.3390/ijms23052625.

Nathani, R. R., Sodhani, S., Vadakekut, E. S. (2025). Irritable bowel syndrome. StatPearls Publishing. Available from: https://www.ncbi.nlm.nih.gov/books/NBK534810/.

Ng, Q. X., Peters, C., Ho, C. Y. X., Lim, D. Y., & Yeo, W. S. (2018). A meta-analysis of the use of probiotics to alleviate depressive symptoms. *Journal of Affective Disorders*, 228, 13–19. https://doi.org/10.1016/j.jad.2017.12.073.

Obadia, N., Estato, V., & Neto, H. C. D. C. F. (2024). The enteric nervous system's (ENS) impact on the neurobiology of infectious diseases. *Neurobiology of Infectious Diseases*, 31–39. https://doi.org/10.1016/B978-0-443-19130-5.00002-2.

Olanow, C. W., Wakeman, D. R., & Kordower, J. H. (2014). Peripheral alpha-synuclein and Parkinson's disease. *Movement Disorders*, 29(8), 963–966. https://doi.org/10.1002/mds.25966.

Oroojzadeh, P., Bostanabad, S. Y., & Lotfi, H. (2022). Psychobiotics: The influence of gut microbiota on the gut-brain axis in neurological disorders. *Journal of Molecular Neuroscience: MN*, 72(9), 1952–1964. https://doi.org/10.1007/s12031-022-02053-3.

Pellissier, S., Dantzer, C., Mondillon, L., Trocme, C., Gauchez, A. S., Ducros, V., Mathieu, N., Toussaint, B., Fournier, A., Canini, F., & Bonaz, B. (2014). Relationship between vagal tone, cortisol, TNF-alpha, epinephrine, and negative affects in Crohn's disease and irritable bowel syndrome. *PLOS ONE*, 9(9), e105328. https://doi.org/10.1371/journal.pone.0105328.

Qin, J., Li, R., Raes, J. et al. (2010). A human gut microbial gene catalogue established by metagenomic sequencing. *Nature*, 464(7285), 59–65. https://doi.org/10.1038/nature08821.

Raskov, H., Burcharth, J., Pommergaard, H. C., & Rosenberg, J. (2016). Irritable bowel syndrome, the microbiota, and the gut-brain axis. *Gut Microbes*, 7(5), 365–383. https://doi.org/10.1080/19490976.2016.1218585.

Rosin, S., Xia, K., Azcarate-Peril, M. A., Stuebe, A. M., Grewen, K. M., Knickmeyer, R. C., & Thompson, A. L. (2021). A preliminary study of gut microbiome variation and HPA axis reactivity in healthy infants. *Psychoneuroendocrinology*, 124, 105046. https://doi.org/10.1016/j.psyneuen.2020.105046.

Rusch, J. A., Layden, B. T., & Dugas, L. R. (2023). Signaling cognition: The gut microbiota and hypothalamic-pituitary-adrenal axis. *Frontiers in Endocrinology (Lausanne)*, 14, 1130689. https://doi.org/10.3389/fendo.2023.1130689.

Rutsch, A., Kantsjö, J. B., & Ronchi, F. (2020). The gut-brain axis: How microbiota and host inflammasome influence brain physiology and pathology. *Frontiers in Immunology*, 11, 604179. https://doi.org/10.3389/fimmu.2020.604179.

Sam, C., & Bordoni, B. (2023). *Acetylcholine*. StatPearls Publishing. Available from: https://www.ncbi.nlm.nih.gov/books/NBK557825/.

Samanta, M. K., Wilson, B., Santhi, K., Kumar, K. P., & Suresh, B. (2006). Alzheimer disease and its management: A review. *American Journal of Therapeutics*, 13(6), 516–526. https://doi.org/10.1097/01.mjt.0000208274.80496.f1.

Sarkar, A., Harty, S., Lehto, S. M., Moeller, A. H., Dinan, T. G., Cryan, J. F., & Burnet, P. W. (2020). The microbiome in psychology and cognitive neuroscience. *Trends in Cognitive Sciences*, 24(8), 635–648. https://doi.org/10.1016/j.tics.2020.05.007.

Sawamura, H., Taniguchi, K., Ikeda, Y., Tsuji, A., Kitagishi, Y., & Matsuda, S. (2022). Gut microbiota could modulate the effects of neuro-immune responses and memory traces via the gut-brain-immune axis in schizophrenia. *Exploratory Neuroprotective Therapy*, 2, 74–86. https://doi.org/10.37349/ent.2022.00019.

Schwasinger-Schmidt, T. E., & Macaluso, M. (2019). Other antidepressants. In: Hammond D. L. and Bowery N. G. (Eds.), *Handbook of Experimental Pharmacology* (Vol. 250, pp. 325–355). Springer. https://doi.org/10.1007/164_2018_167.

Severance, E. G., Prandovszky, E., Castiglione, J., & Yolken, R. H. (2015). Gastroenterology issues in schizophrenia: Why the gut matters. *Current Psychiatry Reports*, 17(5), 27. https://doi.org/10.1007/s11920-015-0574-0.

Soussou, S., Jablaoui, A., Mariaule, V., Kriaa, A., Boudaya, H., Wysocka, M., Amouri, A., Gargouri, A., Lesner, A., Maguin, E., & Rhimi, M. (2023). Serine proteases and metalloproteases are highly increased in irritable bowel syndrome Tunisian patients. *Scientific Reports*, 13(1), 1–10. https://doi.org/10.1038/s41598-023-44454-3.

Sternini, C. (1997). Organization of the peripheral nervous system: Autonomic and sensory ganglia. *Journal of Investigative Dermatology Symposium Proceedings*, 2(1), 1–7. https://doi.org/10.1038/jidsymp.1997.2.

Stokholm, M. G., Danielsen, E. H., Hamilton-Dutoit, S. J., & Borghammer, P. (2016). Pathological α-synuclein in gastrointestinal tissues from prodromal Parkinson disease patients. *Annals of Neurology*, 79(6), 940–949. https://doi.org/10.1002/ana.24648.

Suganya, K., & Koo, B.-S. (2020). Gut–brain axis: Role of gut microbiota on neurological disorders and how probiotics/prebiotics beneficially modulate microbial and immune pathways to improve brain functions. *International Journal of Molecular Sciences*, 21(20), 7551. https://doi.org/10.3390/ijms21207551.

Tan, A. H., Chuah, K. H., Beh, Y. Y., Schee, J. P., Mahadeva, S., & Lim, S. Y. (2023). Gastrointestinal dysfunction in Parkinson's Disease: Neuro-gastroenterology perspectives on a multifaceted problem. *Journal of Movement Disorders*, 16(2), 138–151. https://doi.org/10.14802/jmd.22220.

Wang, Y. B., de Lartigue, G., & Page, A. J. (2020). Dissecting the role of subtypes of gastrointestinal vagal afferents. *Frontiers in Physiology*, 11, 643. https://doi.org/10.3389/fphys.2020.00643.

Xiong, R. G., Li, J., Cheng, J., Zhou, D.-D., Wu, S.-X., Huang, S.-Y., Saimaiti, A., Yang, Z.-J., Gan, R.-Y., & Li, H.-B. (2023). The role of gut microbiota in anxiety, depression, and other mental disorders as well as the protective effects of dietary components. *Nutrients*, 15(14), 3258. https://doi.org/10.3390/nu15143258.

Yano, J. M., Yu, K., Donaldson, G. P., Shastri, G. G., Ann, P., Ma, L., Nagler, C. R., Ismagilov, R. F., Mazmanian, S. K., & Hsiao, E. Y. (2015). Indigenous bacteria from the gut microbiota regulate host serotonin biosynthesis. *Cell*, 161(2), 264–276. https://doi.org/10.1016/j.cell.2015.02.047.

Zhai, L., Xiao, H., Lin, C. et al. (2023). Gut microbiota-derived tryptamine and phenethylamine impair insulin sensitivity in metabolic syndrome and irritable bowel syndrome. *Nature Communications*, 14(1), 1–14. https://doi.org/10.1038/s41467-023-40552-y.

5 Brain and Neural System

Clinical Implications and Mental Health

Aarti and Kamal Singh

5.1 INTRODUCTION

Human beings possess the nervous system as their most developed communication system, which controls the functions of the whole body. We receive environmental and internal perceptions through this system, which leads to the generation of thought, the regulation of emotions, and the coordination of movement. At the center of this network lies the brain, which integrates sensory input, interprets experiences, and controls/guides the responses that define human behavior and personality. It is all made up of neurons with some supporting neuroglial cells, but it is really amazing that only one cell has the capacity to receive and send impulses for the control of the human body.

5.1.1 Significance of Brain Structure in Cognition, Emotion, and Mental Health

- *Cognition*: The cerebral cortex, the prefrontal lobes in particular, enables reasoning, problem-solving, and supports memory storage as well as planning. Damage in these areas may lead to executive dysfunction, disorganized thinking, or memory loss (Friedman and Robbins 2021).
- *Emotion*: The limbic system including the hippocampus, amygdala, and cingulate gyrus functions as an emotional control center that regulates fear responses, pleasure reactions, motivation, and emotional memory processing (Rolls 2013). Dysfunctions in these regions result in anxiety disorders, depression, and post-traumatic stress disorder (PTSD) (LeDoux 2000).
- *Mental health*: Structural abnormalities in neural pathways, altered neurotransmission, and developmental anomalies contribute to psychiatric disorders such as schizophrenia, bipolar disorder, and dementia (Purves et al. 2018) (Table 5.1).

TABLE 5.1
Clinical Insight Box: Effects on Cognition

Brain Region	Typical Function	Effect of Damage	Example/Clinical Condition
Prefrontal cortex	Planning, judgment, impulse control, decision-making	Executive dysfunction: Poor planning, impulsivity, loss of social inhibition	Traumatic brain injury; frontal lobe stroke (Szczepanski and Knight 2014)
Association areas	Integration of thoughts, attention, logical reasoning	Disorganized thinking—fragmented or illogical thought processes	Schizophrenia; diffuse cortical injury
Hippocampus	Learning and memory formation	Memory loss: Anterograde amnesia: inability to form new memories or Retrograde amnesia: recall past ones	Alzheimer's disease; temporal lobe epilepsy

DOI: 10.1201/9781003610274-6

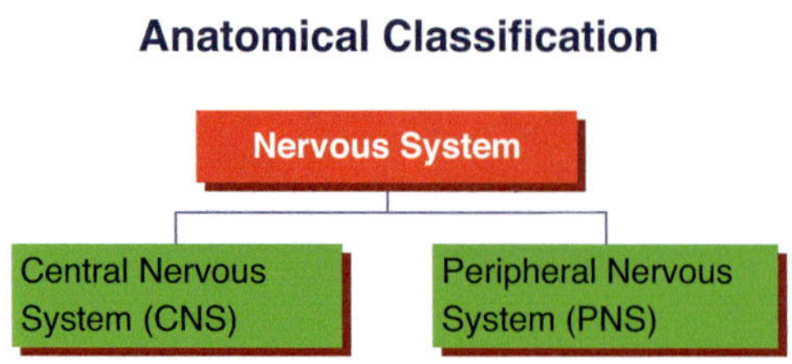

FIGURE 5.1 Structural and functional classification of the nervous system based on location of the organs.

5.1.2 Overview of the Nervous System

The nervous system may be classified into two complementary ways: structurally (based on anatomy) and functionally (based on roles) (Figure 5.1).

5.2 THE AUTONOMIC NERVOUS SYSTEM

The Autonomic Nervous System (ANS) controls vital involuntary functions including circulation, digestion, thermoregulation, and glandular activity. It consists of the sympathetic and parasympathetic nervous systems.

5.2.1 Functional Integration of the Sympathetic and Parasympathetic Nervous Systems

- *Sympathetic system*: Mobilizes energy and prepares body for "fight-or-flight."
- *Parasympathetic system*: Conserves energy and supports "rest-and-digest."
- *Dual innervation*: Most organs receive both inputs, which ensure balanced physiological regulation (Gann 1988).

5.2.2 Modern Network Neuroscience

Contemporary neuroscience emphasizes that cortical functions emerge from dynamic networks, not isolated centers:

- *Default mode network (DMN)*: Medial prefrontal cortex, posterior cingulate, parietal cortex; active in introspection, memory, and self-referential thought (Raichle 2015).
- *Salience network*: Anterior insula and anterior cingulate; detects and filters relevant stimuli, switches between DMN and executive control (Menon 2011; Seeley et al. 2007)
- *Executive control network*: Dorsolateral prefrontal and parietal cortex; supports working memory, attention, and goal-directed behavior (Miller and Cohen 2001).
- *Dual-stream models*:
  - *Visual system*: Dorsal stream ("where pathway," parietal) vs. ventral stream ("what pathway," temporal).
  - *Language system*: Dorsal stream (phonological–motor mapping) vs. ventral stream (semantic comprehension) (Raichle 2015; Miller and Cohen 2001)

5.2.3 Clinical Correlations

- *Frontal lobe lesions*: Hemiplegia, expressive aphasia, personality change.
- *Parietal lobe lesions*: Astereognosis, neglect, Gerstmann's syndrome.

- *Temporal lobe lesions*: Receptive aphasia, prosopagnosia, seizures.
- *Occipital lobe lesions*: Contralateral homonymous hemianopia, visual agnosia.
- *Insular lesions*: Loss of taste, autonomic dysfunction, altered self-awareness.

Expansion of the association cortex (especially prefrontal and parietal) underlies human capacities for abstract reasoning, language, and social behavior. Cortical plasticity ensures that networks can reorganize after injury, forming the basis of rehabilitation and learning. The cerebral cortex is not a mere collection of gyri and sulci but a dynamic, networked structure. Motor areas initiate action; sensory areas decode the world, and association areas integrate them with memory, language, and emotion. Its laminar histology provides the substrate for hierarchical processing, while cortical networks enable the integration of perception, action, and cognition, which makes the cortex the anatomical basis of human thought and behavior (Standring 2016).

5.3 LIMBIC SYSTEM

The limbic system is a set of interconnected cortical and subcortical structures that mediate emotion, motivation, memory, and behavior. It forms a bridge between the primitive drives of the hypothalamus and the higher cognitive functions of the neocortex. Traditionally, it was described as the "visceral brain" or "emotional brain," but modern neuroscience recognizes it as part of broader distributed neural networks (Dalkhsuren 2020; Kandel 2013; Standring 2016)

5.3.1 Components of the Limbic System

- *Cortical Structures*: The cingulate gyrus has a role in processing emotions, focusing attention, and motivating behavior. The parahippocampal gyrus and uncus help with memory consolidation and smell. The hippocampal formation includes the hippocampus, dentate gyrus, subiculum, gyrus fasciolaris, and indusium griseum.
- *Subcortical Structures*: The amygdaloid nuclear complex is the center for fear, aggression, and emotional salience. The septal area is associated with pleasure, reinforcement, and reward. The hypothalamus regulates autonomic and endocrine responses to emotional states. The anterior nucleus of the thalamus relays limbic signals to the cingulate cortex. Habenular nuclei and reticular formation modulate behavior and autonomic responses.

The limbic system functions in the integration of emotion with cognition, attention, and decision-making (Dalkhsuren 2020; Kandel 2013; Standring 2016)

5.3.2 Papez Circuit (Classical Hippocampal Circuit)

The Papez circuit was described by James Papez in 1937. It was the first anatomical model that proposed to explain the neuronal basis of emotion and memory. Originally, Papez proposed this loop as the central "emotional circuit" of the brain, linking cortical and subcortical regions in the regulation of affective states (Kandel 2013). Later, it has been observed that it has its principal role in the consolidation of short-term memory into long-term memory, rather than being central to emotion alone. Despite its historical importance, the Papez model has notable limitations. It overemphasized the hippocampus as the center of emotional life and omitted the amygdala, which modern evidence shows to be essential for processing fear, threat, and social signals (LeDoux 2000). Moreover, the model did not account for the critical role of the prefrontal cortex in regulating and modulating emotions (Kandel 2013). Finally, it presented the limbic system as a closed loop. It overlooked the reality that emotional and memory processes emerge from distributed, multinodal networks, which span the cortex, subcortex, and brainstem. The Papez circuit has been considered a foundational but incomplete model, valuable more as a historical milestone than a comprehensive explanation of limbic function (Rolls 2013).

Contemporary neuroscience highlights that motivational states and emotions do not arise from a single closed circuit, but rather from distributed and dynamic networks. The salience network detects and prioritizes biologically relevant stimuli. Its major areas include anterior insula, cingulate cortex, and amygdala. It detects relevant stimuli (threats, pain, rewards) and shifts activity between default mode and executive control networks. Its dysfunctions can be involved in schizophrenia, anxiety, and mood disorders (Menon 2011; Seeley et al. 2007). The fear circuit, which is anchored in the amygdala and its brainstem connections, orchestrates defensive responses. Basolateral amygdala integrates sensory input, and the central nucleus controls autonomic and motor fear responses via the hypothalamus and brainstem (Rolls 2013). The prefrontal cortex exerts inhibitory control (fear extinction) (Kandel 2013). This clinically correlates well with hyperactivity in PTSD, phobias, and generalized anxiety disorder. Similarly, the reward pathway links the ventral tegmental area, nucleus accumbens, and prefrontal cortex. Nucleus accumbens receives dopaminergic input from ventral tegmental area (VTA) and integrates hippocampal context, amygdala valence, and prefrontal decision-making. Thus, it established a clinical link between addiction, compulsive behavior, and anhedonia in depression. These modern models highlight that emotion and memory are products of parallel but interacting systems. The amygdala and hippocampus process emotional significance and contextual details, while the prefrontal cortex provides higher-order regulation (Rolls 2013; LeDoux 2000).

5.3.3 Clinical Correlations of the Limbic System

Disorders of the limbic system manifest in both neurological syndromes and psychiatric conditions, reflecting its central role in memory, emotion, and motivation. Neurologically, one of the classic syndromes is Klüver–Bucy syndrome, which arises from bilateral lesions of the temporal lobes, particularly the amygdala. Patients develop striking behavioral changes including placidity, hypersexuality, a tendency to explore objects orally, and visual agnosia (inability to recognize familiar objects despite intact vision). Amnesia represents another important consequence of limbic injury: damage to the hippocampus, fornix, or mammillothalamic tract disrupts the consolidation of new memories, leaving patients unable to retain recent experiences even though remote memory may remain intact. In addition, the hippocampus and amygdala are frequent foci for temporal lobe epilepsy, where seizures are often preceded by emotional auras such as fear or déjà vu and may be accompanied by stereotyped automatisms (Papez 1995; Raichle 2015). From a psychiatric perspective, disturbances of limbic connectivity underpin a wide spectrum of major mental illnesses.

In schizophrenia, interactions between the hippocampus, prefrontal cortex, and nucleus accumbens go dysfunctional and produce abnormal salience attribution and psychotic symptoms (Anticevic et al. 2013). In addition, the mesolimbic dopamine pathway (ventral tegmental area to nucleus accumbens) becomes pathologically overactive, and the weakened prefrontal control fails to suppress compulsive drug-seeking (Volkow, Koob, and McLellan 2016). Similarly, in obsessive–compulsive disorder (OCD), there is a malfunction within cortico-striato-thalamo-cortical loops, especially the circuits that link the cingulate gyrus and orbitofrontal cortex. This initiates intrusive thoughts and repetitive behavior (Paul's et al. 2014).

Depression is increasingly understood as a disorder of dysfunctional limbic–cortical connectivity. Neuroimaging studies consistently demonstrate overactivity of the subgenual anterior cingulate cortex (Brodmann area 25), a region strongly interconnected with the amygdala, hippocampus, hypothalamus, and brainstem autonomic centers. This hyperactivity is thought to generate persistent feelings of sadness, hopelessness, and negative self-referential thought. At the same time, the amygdala shows exaggerated responses to emotionally laden stimuli, but the regulatory influence of the prefrontal cortex, especially the dorsolateral and ventromedial regions, is diminished. The characteristic features of major depressive disorder (MDD) result from this imbalance in the form of impaired emotional regulation, ruminative thinking, and poor cognitive flexibility.

In anxiety disorders and PTSD, the amygdala plays a central role as the "fear hub." Patients demonstrate increased activation of amygdala in response to threat-related cues, even when the stimuli are objectively safe (Etkin and Wager 2007). Normally, when amygdala is hypersensitive, there is top-down inhibition from the medial and dorsolateral prefrontal cortex, which normally dampens amygdala output once a stimulus has been evaluated as nonthreatening. In anxiety, this hyperreactivity is intensified because of the relative failure of the said mechanism. As a result, patients experience exaggerated threat perception, hypervigilance, and autonomic arousal. In PTSD specifically, this circuitry interacts with the hippocampus and leads to intrusive memories, flashbacks, and impaired contextual discrimination, i.e., the inability to distinguish between past trauma cues and present safe contexts (Hayes, Hayes, and Mikedis 2012).

Functional magnetic resonance imaging (MRI) studies demonstrate that in people who face trauma there is abnormal synchrony between the amygdala and hippocampus. The amygdala and hippocampus tend to fire in sync when they shouldn't. The amygdala, which tags experiences with strong emotion, keeps on sending alarm signals, while the hippocampus, which normally helps put memories in context and time, can't calm it down. This dysregulated connectivity amplifies emotional salience while impairing the contextual integration of memory. This makes traumatic memories feel as if they are happening "right now," thereby sustaining maladaptive recall patterns seen in PTSD (Rauch, Shin, and Phelps 2006).

Together, these patterns highlight that the amygdala and cingulate cortex act as overactive emotional amplifiers, while the prefrontal cortex fails to provide adequate regulatory "brakes." This failure of emotional homeostasis explains why treatments that enhance prefrontal function (e.g., selective serotonin reuptake inhibitors (SSRIs), CBT, and mindfulness-based therapies) or dampen limbic overdrive (e.g., anxiolytics and exposure therapy) remain central to clinical management.

5.4 THALAMUS AND BASAL GANGLIA

5.4.1 Thalamus: Relay and Modulation

The thalamus is the principal relay and modulator of sensory, motor, and limbic information en route to the cortex. Almost all sensory modalities (except olfaction) pass through thalamic nuclei, which both transmit and filter inputs via corticothalamic feedback and the thalamic reticular nucleus (TRN) (Purves et al. 2018; Standring 2016; Kandel 2013). The thalamus is not a passive switchboard but an active gatekeeper, shaping how sensory, motor, and limbic inputs reach conscious awareness. Corticothalamic feedback and inhibitory neurons called the TRN play a central role in this selective "filtering" (Kandel 2013).

5.4.2 Basal Ganglia: Circuits for Action, Habits, and Motivation

The basal ganglia comprise four major components, i.e., striatum (caudate nucleus and putamen), pallidum (globus pallidus externa (GPe) and interna (GPi)), subthalamic nucleus (STN), and substantia nigra, which is further subdivided into pars compacta (SNc, dopaminergic) and pars reticulata (SNr, GABAergic). Core functions include motor control, procedural learning, and habit formation along with rewards and motivation.

A delicate balance between excitation (driven by glutamatergic inputs) and inhibition (mediated by GABAergic outputs) is maintained by these interconnected loops. This dynamic equilibrium is responsible for the purposeful yet flexible human behavior. The basal ganglia ensure that action and restraint are finely tuned whether in the execution of a precise voluntary movement, consolidation of a habitual routine, or adaptation of choices in response to shifting rewards and motivations (Standring 2016; Galvan and Wichmann 2008).

5.4.3 Clinical Implications

- *Parkinsonism*
  - Degeneration of *SNc dopaminergic neurons* → ↓D1 (direct) activity + ↑relative D2 (indirect) activity → *increased GPi/SNr inhibition of VA/VL* → reduced thalamocortical drive → *bradykinesia, rigidity, resting tremor, postural instability.*
  - *Therapeutic note:* Deep brain stimulation of *STN or GPi* helps rebalance abnormal output (Galvan and Wichmann 2008).
- *Huntington's disease*
  - Early loss of *indirect-pathway (D2) medium spiny neurons in the striatum* → thalamic disinhibition → *choreiform movements.*
  - Later progression includes psychiatric disturbances and cognitive decline, with marked *caudate atrophy.*
- *Hemiballismus*
  - *STN lesion* → ↓excitatory drive to GPi → disinhibited thalamus → violent, flinging proximal limb movements.
- *Other basal ganglia disorders*
  - *Dystonia*: Abnormal co-contraction of agonist–antagonist muscles due to BG dysfunction.
  - *Obsessive–compulsive disorder (OCD)*: Linked to limbic basal ganglia–thalamocortical loops.
- *Thalamic strokes*
  - *VPL/VPM infarcts*: Contralateral hemianesthesia ± *Dejerine–Roussy syndrome (thalamic pain).*
  - *VL involvement*: May produce cerebellar outflow interruption with ataxia and dysmetria.

5.5 BRAINSTEM

The brainstem is the stalk-like portion of the brain that connects the cerebrum with the spinal cord. It is composed of three major subdivisions: midbrain, pons, and medulla oblongata. It serves as a vital conduit for ascending and descending tracts and has the nuclei of cranial nerves III–XII along with the reticular formation, which governs consciousness, arousal, and autonomic regulation (Dalkhsuren 2020; Standring 2016; Kandel 2013). The reticular formation, a network of interconnected neurons, extends throughout the brainstem.

- *Ascending reticular activating system (ARAS)* maintains consciousness; lesions result in coma.
- *Descending reticulospinal pathways* modulate posture, reflexes, and autonomic output.
- *Autonomic centers* regulate cardiovascular, respiratory, and gastrointestinal functions (Table 5.2).

5.6 PROTECTIVE MECHANISMS OF THE BRAIN

The brain is one of the most delicate organs in the body, yet it is remarkably well protected in the cranial cavity. The nerves emerging from the brain leave the cranial cavity through various foramina. In the cranial cavity, there are three major protective mechanisms: the meninges, the cerebrospinal fluid (CSF), and the blood–brain barrier (BBB). These mechanisms work together to protect the nervous tissue from mechanical injury, pathogens, and toxins along with the maintenance of biochemical stability.

TABLE 5.2
Functional and Clinical Correlations of the Brain Stem

Region	Key Functions	Clinical Correlations
Midbrain	Eye movement, motor control, visual and auditory reflexes	CN III/IV palsy, impaired consciousness
Pons	Cerebellar relay Facial cranial nerves Respiratory rhythm regulation	Locked-in syndrome: ventral pons lesion leads to quadriplegia with preserved consciousness because the ARAS remains intact.
Medulla	Vital centers, CN IX–XII, pyramidal decussation	Respiratory/cardiac failure Lateral medullary (Wallenberg) syndrome: Involves nuclei of IX, X and leads to dysphagia, hoarseness, ipsilateral facial sensory loss.
Reticular formation	Consciousness, arousal, autonomic regulation, motor integration	Coma, autonomic instability (Dalkhsuren 2020; Kandel 2013)

5.6.1 Clinical Correlation

- *Meningitis*: Inflammation of the meninges, often due to bacterial or viral infection, presents with fever, neck rigidity, and altered sensorium; can be fatal if untreated.
- *Subdural hematoma*: Tearing of bridging veins causes venous blood accumulation between dura and arachnoid, often after trauma.
- *Subarachnoid hemorrhage*: Rupture of cerebral aneurysms leads to bleeding into the CSF space, producing sudden, severe headache ("thunderclap headache") (Purves et al. 2018).
- *Disruption of BBB occurs in infections and inflammation* like meningitis and encephalitis, which allows the entry of immune cells and toxins and worsens the brain damage. *Neurodegenerative diseases* such as multiple sclerosis and Alzheimer's disease are associated with BBB dysfunction, which contributes to disease progression. Many therapeutic agents (e.g., antibiotics and anticancer drugs) cannot cross the intact BBB. This drug delivery challenge limits the treatment options for CNS disorders.

5.7 NEURONS, SYNAPSES, AND NEUROTRANSMISSION

The basic structural and functional unit of the nervous system is the neuron. It is a specialized cell, which has the ability to conduct nerve impulses and control the functions of the body, i.e., from reflexes to cognition and emotion. Communication between neurons occurs via synapses, where electrical signals are converted into chemical messages in the form of neurotransmitters. Alterations in these processes underlie many neurological and psychiatric disorders.

5.7.1 Major Neurotransmitters and Mental Health

Neurotransmitters are essential for sensory and motor activity but also for higher functions such as emotion, cognition, motivation, and behavior (Russo and Nestler 2013). Dysregulation of neurotransmitter systems can be involved in the causation of several neurological and psychiatric conditions.

- *Dopamine* is a neurotransmitter for movement regulation, reward, pleasure, and motivation. Its *hyperactivity in mesolimbic pathways* leads to psychosis and positive symptoms of schizophrenia. Its *deficiency in nigrostriatal pathways* is associated with Parkinson's disease with tremor, rigidity, and bradykinesia.

- *Serotonin (5-HT)* is the modulator for mood, appetite, sleep, and pain perception. Depression, anxiety, and OCD are associated with its deficiency. Its receptors are the target for *SSRIs* in depression treatment.
- *GABA (gamma-aminobutyric acid)* is the principal *inhibitory neurotransmitter* in the CNS. It reduces neuronal excitability and maintains balance against excitation. Its *deficits* are associated with anxiety disorders, epilepsy, and insomnia. Drugs like benzodiazepines enhance GABAergic activity for the treatment of anxiety disorders.
- *Glutamate* is the major *excitatory neurotransmitter* and has an important role in synaptic plasticity, memory, and learning. Its *excessive activity* leads to excitotoxicity and contributes to stroke, traumatic brain injury, and neurodegenerative diseases (e.g., Amyotrophic lateral sclerosis (ALS) and Alzheimer's). *Hypofunction of NMDA receptors* is implicated in schizophrenia (cognitive and negative symptoms).

These neurotransmitter systems are targets of most psychotropic drugs, from antidepressants to antipsychotics and anxiolytics.

5.8 BRAIN DEVELOPMENT AND PLASTICITY

The process of brain development starts from early embryonic, and the process remains active throughout the entire lifespan. Neurobiological events in a specific order direct the formation of nervous system structure and function. The genetic blueprint serves as the foundation, but environmental factors including nutrition and sensory input and psychosocial experiences determine the final developmental results.

5.8.1 Neurodevelopmental Stages

- *Neurogenesis*: It starts from the ectoderm, which differentiates to form neuroectoderm and gives rise to the neural plate. Then it undergoes dorsal folding to form the neural tube. First, there is closure of the neural tube from the middle portion; cranial and caudal ends remain open as cranial and caudal neuropores. Caudal neuropore closes on Day 25 of intrauterine life, prior to cranial neuropore on Day 27 of intrauterine life (IUL). After the neural tube formation, there is the formation of neural progenitor cells in the wall of the neural tube. Here, neural progenitor cells rapidly divide and form the adult nervous system. Although the majority of neurons are formed prenatally, restricted neurogenesis persists into adulthood, particularly in the hippocampus and subventricular zone. This adult neurogenesis supports learning, memory, and emotional regulation (T.W. Sadler 2018).
- *Neuronal migration*: Newly born neurons migrate along radial glial cells to reach their appropriate cortical and subcortical destinations. Cortical malformations such as lissencephaly and heterotopias, often associated with epilepsy, can be the result of the disrupted migration process.
- *Differentiation and synaptogenesis*: Neurons begin the process of *differentiation*, only when they reach their designated locations. Each neuron forms the *axons* for transmission of information, and *dendrites*, for sensory input. Along with this, there is also development of intricate networks of connections called *synapses,* and the process is called synaptogenesis.
- *Synaptogenesis* is especially active during *late gestation and the early postnatal years*, when the infant brain has been forming trillions of synaptic contacts. This abundance offers the flexibility needed for rapid *sensory, motor, and cognitive development*. These connections are shaped with experience and time. Frequently used pathways are strengthened, while unused ones are pruned away.
- *Myelination*: It is the process by which myelin sheath is formed around the axons. Myelin is a phospholipid-rich insulating sheath formed by oligodendrocytes in the central nervous

system (CNS) and Schwann cells in the peripheral nervous system. It wraps around axons much like insulation around an electrical wire for the impulses to travel rapidly. This process begins in the *late fetal stage* and continues up to the *third decade of life*. Basic sensory and motor pathways are myelinated first, which makes the infants respond to stimuli and begin motor exploration. Association areas such as the *prefrontal cortex,* which is responsible for judgment, planning, and impulse control, undergo myelination much later. This delayed maturation is responsible for the *protracted development of executive functions.* It is for this reason adolescents often show increased risk-taking and impulsive behavior before these circuits are fully refined.

5.8.2 Critical Periods and Synaptic Pruning

Brain development is not a uniform and steady process and is punctuated by critical periods. These are specific windows of heightened plasticity when developing neural circuits to have maximum sensitivity to environmental input. For example, the development of the visual cortex requires adequate visual stimulation during the first months of life. Its deprivation (e.g., congenital cataract) results in irreversible deficits if uncorrected (Stiles and Jernigan 2010).

Following the exuberant phase of synaptogenesis in infancy and childhood, the nervous system undergoes synaptic pruning. It's a process of refinement in which redundant or weak synapses are eliminated while active connections are strengthened. Pruning peaks during adolescence, and it coincides with the maturation of association cortices that underlie higher-order cognition, emotional regulation, and social behavior.

Aberrations in synaptic pruning are recognized as a basis for neurodevelopmental and psychiatric disorders. Schizophrenia, a condition in which there is impaired cortical connectivity and cognitive integration, has been linked to excessive pruning during adolescence. Whereas reduced or atypical pruning in autism spectrum disorders (ASD) results in an overabundance of synaptic connections and is associated with sensory hypersensitivity (Kolb and Gibb 2011; Watson and Tang 2022).

5.8.3 Neuroplasticity and Adaptation

The human brain has the ability of structural as well as functional reorganization and adaptation in response to experience, learning, or injury. This phenomenon is called neuroplasticity and adaptation. It is an ongoing process that occurs in various forms and at various stages of life:

- *Developmental plasticity*: It is most noticeable in infancy and childhood and facilitates the acquisition of language, motor skills, and sensory integration. For example, when infants have exposure to multiple languages in the early years of life, they attain fluency with native-like proficiency. Conversely, the lack of input during this stage leads to permanent deficits. For example, in cases of congenital deafness, there is a lack of perception of stimulus since birth, and it can result in underdeveloped auditory and language circuits if not treated timely. But the early interventions such as *cochlear implantation* within the critical period have a positive role in partial restoration of the circuits (Markman et al. 2011).
- *Experience-dependent plasticity*: Continuous structural and functional adaptation is possible in the human brain across the lifespan. This form of plasticity is the result of the influence of unique experiences, learning, and skill practice. For instance, violinists have enlarged cortical maps in the somatosensory cortex corresponding to the fingers of their left hand, which are used for fingering the strings. This means repeated fine motor practice actually expands the brain's representation of that movement. Experienced London taxi drivers had an enlarged posterior hippocampus compared to nondrivers. The hippocampus is crucial for spatial navigation and memory, and this enlargement was linked to the intensive spatial mapping skills required to navigate London's complex streets. This form

of plasticity demonstrates that the brain remains "sculpted by experience" and constantly reorganizes synaptic connections according to use.

- *Functional plasticity*: This describes the brain's ability to reassign functions to different areas following injury or disease. For example, if a stroke damages language centers in the left cerebral hemisphere in childhood, homologous regions of the right hemisphere can often assume language function with a near-complete recovery. In adults, although the capacity is reduced but the rehabilitation therapies exploit functional plasticity by engaging adjacent or contralateral brain regions to compensate for lost functions. Much of stroke rehabilitation, occupational therapy, and constraint-induced movement therapy are based on this principle (Martin, Ketchabaw, and Turkeltaub 2022).
- *Structural plasticity*: Long-term potentiation and long-term depression support learning and memory at the cellular level as they either strengthen or weaken the synapses based on activity. Although such changes are critical for positive adaptation, they can also give rise to maladaptive plasticity. For example, aberrant sprouting in pain pathways may contribute to chronic neuropathic pain. Similarly, after limb amputation, the cortical area previously devoted to the missing limb is "invaded" by adjacent representations (e.g., face or trunk). This leads to *phantom limb sensations,* and the patients continue to feel the presence of the amputated limb, sometimes accompanied by pain (Marzola et al. 2023).

5.8.4 Impact of Early Life Trauma and Chronic Stress

Brain development is also modulated by early psychosocial experiences and has profound effects on both structure and function.

- *Stress and the HPA axis*: The hypothalamic–pituitary–adrenal (HPA) axis remains persistently activated and increases cortisol secretion due to exposure of chronic early stress. Prolonged cortisol exposure exerts widespread neurotoxic effects. It suppresses hippocampal neurogenesis, so there is weakened memory formation and stress-buffering capacity. It also disrupts prefrontal cortical regulation and reduces executive control and inhibitory processes. There is amplification of amygdala reactivity, which increases fear and emotional salience. This triad predisposes individuals to anxiety disorders, depression, impulsivity, and long-term impairments in cognitive control.
- *Epigenetic modifications*: Long-lasting epigenetic alterations can be induced by adverse experiences in early life such as neglect, abuse, or malnutrition. Silencing or activation of genes responsible for stress regulation, emotional control, and synaptic plasticity can be done by DNA methylation and histone modification. Such changes may not limit this generation itself but may be intergenerational and may amplify subsequent generations.
- *Neurodevelopmental consequences*: Neuroimaging studies in adults with the history of childhood trauma have revealed reduced hippocampal volume, altered integrity of corpus callosum, and functional changes in amygdala reactivity. Persistent deficits in the development of memory, executive functions, emotional regulation, and social skills are linked to these neurobiological changes. All this highlights how early trauma or adverse experiences leave imprints on long-term developmental trajectories (Kaufman et al. 2000).

5.8.5 Lifetime Perspective of Neuroplasticity

- *Infancy and childhood*: There is rapid synaptogenesis, exuberant connections, and the development of sensory integration and motor milestones. Environmental influences during this period profoundly modify cognitive, social, and emotional outcomes.
- *Adolescence*: A second wave of remodeling occurs, with extensive synaptic pruning and myelination in association with the cortices. This helps in attainment as well as enhancement

of higher cognitive functions. But this simultaneously increases the vulnerability to psychiatric disorders such as schizophrenia, bipolar disorder, and substance use disorders. Modulation of brain circuits during this sensitive period is further influenced by hormonal changes and social stressors or peer pressures.

- *Adulthood*: This phase observes ongoing experience-dependent plasticity, which allows for learning, memory consolidation, and skill refinement. During this phase, relative stability is achieved, but the neural circuits keep on sculpting as per the occupational demands, relationships, and lifestyle factors. Chronic stress or trauma can still dysregulate neural networks even though resilience is generally stronger.
- *Aging*: Synaptic density, dendritic branching, and overall plasticity show a gradual decline during this stage of life. Despite this biological reduction, cognitive stimulation, regular physical activity, and supportive environment play a protective role in sustaining neurocognitive function. It is further emphasized that functional resilience can be preserved at old age by the brain's persistent capacity for adaptation in regions such as the hippocampus where lifelong neurogenesis is possible (Navakkode and Kennedy 2024).

5.8.6 Brain–Body Interactions

The brain is the organ that is not meant to function in isolation. It is in constant communication with the whole body through various neural, endocrine, immune, and metabolic pathways. The brain is able to respond to various physiological and environmental changes because of this continuous communication. With this, the brain can monitor the body's internal state and, in turn, modulate behavior, mood, and cognition. The gut–brain axis (GBA) and the HPA axis are two most prominent examples of such bidirectional signaling.

The GBA represents a complex crosstalk system, which links the CNS with the gastrointestinal tract through the vagus nerve, hormonal messengers, immune molecules, and the gut microbiota. Signals can originate in the gut and may range from microbial metabolites to inflammatory cytokines. These signals have the capacity to directly influence brain function, shape emotional as well as stress responses, and even higher-order cognition. Similarly, psychological stress and altered brain activity can remodel gut physiology, motility, and microbial balance, which emphasizes the reciprocal nature of this pathway (Navakkode and Kennedy 2024).

The HPA axis, in contrast, is the body's quintessential stress-response system, which translates psychological and physical stressors into hormonal cascades. Its activation begins in the hypothalamus, which releases corticotropin-releasing hormone, thereby act on pituitary and causes secretion of adrenocorticotropic hormone (ACTH), and culminating in the release of cortisol from the adrenal cortex. While short-term activation of this system promotes vigilance, energy mobilization, and survival, chronic dysregulation disrupts neuronal health, especially in stress-sensitive regions such as the hippocampus, prefrontal cortex, and amygdala. All this increases the vulnerability to psychiatric disorders.

With the help of these two, body-brain loops maintain homeostasis under normal circumstances. So, the disturbances in these two axes, in which there is exaggerated HPA axis response, or faulty immune signaling or disruptions in the gut microbiota, are now considered common threads in conditions such as depression, anxiety, post-traumatic stress disorder, ASD, and functional gastrointestinal syndromes. Thus, the study of brain-body interactions has moved beyond observation of mental illness as a purely cerebral phenomenon, toward an integrative model that embraces the intimate interplay of neural circuits with peripheral physiology (Carabotti et al. 2015).

5.9 GUT–BRAIN AXIS

The GBA is a complex, bidirectional communication network, which connects the CNS with the enteric nervous system, the vagus nerve, and the immune system along with the gut microbiota. This intricate system is essential for the preservation of homeostasis and the regulation of physiological processes such as digestion, mood, cognition, and immune responses. The brain and the gut are not separate systems; rather, they are intricately linked and constitute what is commonly called the "second brain" in the gastrointestinal system (Carabotti et al. 2015; Wu et al. 2025).

5.9.1 Role of Gut Microbiota and the Vagus Nerve

The gut microbiota is a diverse ecosystem of trillions of microorganisms residing in the gastrointestinal tract, which serves as a biochemical factory that produces a wide array of metabolites and signaling molecules. These include:

- Neurotransmitter precursors such as *tryptophan*, which is essential for serotonin synthesis.
- Short-chain fatty acids like *butyrate*, *propionate*, and *acetate*, which influence neuroinflammation and blood–brain barrier integrity.
- Immune mediators modulate immune responses.

The vagus nerve acts as a primary neural connection between the gut and the brain. It is the longest cranial nerve, which starts from the brainstem, supplies in the head, neck, and thorax, and reaches up to the proximal two-thirds of the transverse colon. It transmits sensory information from the gut to the brain and vice versa to facilitate rapid communication. It has been observed in experimental studies on germ-free animals, i.e., the animals which are raised without exposure to any microorganisms, that there are profound alterations in:

- Neurotransmitter turnover (e.g., dopamine, serotonin),
- Stress reactivity, with heightened HPA axis responses, and
- Hippocampal plasticity is crucial for learning and memory.

The essential role of gut microbiota in neurodevelopment and emotional regulation is further emphasized because these abnormalities tend to normalize after microbial colonization (Mhanna et al. 2024).

5.9.2 Impact on Mood and Cognition

When the normal balance and diversity of gut microbiota is disturbed, this condition is called dysbiosis. Evidence suggests that imbalances of gut microbiota can contribute to the onset and progression of several psychiatric and developmental conditions:

- *Depression and anxiety*: There has been increased systemic inflammation induced by dysbiosis, which adversely affects neurochemical pathways. Reduced availability of precursors like tryptophan due to changes in serotonin metabolism impairs mood regulation. Exaggerated HPA axis responses lead to impairment of stress resilience as well as emotional coping.
- *Autism spectrum disorders (ASD)*: Individuals with ASD frequently experience gastrointestinal symptoms such as constipation, diarrhea, and abdominal discomfort. Alterations in gut microbiota, which is characterized by imbalanced bacterial ratios and reduced microbial diversity, have been strongly associated with ASD. An increase in behavioral symptoms such as irritability, social disengagement, and repetitive behaviors is a sign of gut-mediated influence on neurodevelopment.
- *Stress-related disorders*: Chronic stress leads to a breech in the integrity of the intestinal epithelial barrier. The result is increased intestinal permeability, commonly referred to as "leaky gut," which facilitates the translocation of microbial antigens and endotoxins into

systemic circulation. This process gives rise to a state of chronic low-grade inflammation as it activates innate immune responses. The inflammatory milieu further modulates the gut microbial ecosystem and promotes dysbiosis. All of this exacerbates both neuropsychiatric and gastrointestinal symptoms, demonstrating the GBA reciprocal nature. The cyclical nature of stress-induced immune activation, barrier dysfunction, and microbial imbalance creates a self-reinforcing loop in which microbial dysregulation and psychological distress reinforce each other (Cryan and Dinan 2012).

- *Therapeutic implications*: The therapeutic potential of microbiota-targeted interventions is rapidly evolving and seems promising. Some of the strategies have shown promise in the modulation of mood and reduction of stress reactivity such as
  - Probiotics (live beneficial bacteria),
  - Prebiotics (compounds that promote the growth of beneficial microbes),
  - Fecal microbiota transplantation. For example, certain strains of *Lactobacillus* and *Bifidobacterium* have demonstrated anxiolytic and antidepressant-like effects in both animal models and preliminary human trials.

Although preclinical evidence from research is strong but the translation of these findings into human research is still in its formative stages. A major obstacle is the heterogeneity of gut microbiota profiles across individuals, which is influenced by host genetics, dietary patterns, environmental exposures, and lifestyle behaviors. These variables make it difficult to establish standardized therapeutic protocols or provide universally applicable and effective microbial interventions (Ng et al. 2017).

Nevertheless, the GBA continues to represent a promising frontier for personalized medicine. Integration of microbial profiling with genetic, dietary, and psychosocial data; future research may give tailored interventions that optimize mental health outcomes. This approach is particularly relevant in the context of neuropsychiatric and neurodevelopmental disorders, where conventional therapies often fall short and there is an urgent need for novel adjunctive strategies (Bistas and Tabet 2023).

5.9.3 Hormonal Response and Stress

The HPA axis serves as the central neuroendocrine mechanism, which regulates the body's response to stress. When there is any exposure to stress either physical or emotional, the hypothalamus secretes corticotropin-releasing factor, which stimulates the anterior pituitary to release ACTH. ACTH then stimulates the adrenal cortex to release cortisol, which is the primary glucocorticoid required for stress adaptation. The amygdala develops dendritic hypertrophy and hyperreactivity when exposed to chronic stress while its volume tends to increase. These changes lead to increased anxiety and stronger fear responses and worse emotional control (Bistas and Tabet 2023). Stress-related conditions like depression, PTSD, and generalized anxiety disorder are associated with the triad of amygdala hyperactivity, prefrontal cortical dysfunction, and hippocampus atrophy.

5.9.4 HPA Axis Dysregulation in Psychiatric Disorders

The HPA axis dysregulation leads to the development of various psychiatric disorders that result from stress. The main characteristic of MDD is the prolonged elevation of hypercortisolemia. The prolonged elevation of cortisol causes hippocampal atrophy and impairs neurogenesis, which results in the cognitive deficits, memory impairments, and emotional instability that are characteristic of depression. The continuous activation of the HPA axis system reduces glucocorticoid receptor sensitivity and weakens the negative feedback mechanism, which further perpetuates the cycle of cortisol production.

The HPA axis dysregulation affects amygdala-prefrontal interactions, which results in changes to these brain regions. The amygdala shows excessive activity, which combines with insufficient prefrontal cortex inhibition to produce intense fear responses, heightened alertness, and poor emotional control. The imbalance in PTSD strengthens the process of traumatic memory encoding and their intrusive re-experiencing while anxiety disorders maintain elevated arousal and fear anticipation (P and Vellapandian 2024).

5.10 CONCLUSION

The HPA axis and the GBA interact dynamically and are not independent. For instance, stress hormones change gut permeability and microbial composition, while gut microbiota controls HPA axis set points. Therefore, a self-reinforcing loop linked to depression, anxiety, ASD, and functional gastrointestinal disorders can occur when a dysregulated microbiome intensifies stress responses and chronic stress reinforces dysbiosis.

5.11 DECLARATION OF GENERATIVE AI AND AI-ASSISTED TECHNOLOGIES IN THE WRITING PROCESS

During the preparation of this work, the author(s) used AI assistance to improve the readability and language of the manuscript. The authors reviewed and edited the final content and took full responsibility for the content of the published article.

REFERENCES

Anticevic, Alan, Michael W. Cole, Grega Repovs, Aleksandar Savic, Naomi R. Driesen, Genevieve Yang, Youngsun T. Cho et al. 2013. "Connectivity, pharmacology, and computation: toward a mechanistic understanding of neural system dysfunction in Schizophrenia." *Frontiers in Psychiatry* 4 (January): 169. https://doi.org/10.3389/fpsyt.2013.00169.

Bistas, Karlyle G, and Jean Paul Tabet. 2023. "The benefits of prebiotics and probiotics on mental health." *Cureus* 15 (8): e43217. https://doi.org/10.7759/cureus.43217.

Carabotti, Marilia, Annunziata Scirocco, Maria Antonietta Maselli, and Carola Severi. 2015. "The gut-brain axis: interactions between enteric microbiota, central and enteric nervous systems." June 1, 2015. https://pmc.ncbi.nlm.nih.gov/articles/PMC4367209/.

Cryan, John F., and Timothy G. Dinan. 2012. "Mind-altering microorganisms: the impact of the gut microbiota on brain and behaviour." *Nature Reviews. Neuroscience* 13 (10): 701–12. https://doi.org/10.1038/nrn3346.

Dalkhsuren, Shine-Od. 2020. "Snell's clinical neuroanatomy." *Central Asian Journal of Medical Sciences* 6 (3): 208–9. https://doi.org/10.24079/cajms.2020.09.013.

Etkin, Amit, and Tor D. Wager. 2007. "Functional neuroimaging of anxiety: a meta-analysis of emotional processing in PTSD, social anxiety disorder, and specific phobia." *American Journal of Psychiatry* 164 (10): 1476–88. https://doi.org/10.1176/appi.ajp.2007.07030504.

Friedman, Naomi P., and Trevor W. Robbins. 2021. "The role of prefrontal cortex in cognitive control and executive function." *Neuropsychopharmacology* 47 (1): 72–89. https://doi.org/10.1038/s41386-021-01132-0.

Galvan, Adriana, and Thomas Wichmann. 2008. "Pathophysiology of parkinsonism." *Clinical Neurophysiology* 119 (7): 1459–74. https://doi.org/10.1016/j.clinph.2008.03.017.

Gann, Myrna. 1988. "Cunningham's manual of practical anatomy (15th ed.) upper and lower limbs (vol. 1) thorax and abdomen (vol. 2) head and neck and brain (vol. 3)." *American Journal of Occupational Therapy* 42 (3): 199–200. https://doi.org/10.5014/ajot.42.3.199c.

Hayes, Jasmeet P., Scott M. Hayes, and Amanda M. Mikedis. 2012. "Quantitative meta-analysis of neural activity in posttraumatic stress disorder." *Biology of Mood & Anxiety Disorders* 2 (1): 9. https://doi.org/10.1186/2045-5380-2-9.

Kandel, Eric R. 2013. *Principles of Neural Science,* 5th ed., McGraw Hill.

Kaufman, Joan, Paul M. Plotsky, Charles B. Nemeroff, and Dennis S. Charney. 2000. "Effects of early adverse experiences on brain structure and function: clinical implications." *Biological Psychiatry* 48 (8): 778–90. https://doi.org/10.1016/s0006-3223(00)00998-7.

Kolb, Bryan, and Robbin Gibb. 2011. "Brain plasticity and behaviour in the developing brain." *Journal of the Canadian Academy of Child and Adolescent Psychiatry* 20 (4): 265. https://pmc.ncbi.nlm.nih.gov/articles/PMC3222570/.

LeDoux, Joseph E. 2000. "Emotion circuits in the brain." *Annual Review of Neuroscience* 23 (1): 155–84. https://doi.org/10.1146/annurev.neuro.23.1.155.

Markman, Timothy M., Alexandra L. Quittner, Laurie S. Eisenberg, Emily A. Tobey, Donna Thal, John K. Niparko, and Nae-Yuh Wang. 2011. "Language development after cochlear implantation: an epigenetic model." *Journal of Neurodevelopmental Disorders* 3 (4): 388–404. https://doi.org/10.1007/s11689-011-9098-z.

Martin, Kelly C., W. Tyler Ketchabaw, and Peter E. Turkeltaub. 2022. "Plasticity of the language system in children and adults." *Handbook of Clinical Neurology* 184 (January): 397–414. https://doi.org/10.1016/b978-0-12-819410-2.00021-7.

Marzola, Patrícia, Thayza Melzer, Eloisa Pavesi, Joana Gil-Mohapel, and Patricia S. Brocardo. 2023. "Exploring the role of neuroplasticity in development, aging, and neurodegeneration." *Brain Sciences* 13 (12): 1610. https://doi.org/10.3390/brainsci13121610.

Menon, Vinod. 2011. "Large-scale brain networks and psychopathology: a unifying triple network model." *Trends in Cognitive Sciences* 15 (10): 483–506. https://doi.org/10.1016/j.tics.2011.08.003.

Mhanna, Amjad, Nafiza Martini, Ghefar Hmaydoosh, George Hamwi, Mulham Jarjanazi, Ghaith Zaifah, Reem Kazzazo, Aya Haji Mohamad, and Zuheir Alshehabi. 2024. "The correlation between gut microbiota and both neurotransmitters and mental disorders: a narrative review." *Medicine* 103 (5): e37114. https://doi.org/10.1097/md.0000000000037114.

Miller, Earl K., and Jonathan D. Cohen. 2001. "An integrative theory of prefrontal cortex function." *Annual Review of Neuroscience* 24 (1): 167–202. https://doi.org/10.1146/annurev.neuro.24.1.167.

Navakkode, Sheeja, and Brian K. Kennedy. 2024. "Neural ageing and synaptic plasticity: prioritizing brain health in healthy longevity." *Frontiers in Aging Neuroscience* 16 (August): 1428244. https://doi.org/10.3389/fnagi.2024.1428244.

Ng, Qin Xiang, Christina Peters, Collin Yih Xian Ho, Donovan Yutong Lim, and Wee-Song Yeo. 2017. "A meta-analysis of the use of probiotics to alleviate depressive symptoms." *Journal of Affective Disorders* 228 (November): 13–19. https://doi.org/10.1016/j.jad.2017.11.063.

Sharan, P., and Chitra Vellapandian. 2024. "Hypothalamic-Pituitary-Adrenal (HPA) Axis: unveiling the potential mechanisms involved in Stress-Induced Alzheimer's Disease and depression." *Cureus* 16 (8): e67595. https://doi.org/10.7759/cureus.67595.

Papez, James W. 1995. "A proposed mechanism of emotion. 1937 [classical article]." *Journal of Neuropsychiatry* 7 (1): 103–12. https://doi.org/10.1176/jnp.7.1.103.

Pauls, David L., Amitai Abramovitch, Scott L. Rauch, and Daniel A. Geller. 2014. "Obsessive–compulsive disorder: an integrative genetic and neurobiological perspective." *Nature Reviews. Neuroscience* 15 (6): 410–24. https://doi.org/10.1038/nrn3746.

Purves, Dale, George Augustine, David Fitzpatrick, William Hall, Anthony LaMantia, Leonard White, Richard Mooney, and Michael Platt. 2018. *Neuroscience*.

Raichle, Marcus E. 2015. "The brain's default mode network." *Annual Review of Neuroscience* 38 (1): 433–47. https://doi.org/10.1146/annurev-neuro-071013-014030.

Rauch, Scott L., Lisa M. Shin, and Elizabeth A. Phelps. 2006. "Neurocircuitry models of posttraumatic stress disorder and extinction: human neuroimaging research—past, present, and future." *Biological Psychiatry* 60 (4): 376–82. https://doi.org/10.1016/j.biopsych.2006.06.004.

Rolls, Edmund T. 2013. "Limbic systems for emotion and for memory, but no single limbic system." *Cortex* 62 (December): 119–57. https://doi.org/10.1016/j.cortex.2013.12.005.

Russo, Scott J., and Eric J. Nestler. 2013. "The brain reward circuitry in mood disorders." *Nature Reviews. Neuroscience* 14 (9): 609–25. https://doi.org/10.1038/nrn3381.

Seeley, William W., Vinod Menon, Alan F. Schatzberg, Jennifer Keller, Gary H. Glover, Heather Kenna, Allan L. Reiss, and Michael D. Greicius. 2007. "Dissociable intrinsic connectivity networks for salience processing and executive control." *Journal of Neuroscience* 27 (9): 2349–56. https://doi.org/10.1523/jneurosci.5587-06.2007.

Standring, Susan. 2016. "The Brain." In *Gray's Anatomy: The Anatomical Basis of Clinical Practice*, edited by Susan Standring, Neel Anand, Rolfe Birch, Patricia Collins, Alan R. Crossman, Michael Gleeson, Ariana L. Smith, Jonathan D. Spratt, Mark D. Stringer, R. Shane Tubbs, Alan J. Wein, and Caroline B. Wigley, 41st ed., 323–81. Elsevier.

Stiles, Joan, and Terry L. Jernigan. 2010. "The basics of brain development." *Neuropsychology Review* 20 (4): 327–48. https://doi.org/10.1007/s11065-010-9148-4.

Szczepanski, Sara M., and Robert T. Knight. 2014. "Insights into human behavior from lesions to the prefrontal cortex." *Neuron* 83 (5): 1002–18. https://doi.org/10.1016/j.neuron.2014.08.011.

Sadler, Thomas W. 2018. "Central nervous system." *Langman's Medical Embryology*, 13: 90–96. ISBN/ISSN: 9781496383907 (https://www.mendeley.com/catalogue/e4b96bf0-b71f-3ffc-8502-8ece7e539db1/).

Volkow, Nora D., George F. Koob, and A. Thomas McLellan. 2016. "Neurobiologic advances from the brain disease model of addiction." *New England Journal of Medicine* 374 (4): 363–71. https://doi.org/10.1056/nejmra1511480.

Watson, Zachary, and Shao-Jun Tang. 2022. "Aberrant synaptic pruning in CNS diseases: a critical player in HIV-Associated neurological dysfunction?" *Cells* 11 (12): 1943. https://doi.org/10.3390/cells11121943.

Wu, Dezhi, Hui Wang, Yueqiong Ni, Yurun Lu, Yong Wang, Huating Li, and Luonan Chen. 2025. "Decoding the gut–brain axis: toward AI-driven integration of neuroimaging and gut microbiota in human health." *The Visual Computer* 41 (12): 9793–9804. https://doi.org/10.1007/s00371-025-04000-w.

Section B

Aging and Neurological Disorders

6 Aging
Overview and In-Depth Analysis

Vrushali Bhalchim, Vrushali Neve, Namrata Sarma, Mohini Rana, and Muthuraju Sangu

6.1 INTRODUCTION

Aging is a complex biological process involving declining physiological functions and increased disease vulnerability, with interconnected mechanisms at molecular, cellular, and systemic levels (López-Gil et al. 2023). Research has revealed aging's causes, including genomic instability, epigenetic alterations, cellular senescence, stem cell exhaustion, and dysregulated nutrient-sensing pathways (Wang et al. 2022). The field has progressed from descriptive to molecular biology-based studies, with aging now recognized as malleable through genetic and environmental factors (Dhar et al. 2022). Genomic instability, a fundamental aging hallmark, occurs as cells accumulate DNA damage from various sources, causing cellular dysfunction. Epigenetic alterations affect cellular function and contribute to aging (Figure 6.1) (Dhar et al. 2022). Cellular senescence, responding to stressors, acts in tumor suppression but causes inflammation through the senescence-associated secretory phenotype (SASP) (Zhang et al. 2022). Stem cell exhaustion reduces tissue regeneration capacity (Oh et al. 2014). Dysregulation of nutrient-sensing pathways, including insulin/IGF-1 signaling, mTOR, and sirtuins, affects cellular metabolism and aging (Tomtheelnganbee et al. 2022). Interventions like caloric restriction and rapamycin show promise in extending lifespan. These mechanisms are interconnected, with genomic instability affecting cellular senescence and stem cell function, while nutrient-sensing pathways impact metabolism and genomic stability (Fernandes and Demetriades 2021).

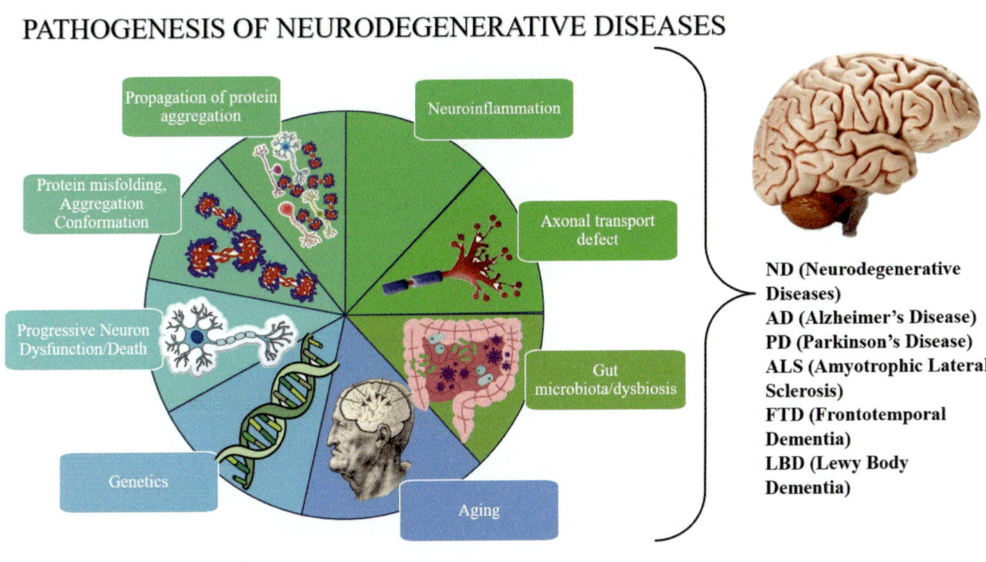

FIGURE 6.1 Common factors for neurodegenerative diseases.

DOI: 10.1201/9781003610274-8

6.2 HALLMARKS OF AGING

6.2.1 Genomic Instability

Genomic instability, a hallmark of aging, involves accumulated genetic damage including DNA alterations and telomere attrition (López-Gil et al. 2023). Aging cells show reduced genomic maintenance, leading to DNA lesions and repair errors that cause cellular dysfunction (Burhans and Weinberger 2007). This accumulation is influenced by oxidative stress and environmental mutagens, with understanding these mechanisms being crucial for developing aging interventions (Maslov and Vijg 2009).

6.2.2 Telomere Attrition

Telomere attrition involves progressive telomere shortening during cell division due to DNA polymerase's incomplete chromosome end replication (Rossiello et al. 2022). When telomeres become critically short, they trigger cellular senescence or apoptosis, causing tissue dysfunction and age-related diseases (Vaiserman and Krasnienkov 2021). Environmental factors accelerate this process, while research explores mechanisms and interventions (Shammas 2011).

6.2.3 Epigenetic Alterations

Epigenetic alterations are a hallmark of aging, involving changes in gene expression without DNA sequence changes. These include DNA methylation, histone modifications, and chromatin remodeling. With age, loss of epigenetic regulation causes aberrant gene expression (la Torre et al. 2023). This epigenetic drift can activate harmful genes or silence beneficial ones. Environmental factors impact epigenetic patterns, suggesting potential interventions (Li and Tollefsbol 2016). Research focuses on understanding these alterations and developing stability strategies (Mangelinck and Mann 2021).

6.2.4 Loss of Proteostasis

Declining protein quality control with age leads to misfolded, aggregated proteins that disrupt cellular functions (Labbadia and Morimoto 2015). The proteostasis network becomes overwhelmed, impairing protein synthesis and degradation. Oxidative stress, mitochondrial dysfunction, and senescence worsen proteostasis loss (Morimoto 2014). Protein aggregates trigger stress responses and inflammation, compromising tissue function. Research focuses on understanding proteostasis decline and developing interventions (Higuchi-Sanabria et al. 2018).

6.2.5 Deregulated Nutrient Sensing

Nutrient-sensing pathways including insulin/IGF-1 signaling (IIS), mTOR, AMPK, and sirtuins affect cellular responses to nutrients. With age, these pathways become less responsive, causing metabolic imbalances (Bettedi and Foukas 2017). Decreased insulin/IGF-1 sensitivity and mTOR dysregulation impair metabolism and protein synthesis, while reduced AMPK and sirtuin activity affect energy homeostasis and stress responses (Miyamoto et al. 2013).

6.2.6 Mitochondrial Dysfunction

Mitochondrial dysfunction contributes to cellular aging and age-related diseases. As cells age, mitochondrial efficiency declines (Guo et al. 2023), with decreased ATP production, increased reactive oxygen species (ROS), and DNA mutations. ROS further damages mitochondrial DNA and proteins (Yapa et al. 2021). Mitochondrial dysfunction affects metabolism, calcium homeostasis, and apoptosis. Impaired quality control leads to damaged mitochondria accumulation (Bakula and Scheibye-Knudsen 2020), causing reduced energy, oxidative stress, and accelerated aging (Miwa et al. 2022).

6.2.7 Cellular Senescence

Cellular senescence involves irreversible cell division arrest and phenotypic changes, affecting aging, tumor suppression, and tissue repair. Senescent cells exhibit growth arrest, apoptosis resistance, and an SASP (Herranz and Gil 2018). The SASP releases proinflammatory factors affecting neighboring cells. Senescence occurs through telomere shortening, DNA damage, oxidative stress, and oncogene activation (Takasugi et al. 2023). While senescence prevents malignant cell proliferation, accumulated senescent cells impair organ function and cause age-related Takasugi diseases. Research aims to understand senescence mechanisms for aging and regenerative medicine (Dong et al. 2024).

6.2.8 Stem Cell Exhaustion

Stem cell exhaustion impacts aging and age-related diseases. As organisms age, tissue regeneration decreases due to declining stem cell function (Oh et al. 2014). This depletion occurs through senescence, telomere attrition, and DNA damage. Stem cell exhaustion impairs tissue homeostasis (Neves et al. 2017). This decline affects proliferative tissues like blood, skin, and intestinal epithelium, contributing to osteoporosis and cognitive decline (Schumacher et al. 2021). Research aims to rejuvenate stem cells for treating age-related diseases (Liu and Rando 2011).

6.2.9 Altered Intercellular Communication

Altered intercellular communication impacts cellular function and homeostasis. With aging, cell signaling pathways deteriorate, disrupting processes through changes in the extracellular matrix, receptors, and signaling molecules. This impairment affects tissue repair and metabolic regulation. Age-related changes lead to chronic inflammation and metabolic imbalances. Understanding these mechanisms is crucial for developing interventions to promote healthy aging (Statzer et al. 2023).

6.3 MOLECULAR MECHANISMS OF AGING

6.3.1 DNA Damage and Repair

DNA damage and repair are key mechanisms in aging. Cells accumulate damage from oxidative stress and environmental factors as organisms age, leading to genomic instability and dysfunction. DNA repair pathways become less efficient with age (Paul et al. 2011), resulting in lesions, mutations, and cellular senescence. DNA damage triggers response signaling, contributing to inflammation and altered gene expression. Understanding these mechanisms is crucial for reducing age-related genomic instability (Nicolai et al. 2015).

6.3.2 Oxidative Stress and Free Radical Theory

Oxidative stress and free radical theory are keys to understanding aging (Ziada et al. 2020). The theory proposes that damage from ROS and free radicals drives aging. These molecules damage proteins, lipids, and DNA, causing cellular dysfunction. Mitochondria are vulnerable to oxidative stress (Cui et al. 2012). Their declining function increases ROS production, creating damage cycles. Aging weakens antioxidant defenses (Maldonado et al. 2023), linking cellular damage to physiological aging and suggesting interventions (Iakovou and Kourti 2022).

6.3.3 Protein Aggregation and Misfolding

Protein aggregation and misfolding are critical cellular processes with disease implications (Gandhi et al. 2019). When proteins lose their native structure, they form toxic aggregates (Sweeney et al. 2017). These processes are linked to neurodegenerative diseases like Alzheimer's and Parkinson's (Ciechanover and Kwon 2015). Misfolded protein accumulation overwhelms cellular quality control (Ajmal 2023), making understanding these mechanisms essential for therapies (Houck et al. 2012).

6.3.4 Cellular Senescence Pathways

Cellular senescence, marked by permanent cell division arrest, influences aging, tumor suppression, and tissue repair (Herranz and Gil 2018). The p53/p21 pathway is activated by DNA damage, while p16/Rb responds to stressors like oxidative stress (Kumari and Jat 2021). Senescent cells secrete proinflammatory factors (SASP), affecting surrounding tissues and age-related pathologies (Mijit et al. 2020).

6.3.5 Mitochondrial Dysfunction and Energy Metabolism

Mitochondrial dysfunction and altered metabolism contribute to senescence and aging. Damage accumulation reduces ATP and increases ROS, compromising energy homeostasis and increasing oxidative stress. Impaired mitochondria activate stress pathways via p53 and AMPK, triggering senescence. Metabolic reprogramming shifts toward glycolysis to support the SASP. Understanding these is crucial for developing interventions for healthy aging (Salih and Brunet 2008).

6.4 GENETIC AND EPIGENETIC FACTORS

6.4.1 Longevity Genes and Their Role in Aging

Longevity genes regulate aging through mechanisms promoting organismal health. FOXO transcription factors enhance longevity by controlling stress responses and metabolism. Sirtuins (SIRT1 and SIRT6) modulate DNA repair and metabolism (Salih and Brunet 2008). The mTOR pathway integrates nutrient sensing, with reduced activity increasing lifespan (Papadopoli et al. 2019). Autophagy genes like ATG5 and BECN1 recycle damaged components. Understanding these genes' interactions is crucial for aging interventions (Ambrosio and Majello 2020).

6.4.2 Epigenetic Changes Associated with Aging

Epigenetic changes during aging affect DNA methylation, histone modifications, and chromatin remodeling (Jiang and Guo 2020), causing genomic instability and altered gene expression. DNA methylation shows age-related global hypomethylation and site-specific hypermethylation (Hamon and Cossart 2008), with the "epigenetic clock" serving as a biological age biomarker. Histone modifications and noncoding RNA changes (Lu et al. 2006) affect cellular processes, contributing to aging phenotype and age-related diseases (Wang et al. 2022).

6.4.3 Role of Noncoding RNAs in Aging Processes

Noncoding RNAs influence aging through cellular mechanisms. These include microRNAs (miRNAs), long noncoding RNAs (lncRNAs), and circular RNAs (circRNAs). miRNAs regulate gene expression and cellular senescence. lncRNAs affect chromatin structure and gene transcription. circRNAs accumulate with age as biomarkers. Their dysregulation alters gene expression, contributing to aging disorders (Kour and Rath 2016).

6.5 CELLULAR AND TISSUE-SPECIFIC AGING

6.5.1 Stem Cell Aging and Tissue Regeneration

Stem cell aging impacts tissue regeneration and organismal aging. Aging stem cells show reduced self-renewal and differentiation capabilities, limiting tissue repair (Rossi et al. 2007). This results from DNA damage, epigenetic changes, and niche alterations. Aging affects multiple stem cell types, reducing tissue homeostasis (Ruegsegger and Saxena 2016). Understanding aging mechanisms and developing restoration strategies are crucial for addressing age-related disorders (Scadden 2014).

6.5.2 Immunosenescence and Inflammaging

Immunosenescence and inflammaging contribute to age-related immune decline. Immunosenescence is immune system deterioration with age, reducing antigen responses and immune memory (Liu et al. 2023). This affects immunity, increasing disease susceptibility. Inflammaging is chronic inflammation with age (Coppack 2001), marked by proinflammatory cytokines that drive age-related diseases. These processes interact, creating a cycle where weakened immunity worsens inflammation, accelerating aging pathologies (Wadden et al. 2007).

6.5.3 Neurodegeneration and Brain Aging

Neurodegeneration and brain aging impact cognitive function in older individuals. The aging brain shows reduced volume, synaptic density, and altered neurotransmitter systems (Scheibel 1992), facilitating neuronal loss. Neurodegenerative disorders like Alzheimer's and Parkinson's increase with age, accelerating cognitive decline. Oxidative stress and protein aggregation contribute to these disorders while affecting brain aging (Chaudhary et al. 2022). Blood–brain barrier changes affect nutrient delivery and waste removal, promoting neurodegeneration. Understanding this aging–neurodegeneration relationship is crucial for maintaining cognitive health (Kaur et al. 2021).

6.5.4 Cardiovascular Aging

Cardiovascular aging causes changes in the heart and blood vessels. Arterial walls stiffen, increasing blood pressure and cardiac strain. The left ventricle thickens, reducing filling efficiency. Endothelial dysfunction impairs vascular regulation (Stenvinkel 2001). Age affects heart rate and blood pressure through autonomic changes. Atherosclerotic plaque accumulation narrows vessels, increasing cardiovascular risk. These changes reduce function and disease resistance in older adults (Menghini et al. 2014).

6.5.5 Musculoskeletal Aging

Musculoskeletal aging impacts body structure and function. Age-related muscle mass decline (sarcopenia) reduces mobility and increases falls. Bone density loss causes osteoporosis and fractures, particularly in postmenopausal women (Laurent et al. 2019). Joint degeneration leads to osteoarthritis. Less elastic connective tissues affect posture and balance. Reduced regeneration slows injury healing, impacting life quality and independence (Kwan et al. 2023).

6.6 SYSTEMIC AND ORGANISMAL AGING

6.6.1 Endocrine and Metabolic Changes

Aging causes decreased hormone production and sensitivity in the hypothalamic-pituitary-adrenal and growth hormone/insulin-like growth factor-1 axes (Sherlock and Toogood 2007). These changes lead to increased fat and decreased muscle mass. Metabolic efficiency declines with age, reducing basal metabolism and glucose processing. Thyroid function and sex hormones decrease, affecting metabolism and bone density (McAninch and Bianco 2014). These changes contribute to type 2 diabetes and osteoporosis, making their understanding crucial for developing aging interventions (Sletten et al. 2020).

6.6.2 Circadian Rhythm Disruption

Circadian rhythm disruption occurs when the body's internal clock misaligns with environmental light-dark cycles. This disruption stems from shift work, jet lag, and nighttime light exposure. Chronic disruption is linked to metabolic disorders, cardiovascular diseases, cognitive issues, cancer risk, and affects mood. Maintaining consistent sleep schedules and reducing evening light exposure are essential for preserving circadian health (Bishehsari et al. 2020).

6.6.3 Gut Microbiome Alterations

Gut microbiome alterations significantly result from circadian rhythm disruption, showing the link between the internal clock and gut microbial communities (Figure 6.2). The microbiome exhibits diurnal fluctuations synchronized with host circadian rhythms (Heddes et al. 2022). Circadian disruption causes microbial dysbiosis, affecting health. Studies demonstrate that circadian misalignment alters microbial diversity and metabolism (Zhang et al. 2025). These changes contribute to metabolic disorders and compromised intestinal function. The bidirectional gut microbiome-circadian communication creates a feedback loop affecting host health (Giovannini et al. 2011).

6.6.4 Chronic Low-Grade Inflammation

Chronic low-grade inflammation, a persistent mild immune activation, has significant health impacts. Unlike acute inflammation, it persists at low levels without symptoms (Bachmann et al. 2020).

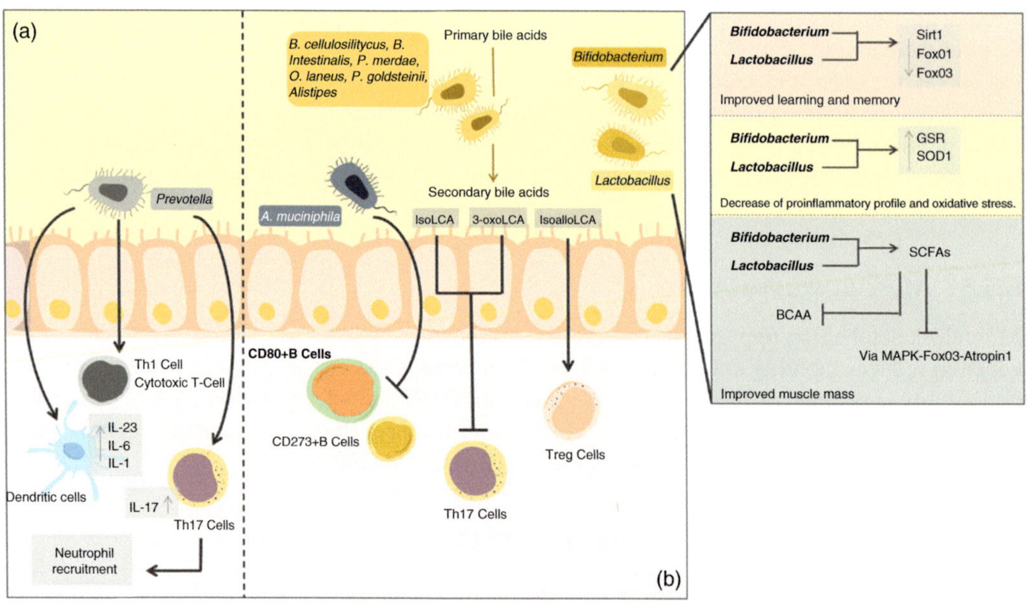

FIGURE 6.2 Gut microbiome and health aging (Salazar et al. 2023). (Published by MDPI in open access under common creative license.)

It features elevated proinflammatory markers like C-reactive protein (CRP), interleukin (IL-6), and tumor necrosis factor alpha (TNF-α), linked to cardiovascular disorders, type 2 diabetes, obesity, and certain cancers (Li et al. 2024), from poor diet, sedentary lifestyle, stress, and toxins. Inflammatory mediators cause tissue damage and cellular dysfunction, contributing to age-related diseases. This is a key focus in preventive medicine and aging strategies (Madison et al. 2021).

6.7 ENVIRONMENTAL AND LIFESTYLE FACTORS

6.7.1 Dietary Influences on Aging

Diet significantly influences aging rate and quality. A balanced diet can mitigate age-related decline and promote longevity (Tessier et al. 2025). Antioxidant-rich foods combat oxidative stress, while omega-3 fatty acids protect against age-related diseases. Protein supports muscle maintenance during aging (Spadaro et al. 2022). Processed foods and saturated fats accelerate aging through inflammation. Caloric restriction extends animal lifespans, though human effects remain under study. A fiber-rich diet maintains gut health and may enhance longevity (Guarnera et al. 2023). The Mediterranean diet reduces age-related disease risk and increases lifespan.

6.7.2 Exercise and Physical Activity

Exercise and physical activity are vital for healthy aging. Regular activity maintains muscle mass and bone density, reducing fall risks. Cardiovascular exercise improves heart health and lowers age-related cardiovascular diseases. Strength training builds muscle and aids weight management. Physical activity enhances cognitive function and may delay neurodegenerative diseases. Exercise reduces depression and anxiety symptoms, while group activities promote social interaction. The World Health Organization recommends 150 minutes of moderate or 75 minutes of vigorous aerobic activity weekly, with muscle-strengthening twice weekly (Geiger et al. 2016).

6.7.3 Stress and Psychological Factors

Stress and psychological factors influence aging and health. Chronic stress accelerates cellular aging through inflammation and oxidative stress, leading to age-related diseases. Depression and isolation can impair cognition and increase dementia risk (Carrier et al. 2021).. However, positive outlook and social connections promote cognitive resilience. Research shows that psychological well-being correlates with lower age-related diseases (Newbury et al. 2021).

6.7.4 Environmental Toxins and Pollutants

Environmental toxins and pollutants threaten human health and ecosystems globally. Heavy metals, pesticides, and air pollutants accumulate in organisms, causing respiratory problems, neurological disorders, and cancers (Shetty et al. 2023). These pollutants disrupt ecological balances, particularly in industrial zones. Growing awareness has led to stricter regulations and cleaner technologies to protect health and the environment (Trepanowski et al. 2011).

6.8 INTERVENTIONS AND POTENTIAL THERAPIES

6.8.1 Caloric Restriction and Dietary Interventions

Caloric restriction and dietary interventions promote health and longevity. Caloric restriction reduces intake by 20%–40% while maintaining nutrition, extending lifespan in animals, and improving insulin sensitivity. Dietary interventions include plant-based and Mediterranean diets, showing potential metabolic and cardiovascular benefits. Their long-term human effects remain under study. Dietary changes should be guided by healthcare professionals (Hoang et al. 2022).

6.8.2 Stem Cell Therapies and Regenerative Medicine

Stem cell therapies and regenerative medicine treat diseases by harnessing the body's healing capabilities. Stem cells differentiate into multiple cell types to repair damaged tissues. Regenerative medicine includes tissue engineering and gene therapy (Cerneckis et al. 2024). These therapies show promise for heart disease and neurological disorders. The development of induced pluripotent stem cells from adult cells enables personalized treatments while addressing ethical concerns. These therapies offer hope for previously incurable conditions (Uddin et al. 2020).

6.8.3 Genetic and Epigenetic Interventions

Genetic and epigenetic interventions are promising approaches in personalized medicine. Genetic interventions modify DNA sequences through techniques like Clustered Regularly Interspaced Short Palindromic Repeats and CRISPR-associated protein 9 (CRISPR-Cas9) (Castro-Muñoz et al. 2023), while epigenetic interventions alter gene expression through DNA methylation and histone modifications (López-Otín et al. 2013). These interventions target genetic disorders and cancer but require evaluation of long-term effects (López-Otín et al. 2013).

6.9 AGING AND AGE-RELATED DISEASES

6.9.1 Connections between Aging and Chronic Diseases

Aging is a major risk factor for chronic diseases due to accumulated cellular and molecular changes. These changes include genomic instability, telomere attrition, cellular senescence, and mitochondrial dysfunction, leading to cardiovascular disease, diabetes, cancer, and neurodegeneration (Ferrucci and Fabbri 2018). Age-related inflammation accelerates disease progression, while lifestyle and genetic factors influence disease onset in older adults (Dong et al. 2024).

6.9.2 Cancer and Aging

Cancer and aging are interconnected processes affecting health. Cancer risk increases with age due to mutations and declining immunity. Aging hallmarks like genomic instability and cellular senescence contribute to cancer development (Sprenger et al. 2010). Senescent cells promote tumors while age-related changes create conditions for cancer progression. Understanding this relationship is crucial for reducing cancer in older populations (Roth et al. 2020).

6.9.3 Cardiovascular Diseases

Cardiovascular diseases affect the heart and blood vessels, representing a major health burden. These include coronary disease, heart failure, stroke, and hypertension. Risk factors include poor diet, inactivity, tobacco, alcohol, age, genetics, diabetes, and obesity (Mendis et al. 2022). Prevention focuses on lifestyle changes and risk management. Medical advances in drugs, surgery, and technology have improved outcomes, though research remains vital (Misrani et al. 2021).

6.9.4 Neurodegenerative Disorders

Neurodegenerative disorders progressively deteriorate neurons in the brain and nervous system. These disorders, including Alzheimer's and Parkinson's, impair cognitive function and motor skills.

The mechanisms involve protein accumulation and mitochondrial dysfunction (Lake et al. 2021). Symptoms include memory loss and impaired movement. Disease progression varies, with treatments focusing on symptoms. Research continues to develop therapies and understand contributing factors (Han and Lean 2015).

6.9.5 Metabolic Disorders

Metabolic disorders disrupt normal metabolic processes, affecting carbohydrate, protein, and lipid breakdown and energy utilization. These disorders stem from genetic mutations, enzyme deficiencies, or environmental factors, leading to metabolite accumulation or compound deficiencies. Common symptoms include fatigue, muscle weakness, and organ dysfunction. Early diagnosis and management through dietary changes, enzyme replacement, or medications prevent complications (Tumurkhuu et al. 2025).

6.10 BIOMARKERS OF AGING

6.10.1 Molecular and Cellular Biomarkers

Molecular and cellular biomarkers are essential for diagnosing metabolic disorders, indicating disease presence and treatment response (Qiu et al. 2023). Key examples include blood glucose for diabetes and phenylalanine for phenylketonuria. Metabolomics has enabled the identification of novel biomarkers (Srikanthan et al. 2016). Cellular biomarkers provide information about affected processes, enabling early detection and targeted therapies (Jeppesen et al. 2021).

6.10.2 Physiological and Functional Biomarkers

Physiological and functional biomarkers indicate health status in metabolic disorders, including heart rate variability, blood pressure, and exercise tolerance. Cardiopulmonary testing reveals aerobic capacity in mitochondrial disorders, while glucose monitoring tracks diabetic fluctuations (Glunde et al. 2011). Functional markers like grip strength show metabolic impacts, and magnetic resonance spectroscopy assesses tissue metabolism. Integration with molecular markers enables targeted treatments (Ruprecht et al. 2024).

6.10.3 Omics Approaches in Aging Research

Omics technologies have revolutionized aging research through molecular analyses. These approaches, including genomics, transcriptomics, proteomics, and metabolomics, provide insights into aging mechanisms (Xie et al. 2025). Genomics identifies longevity-linked variants, while transcriptomics shows age-related expression changes. Proteomics analyzes protein alterations (Rapin et al. 2022), and metabolomics reveals metabolic changes. Multi-omics integration identifies aging biomarkers and therapeutic targets (MacRae et al. 2015).

6.11 COMPARATIVE BIOLOGY OF AGING

6.11.1 Lessons from Long-Lived Species

Long-lived species like whales, tortoises, and naked mole rats have evolved adaptations against aging. Research shows enhanced DNA repair, stress responses, and mitochondrial function, along with reduced inflammation and tumor suppression. Comparing their genomes with shorter-lived relatives reveals key longevity pathways for potential human aging interventions (Le Bihan and Martin 2006).

6.11.2 Evolutionary Perspectives on Aging

Evolutionary perspectives reveal mechanisms of longevity and aging. Natural selection has shaped diverse aging patterns across species. Some organisms show negligible senescence, maintaining reproduction and low mortality throughout life and challenging traditional aging theories. Studies of species with divergent lifespans have identified genetic pathways like DNA repair and telomere maintenance between long-lived bats and mice, providing insights for human healthspan. The disposable soma theory explains aging through resource allocation between reproduction and maintenance (Le Bihan and Martin 2006).

6.12 ETHICAL AND SOCIETAL IMPLICATIONS

6.12.1 Longevity Research and Its Impact on Society

Longevity research has major implications for society, affecting healthcare, economics, and social structures. Scientific advances may extend lifespans and healthspans, shifting demographics. This affects work, retirement, and healthcare systems while raising questions about resources and generational equity. Research may reduce age-related disabilities and healthcare costs but could worsen socioeconomic gaps. Policymakers must ensure equitable access to life-extending technologies (Lucke et al. 2010).

6.12.2 Ethical Considerations in Aging Interventions

Ethical considerations in aging interventions remain complex. Uneven access to life-extending technologies could worsen socioeconomic disparities (Peng et al. 2024). Extended lifespans affect resource allocation and intergenerational fairness. While longevity interventions may reduce age-related disabilities, they impact social and economic structures (Partridge et al. 2009). Policymakers must balance individual autonomy with societal effects and resource allocation between aging interventions and other medical priorities (Saliev and Singh 2024).

6.13 FUTURE DIRECTIONS AND CHALLENGES

6.13.1 Emerging Technologies in Aging Research

Emerging technologies in aging research are revolutionizing our understanding and interventions for healthspan extension, including genetic therapies, senolytic drugs, and regenerative medicine (Huang et al. 2023). AI accelerates drug discovery while nanotechnology enables targeted delivery. CRISPR explores correction of age-related mutations, and epigenetic reprogramming aims to reverse cellular aging (Pereira et al. 2024). Advanced imaging and biomarkers detect age-related changes, though these technologies' long-term effects require careful study (Martinović et al. 2024).

6.13.2 Personalized Approaches to Healthy Aging

Personalized approaches to healthy aging consider genetic, lifestyle, and environmental factors for targeted interventions. Analysis of biological profiles and aging biomarkers helps create customized prevention plans (Austin et al. 2024). This includes personalized nutrition, exercise, and medication based on individual needs. Wearable devices and AI enable monitoring for adjustments, while precision medicine targets specific molecular pathways to improve health outcomes (Lin et al. 2022).

6.13.3 Integration of Multi-Omics Data in Aging Studies

Multi-omics data integration helps unravel aging mechanisms by combining genomics, transcriptomics, proteomics, metabolomics, and epigenomics (Deng et al. 2025). This approach identifies biomarkers and therapeutic targets while exploring interactions between biological layers. Longitudinal studies track age-related changes, enabling predictive models for diseases and interventions. Advanced computational methods enhance understanding of aging disorders (Shen et al. 2024).

6.14 CONCLUSION

Multi-omics data integration advances aging research by combining genomics, transcriptomics, proteomics, metabolomics, and epigenomics to reveal age-related changes. By studying interactions between biological layers, researchers identify biomarkers and therapeutic targets. Longitudinal studies provide insights into aging dynamics and enable predictive models. This approach can revolutionize healthcare through personalized interventions and improved strategies for aging populations.

REFERENCES

Ajmal, Mohammad Rehan. 2023. "Protein Misfolding and Aggregation in Proteinopathies: Causes, Mechanism and Cellular Response." *Diseases* 11 (1). https://doi.org/10.3390/diseases11010030.

Ambrosio, Susanna, and Barbara Majello. 2020. "Autophagy Roles in Genome Maintenance." *Cancers* 12 (7). https://doi.org/10.3390/cancers12071793.

Austin, Charles Adrian, Benjamin Seligman, Sangeetha Shan-Bala et al. 2024. "Aging Precisely: Precision Medicine through the Lens of an Older Adult." *Journal of the American Geriatrics Society* 72 (10). https://doi.org/10.1111/jgs.19036.

Bachmann, María Consuelo, Sofía Bellalta, Roque Basoalto et al. 2020. "The Challenge by Multiple Environmental and Biological Factors Induce Inflammation in Aging: Their Role in the Promotion of Chronic Disease." *Frontiers in Immunology* 11 (October). https://doi.org/10.3389/fimmu.2020.570083.

Bakula, Daniela, and Morten Scheibye-Knudsen. 2020. "MitophAging: Mitophagy in Aging and Disease." *Frontiers in Cell and Developmental Biology* 8 (April). https://doi.org/10.3389/fcell.2020.00239.

Bettedi, Lucia, and Lazaros C. Foukas. 2017. "Growth Factor, Energy and Nutrient Sensing Signalling Pathways in Metabolic Ageing." *Biogerontology* 18 (6). https://doi.org/10.1007/s10522-017-9724-6.

Bihan, Blanche Le, and Claude Martin. 2006. "A Comparative Case Study of Care Systems for Frail Elderly People: Germany, Spain, France, Italy, United Kingdom and Sweden." *Social Policy & Administration* 40 (1). https://doi.org/10.1111/j.1467-9515.2006.00475.x.

Bishehsari, Faraz, Robin M. Voigt, and Ali Keshavarzian. 2020. "Circadian Rhythms and the Gut Microbiota: From the Metabolic Syndrome to Cancer." *Nature Reviews Endocrinology* 16 (12). https://doi.org/10.1038/s41574-020-00427-4.

Burhans, William C., and Marc Weinberger. 2007. "DNA Replication Stress, Genome Instability and Aging." *Nucleic Acids Research* 35 (22). https://doi.org/10.1093/nar/gkm1059.

Carrier, Micaël, Eva Šimončičová, Marie-Kim St-Pierre, Chloe McKee, and Marie-Ève Tremblay. 2021. "Psychological Stress as a Risk Factor for Accelerated Cellular Aging and Cognitive Decline: The Involvement of Microglia-Neuron Crosstalk." *Frontiers in Molecular Neuroscience* 14 (November). https://doi.org/10.3389/fnmol.2021.749737.

Castro-Muñoz, Leonardo, Elenaé Ulloa, Cecilia Sahlgren, Marcela Lizano, Erick De La Cruz-Hernández, and Adriana Contreras-Paredes. 2023. "Modulating Epigenetic Modifications for Cancer Therapy (Review)." *Oncology Reports* 49 (3). https://doi.org/10.3892/or.2023.8496.

Cerneckis, Jonas, Hongxia Cai, and Yanhong Shi. 2024. "Induced Pluripotent Stem Cells (IPSCs): Molecular Mechanisms of Induction and Applications." *Signal Transduction and Targeted Therapy* 9 (1). https://doi.org/10.1038/s41392-024-01809-0.

Chaudhary, Rishabh, Mujeeba Rehman, Vipul Agarwal, Arjun Singh Kaushik, and Vikas Mishra. 2022. "Protein Aggregation in Neurodegenerative Diseases." In: Tabish Qidwai (Ed.), *Neurodegenerative Diseases: Multifactorial Degenerative Processes, Biomarkers and Therapeutic Approaches* (First Edition). Bentham Science Publishers. https://doi.org/10.2174/9789815040913122010005.

Ciechanover, Aaron, and Yong Tae Kwon. 2015. "Degradation of Misfolded Proteins in Neurodegenerative Diseases: Therapeutic Targets and Strategies." *Experimental & Molecular Medicine* 47 (3). https://doi.org/10.1038/emm.2014.117.

Coppack, Simon W. 2001. "Pro-Inflammatory Cytokines and Adipose Tissue." *Proceedings of the Nutrition Society* 60 (3). https://doi.org/10.1079/PNS2001110.

Cui, Hang, Yahui Kong, and Hong Zhang. 2012. "Oxidative Stress, Mitochondrial Dysfunction, and Aging." *Journal of Signal Transduction* 2012 (October). https://doi.org/10.1155/2012/646354.

Deng, Chao, Chenjun Xie, Zixi Li, Jie Mei, and Kewei Wang. 2025. "Multi-Omics Analysis Identifies Diagnostic Circulating Biomarkers and Potential Therapeutic Targets, Revealing IQGAP1 as an Oncogene in Gastric Cancer." *Npj Precision Oncology* 9 (1). https://doi.org/10.1038/s41698-025-00895-9.

Dhar, Poulami, Shailaja S. Moodithaya, and Prakash Patil. 2022. "Epigenetic Alterations—The Silent Indicator for Early Aging and Age-associated Health-risks." *Aging Medicine* 5 (4). https://doi.org/10.1002/agm2.12236.

Dong, Zening, Yahan Luo, Zhangchen Yuan, Yu Tian, Tianqiang Jin, and Feng Xu. 2024. "Cellular Senescence and SASP in Tumor Progression and Therapeutic Opportunities." *Molecular Cancer* 23 (1). https://doi.org/10.1186/s12943-024-02096-7.

Fan, Ting, Jiaman Peng, Huiting Liang, Wenzhi Chen, Junlin Wang, and Renshi Xu. 2025. "Potential Common Pathogenesis of Several Neurodegenerative Diseases." *Neural Regeneration Research* 21 (3): 972–88. https://doi.org/10.4103/nrr.nrr-d-24-01054.

Fernandes, Stephanie A., and Constantinos Demetriades. 2021. "The Multifaceted Role of Nutrient Sensing and MTORC1 Signaling in Physiology and Aging." *Frontiers in Aging* 2 (August). https://doi.org/10.3389/fragi.2021.707372.

Ferrucci, Luigi, and Elisa Fabbri. 2018. "Inflammageing: Chronic Inflammation in Ageing, Cardiovascular Disease, and Frailty." *Nature Reviews Cardiology* 15 (9). https://doi.org/10.1038/s41569-018-0064-2.

Gandhi, Jason, Anthony C. Antonelli, Adil Afridi et al. 2019. "Protein Misfolding and Aggregation in Neurodegenerative Diseases: A Review of Pathogeneses, Novel Detection Strategies, and Potential Therapeutics." *Reviews in the Neurosciences* 30 (4). https://doi.org/10.1515/revneuro-2016-0035.

Geiger, Paul J., Ian A. Boggero, Charles Alex Brake et al. 2016. "Mindfulness-Based Interventions for Older Adults: A Review of the Effects on Physical and Emotional Well-Being." *Mindfulness* 7 (2). https://doi.org/10.1007/s12671-015-0444-1.

Giovannini, Silvia, Graziano Onder, Rosa Liperoti et al. 2011. "Interleukin-6, C-Reactive Protein, and Tumor Necrosis Factor-Alpha as Predictors of Mortality in Frail, Community-Living Elderly Individuals." *Journal of the American Geriatrics Society* 59 (9). https://doi.org/10.1111/j.1532-5415.2011.03570.x.

Glunde, Kristine, Lu Jiang, Siver A. Moestue, and Ingrid S. Gribbestad. 2011. "MRS and MRSI Guidance in Molecular Medicine: Targeting and Monitoring of Choline and Glucose Metabolism in Cancer." *NMR in Biomedicine* 24 (6). https://doi.org/10.1002/nbm.1751.

Guarnera, Jade, Eva Yuen, and Helen Macpherson. 2023. "The Impact of Loneliness and Social Isolation on Cognitive Aging: A Narrative Review." *Journal of Alzheimer's Disease Reports* 7 (1). https://doi.org/10.3233/ADR-230011.

Guo, Ying, Teng Guan, Kashfia Shafiq et al. 2023. "Mitochondrial Dysfunction in Aging." *Ageing Research Reviews* 88 (July). https://doi.org/10.1016/j.arr.2023.101955.

Hamon, Mélanie Anne, and Pascale Cossart. 2008. "Histone Modifications and Chromatin Remodeling during Bacterial Infections." *Cell Host & Microbe* 4 (2). https://doi.org/10.1016/j.chom.2008.07.009.

Han, Thang S., and Michael E. J. Lean. 2015. "Metabolic Syndrome." *Medicine* 43 (2). https://doi.org/10.1016/j.mpmed.2014.11.006.

Heddes, Marjolein, Baraa Altaha, Yunhui Niu et al. 2022. "The Intestinal Clock Drives the Microbiome to Maintain Gastrointestinal Homeostasis." *Nature Communications* 13 (1): 6068. https://doi.org/10.1038/s41467-022-33609-x.

Herranz, Nicolás, and Jesús Gil. 2018. "Mechanisms and Functions of Cellular Senescence." *Journal of Clinical Investigation* 128 (4). https://doi.org/10.1172/JCI95148.

Higuchi-Sanabria, Ryo, Phillip Andrew Frankino, Joseph West Paul, Sarah Uhlein Tronnes, and Andrew Dillin. 2018. "A Futile Battle? Protein Quality Control and the Stress of Aging." *Developmental Cell* 44 (2). https://doi.org/10.1016/j.devcel.2017.12.020.

Hoang, Duc M., Phuong T. Pham, Trung Q. Bach et al. 2022. "Stem Cell-Based Therapy for Human Diseases." *Signal Transduction and Targeted Therapy* 7 (1). https://doi.org/10.1038/s41392-022-01134-4.

Houck, Scott A., Sangita Singh, and Douglas M. Cyr. 2012. "Cellular Responses to Misfolded Proteins and Protein Aggregates." *Methods in Molecular Biology (Clifton, N.J.)* 832. https://doi.org/10.1007/978-1-61779-474-2_32.

Huang, Aiquan, Yingyu Huo, Yong Zhong, and Wenyin Yang. 2023. "AI Technology for Anti-Aging: An Overview." In: *2023 International Conference on Intelligent Supercomputing and BioPharma (ISBP)*, Zhuhai, China, January 6. https://doi.org/10.1109/ISBP57705.2023.10061311.

Iakovou, Evripides, and Malamati Kourti. 2022. "A Comprehensive Overview of the Complex Role of Oxidative Stress in Aging, the Contributing Environmental Stressors and Emerging Antioxidant Therapeutic Interventions." *Frontiers in Aging Neuroscience* 14 (June). https://doi.org/10.3389/fnagi.2022.827900.

Jeppesen, Tina D., Karen L. Madsen, Nanna S. Poulsen, Nicoline Løkken, and John Vissing. 2021. "Exercise Testing, Physical Training and Fatigue in Patients with Mitochondrial Myopathy Related to MtDNA Mutations." *Journal of Clinical Medicine* 10 (8). https://doi.org/10.3390/jcm10081796.

Jiang, Shuang, and Yuchen Guo. 2020. "Epigenetic Clock: DNA Methylation in Aging." *Stem Cells International* 2020 (July). https://doi.org/10.1155/2020/1047896.

Kaur, Jasleen, Lara M. Fahmy, Esmaeil Davoodi-Bojd et al. 2021. "Waste Clearance in the Brain." *Frontiers in Neuroanatomy* 15. https://doi.org/10.3389/fnana.2021.665803.

Kour, Sukhleen, and Pramod C. Rath. 2016. "Long Noncoding RNAs in Aging and Age-Related Diseases." *Ageing Research Reviews* 26 (March). https://doi.org/10.1016/j.arr.2015.12.001.

Kumari, Ruchi, and Parmjit Jat. 2021. "Mechanisms of Cellular Senescence: Cell Cycle Arrest and Senescence Associated Secretory Phenotype." *Frontiers in Cell and Developmental Biology* 9 (March). https://doi.org/10.3389/fcell.2021.645593.

Kwan, Ka Yu Carissa, Ka Wai Kerry Ng, Ying Rao et al. 2023. "Effect of Aging on Tendon Biology, Biomechanics and Implications for Treatment Approaches." *International Journal of Molecular Sciences* 24 (20). https://doi.org/10.3390/ijms242015183.

Labbadia, Johnathan, and Richard I. Morimoto. 2015. "The Biology of Proteostasis in Aging and Disease." *Annual Review of Biochemistry* 84 (1). https://doi.org/10.1146/annurev-biochem-060614-033955.

Lake, Julie, Catherine S. Storm, Mary B. Makarious, and Sara Bandres-Ciga. 2021. "Genetic and Transcriptomic Biomarkers in Neurodegenerative Diseases: Current Situation and the Road Ahead." *Cells* 10 (5). https://doi.org/10.3390/cells10051030.

Laurent, Michaël R., Lenore Dedeyne, Jolan Dupont, Bea Mellaerts, Marian Dejaeger, and Evelien Gielen. 2019. "Age-Related Bone Loss and Sarcopenia in Men." *Maturitas* 122 (April). https://doi.org/10.1016/j.maturitas.2019.01.006.

Li, Ping, Qincheng Qiao, Chenyu Nie et al. 2024. "The Mediating Role of Chronic Low-Grade Inflammation Participation in the Relationship between Obesity and Type 2 Diabetes: Findings from the NHANES." *BMC Endocrine Disorders* 24 (1). https://doi.org/10.1186/s12902-024-01671-9.

Li, Yuanyuan, and Trygve O. Tollefsbol. 2016. "Age-Related Epigenetic Drift and Phenotypic Plasticity Loss: Implications in Prevention of Age-Related Human Diseases." *Epigenomics* 8 (12). https://doi.org/10.2217/epi-2016-0078.

Lin, Biaoyang, Yingying Ma, and ShengJun Wu. 2022. "Multi-Omics and Artificial Intelligence-Guided Data Integration in Chronic Liver Disease: Prospects and Challenges for Precision Medicine." *Omics: A Journal of Integrative Biology* 26 (8). https://doi.org/10.1089/omi.2022.0079.

Liu, Ling, and Thomas A. Rando. 2011. "Manifestations and Mechanisms of Stem Cell Aging." *Journal of Cell Biology* 193 (2). https://doi.org/10.1083/jcb.201010131.

Liu, Zaoqu, Qimeng Liang, Yuqing Ren et al. 2023. "Immunosenescence: Molecular Mechanisms and Diseases." *Signal Transduction and Targeted Therapy* 8 (1). https://doi.org/10.1038/s41392-023-01451-2.

López-Gil, Lucía, Amparo Pascual-Ahuir, and Markus Proft. 2023. "Genomic Instability and Epigenetic Changes during Aging." *International Journal of Molecular Sciences* 24 (18). https://doi.org/10.3390/ijms241814279.

López-Otín, Carlos, Maria A. Blasco, Linda Partridge, Manuel Serrano, and Guido Kroemer. 2013. "The Hallmarks of Aging." *Cell* 153 (6). https://doi.org/10.1016/j.cell.2013.05.039.

Lu, Qianjin, Xiaoqiu Qiu, Nan Hu, Haiyan Wen, Yuan Su, and Bruce C. Richardso. 2006. "Epigenetics, Disease, and Therapeutic Interventions." *Ageing Research Reviews* 5 (4). https://doi.org/10.1016/j.arr.2006.07.001.

Lucke, Jayne C., Danielle Herbert, Brad Partridge, and Wayne D. Hall. 2010. "Anticipating the Use of Life Extension Technologies." *EMBO Reports* 11 (5). https://doi.org/10.1038/embor.2010.48.

MacRae, Sheila L., Matthew McKnight Croken, Robert B. Calder et al. 2015. "DNA Repair in Species with Extreme Lifespan Differences." *Aging* 7 (12). https://doi.org/10.18632/aging.100866.

Madison, Annelise A., Martha A. Belury, Rebecca Andridge et al. 2021. "Omega-3 Supplementation and Stress Reactivity of Cellular Aging Biomarkers: An Ancillary Substudy of a Randomized, Controlled Trial in Midlife Adults." *Molecular Psychiatry* 26 (7). https://doi.org/10.1038/s41380-021-01077-2.

Maldonado, Edio, Sebastián Morales-Pison, Fabiola Urbina, and Aldo Solari. 2023. "Aging Hallmarks and the Role of Oxidative Stress." *Antioxidants (Basel, Switzerland)* 12 (3). https://doi.org/10.3390/antiox12030651.

Mangelinck, Adèle, and Carl Mann. 2021. *DNA Methylation and Histone Variants in Aging and Cancer*. https://doi.org/10.1016/bs.ircmb.2021.06.002.

Martinović, Anđela, Matilde Mantovani, Natalia Trpchevska et al. 2024. "Climbing the Longevity Pyramid: Overview of Evidence-Driven Healthcare Prevention Strategies for Human Longevity." *Frontiers in Aging* 5 (November). https://doi.org/10.3389/fragi.2024.1495029.

Maslov, Alexander Y., and Jan Vijg. 2009. "Genome Instability, Cancer and Aging." *Biochimica et Biophysica Acta (BBA) – General Subjects* 1790 (10). https://doi.org/10.1016/j.bbagen.2009.03.020.

McAninch, Elizabeth A., and Antonio C. Bianco. 2014. "Thyroid Hormone Signaling in Energy Homeostasis and Energy Metabolism." *Annals of the New York Academy of Sciences* 1311 (1). https://doi.org/10.1111/nyas.12374.

Mendis, Shanthi, Ian Graham, and Jagat Narula. 2022. "Addressing the Global Burden of Cardiovascular Diseases; Need for Scalable and Sustainable Frameworks." *Global Heart* 17 (1). https://doi.org/10.5334/gh.1139.

Menghini, Rossella, Robert Stöhr, and Massimo Federici. 2014. "MicroRNAs in Vascular Aging and Atherosclerosis." *Ageing Research Reviews* 17 (September). https://doi.org/10.1016/j.arr.2014.03.005.

Mijit, Mahmut, Valentina Caracciolo, Antonio Melillo, Fernanda Amicarelli, and Antonio Giordano. 2020. "Role of P53 in the Regulation of Cellular Senescence." *Biomolecules* 10 (3). https://doi.org/10.3390/biom10030420.

Misrani, Afzal, Sidra Tabassum, and Li Yang. 2021. "Mitochondrial Dysfunction and Oxidative Stress in Alzheimer's Disease." *Frontiers in Aging Neuroscience* 13 (February). https://doi.org/10.3389/fnagi.2021.617588.

Miwa, Satomi, Sonu Kashyap, Eduardo Chini, and Thomas von Zglinicki. 2022. "Mitochondrial Dysfunction in Cell Senescence and Aging." *Journal of Clinical Investigation* 132 (13). https://doi.org/10.1172/JCI158447.

Miyamoto, Tetsuya, Geraldine Wright, and Hubert Amrein. 2013. "Nutrient Sensors." *Current Biology* 23 (9). https://doi.org/10.1016/j.cub.2013.04.002.

Morimoto, Richard. 2014. "Cell Stress and Proteostasis Networks in Biology, Aging, and Disease." *Biophysical Journal* 106 (2). https://doi.org/10.1016/j.bpj.2013.11.259.

Neves, Joana, Pedro Sousa-Victor, and Heinrich Jasper. 2017. "Rejuvenating Strategies for Stem Cell-Based Therapies in Aging." *Cell Stem Cell* 20 (2). https://doi.org/10.1016/j.stem.2017.01.008.

Newbury, Joanne B., Robert Stewart, Helen L. Fisher et al. 2021. "Association between Air Pollution Exposure and Mental Health Service Use among Individuals with First Presentations of Psychotic and Mood Disorders: Retrospective Cohort Study." *The British Journal of Psychiatry* 219 (6). https://doi.org/10.1192/bjp.2021.119.

Nicolai, Sara, Antonello Rossi, Nicola Di Daniele, Gerry Melino, Margherita Annicchiarico-Petruzzelli, and Giuseppe Raschellà. 2015. "DNA Repair and Aging: The Impact of the P53 Family." *Aging* 7 (12). https://doi.org/10.18632/aging.100858.

Oh, Juhyun, Yang David Lee, and Amy J. Wagers. 2014. "Stem Cell Aging: Mechanisms, Regulators and Therapeutic Opportunities." *Nature Medicine* 20 (8). https://doi.org/10.1038/nm.3651.

Papadopoli, David, Karine Boulay, Lawrence Kazak et al. 2019. "MTOR as a Central Regulator of Lifespan and Aging." *F1000Research* 8 (July). https://doi.org/10.12688/f1000research.17196.1.

Partridge, Brad, Jayne Lucke, Helen Bartlett, and Wayne Hall. 2009. "Ethical, Social, and Personal Implications of Extended Human Lifespan Identified by Members of the Public." *Rejuvenation Research* 12 (5). https://doi.org/10.1089/rej.2009.0907.

Paul, Catriona, Makoto Nagano, and Bernard Robaire. 2011. "Aging Results in Differential Regulation of DNA Repair Pathways in Pachytene Spermatocytes in the Brown Norway Rat1." *Biology of Reproduction* 85 (6). https://doi.org/10.1095/biolreprod.111.094219.

Peng, Yaojin, Lulu Ding, Zhenyu Xiao, Moshi Song, Jianwei Lv, and Guang-Hui Liu. 2024. "Ethical Concerns in Aging Research: Perspectives of Global Frontline Researchers." *Science China Life Sciences* 67 (10). https://doi.org/10.1007/s11427-024-2650-y.

Pereira, Beatriz, Francisca P. Correia, Inês A. Alves et al. 2024. "Epigenetic Reprogramming as a Key to Reverse Ageing and Increase Longevity." *Ageing Research Reviews* 95 (March). https://doi.org/10.1016/j.arr.2024.102204.

Qiu, Shi, Ying Cai, Hong Yao et al. 2023. "Small Molecule Metabolites: Discovery of Biomarkers and Therapeutic Targets." *Signal Transduction and Targeted Therapy* 8 (1). https://doi.org/10.1038/s41392-023-01399-3.

Rapin, Alexis, Maroun Bou Sleiman, and Johan Auwerx. 2022. *Data Integration in Systems Genetics and Aging Research*. arXiv July 7, 10.48550/arXiv.2207.03540.

Rossi, Derrick J., Jun Seita, Agnieszka Czechowicz, Deepta Bhattacharya, David Bryder, and Irving L. Weissman. 2007. "Hematopoietic Stem Cell Quiescence Attenuates DNA Damage Response and Permits DNA Damage Accumulation During Aging." *Cell Cycle* 6 (19). https://doi.org/10.4161/cc.6.19.4759.

Rossiello, Francesca, Diana Jurk, João F. Passos, and Fabrizio d'Adda di Fagagna. 2022. "Telomere Dysfunction in Ageing and Age-Related Diseases." *Nature Cell Biology* 24 (2). https://doi.org/10.1038/s41556-022-00842-x.

Roth, Gregory A., George A. Mensah, Catherine O. Johnson et al. 2020. "Global Burden of Cardiovascular Diseases and Risk Factors, 1990–2019." *Journal of the American College of Cardiology* 76 (25). https://doi.org/10.1016/j.jacc.2020.11.010.

Ruegsegger, Céline, and Smita Saxena. 2016. "Proteostasis Impairment in ALS." *Brain Research* 1648 (October). https://doi.org/10.1016/j.brainres.2016.03.032.

Ruprecht, Nathan A., Sonalika Singhal, Kalli Schaefer, Om Panda, Donald Sens, and Sandeep K. Singhal. 2024. "A Review: Multi-Omics Approach to Studying the Association between Ionizing Radiation Effects on Biological Aging." *Biology* 13 (2). https://doi.org/10.3390/biology13020098.

Salazar, Juan, Pablo Durán, María P. Díaz, Marlene Verónica, Raquel Santeliz, Edgardo Mengual, Emma Virginia et al. 2023. "Exploring the Relationship between the Gut Microbiota and Ageing: A Possible Age Modulator." *International Journal of Environmental Research and Public Health* 20 (10): 5845–45. https://doi.org/10.3390/ijerph20105845.

Saliev, Timur, and Prim B. Singh. 2024. "From Bench to Bedside: Translating Cellular Rejuvenation Therapies into Clinical Applications." *Cells* 13 (24). https://doi.org/10.3390/cells13242052.

Salih, Dervis A. M., and Anne Brunet. 2008. "FoxO Transcription Factors in the Maintenance of Cellular Homeostasis during Aging." *Current Opinion in Cell Biology* 20 (2). https://doi.org/10.1016/j.ceb.2008.02.005.

Scadden, David T. 2014. "Nice Neighborhood: Emerging Concepts of the Stem Cell Niche." *Cell* 157 (1). https://doi.org/10.1016/j.cell.2014.02.013.

Scheibel, Arnold B. 1992. "Structural Changes in the Aging Brain." In: James E. Birren, Gene D. Cohen, Robert Bruce Sloane (Eds.), *Handbook of Mental Health and Aging*. Elsevier. https://doi.org/10.1016/B978-0-12-101277-9.50010-3.

Schumacher, Björn, Joris Pothof, Jan Vijg, and Jan H. J. Hoeijmakers. 2021. "The Central Role of DNA Damage in the Ageing Process." *Nature* 592 (7856). https://doi.org/10.1038/s41586-021-03307-7.

Shammas, Masood A. 2011. "Telomeres, Lifestyle, Cancer, and Aging." *Current Opinion in Clinical Nutrition and Metabolic Care* 14 (1). https://doi.org/10.1097/MCO.0b013e32834121b1.

Shen, Xiaotao, Chuchu Wang, Xin Zhou et al. 2024. "Nonlinear Dynamics of Multi-Omics Profiles during Human Aging." *Nature Aging* 4 (11). https://doi.org/10.1038/s43587-024-00692-2.

Sherlock, Mark, and Andrew A. Toogood. 2007. "Aging and the Growth Hormone/Insulin like Growth Factor-I Axis." *Pituitary* 10 (2). https://doi.org/10.1007/s11102-007-0039-5.

Shetty, Shilpa S., Deepthi D'Souza, Harshitha Somanatha et al. 2023. "Environmental Pollutants and Their Effects on Human Health." *Heliyon* 9 (9). https://doi.org/10.1016/j.heliyon.2023.e19496.

Sletten, Tracey L., Francesco P. Cappuccio, Alec J. Davidson, Eve Van Cauter, Shantha M. W. Rajaratnam, and Frank A. J. L. Scheer. 2020. "Health Consequences of Circadian Disruption." *Sleep* 43 (1). https://doi.org/10.1093/sleep/zsz194.

Spadaro, Olga, Yeonhee Youm, Irina Shchukina et al. 2022. "Caloric Restriction in Humans Reveals Immunometabolic Regulators of Health Span." *Science* 375 (6581). https://doi.org/10.1126/science.abg7292.

Sprenger, Cynthia C., Stephen R. Plymate, and May J. Reed. 2010. "Aging-related Alterations in the Extracellular Matrix Modulate the Microenvironment and Influence Tumor Progression." *International Journal of Cancer* 127 (12). https://doi.org/10.1002/ijc.25615.

Srikanthan, Krithika, Andrew Feyh, Haresh Visweshwar, Joseph I. Shapiro, and Komal Sodhi. 2016. "Systematic Review of Metabolic Syndrome Biomarkers: A Panel for Early Detection, Management, and Risk Stratification in the West Virginian Population." *International Journal of Medical Sciences* 13 (1). https://doi.org/10.7150/ijms.13800.

Statzer, Cyril, Ji Young Cecilia Park, and Collin Y. Ewald. 2023. "Extracellular Matrix Dynamics as an Emerging yet Understudied Hallmark of Aging and Longevity." *Aging and Disease* 14 (3). https://doi.org/10.14336/AD.2022.1116.

Stenvinkel, Peter. 2001. "Endothelial Dysfunction and Inflammation: Is There a Link?" *Nephrology Dialysis Transplantation* 16 (10). https://doi.org/10.1093/ndt/16.10.1968.

Sweeney, Patrick, Hyunsun Park, Marc Baumann et al. 2017. "Protein Misfolding in Neurodegenerative Diseases: Implications and Strategies." *Translational Neurodegeneration* 6 (1). https://doi.org/10.1186/s40035-017-0077-5.

Takasugi, Masaki, Yuya Yoshida, Eiji Hara, and Naoko Ohtani. 2023. "The Role of Cellular Senescence and SASP in Tumour Microenvironment." *The FEBS Journal* 290 (5). https://doi.org/10.1111/febs.16381.

Tessier, Anne-Julie, Fenglei Wang, Andres Ardisson Korat et al. 2025. "Optimal Dietary Patterns for Healthy Aging." *Nature Medicine* 31 (5). https://doi.org/10.1038/s41591-025-03570-5.

Tomtheelnganbee, Elangbam, Puja Sah, and Rohit Sharma. 2022. "Mitochondrial Function and Nutrient Sensing Pathways in Ageing: Enhancing Longevity through Dietary Interventions." *Biogerontology* 23 (6). https://doi.org/10.1007/s10522-022-09978-7.

Torre, Annamariala, Filomena Lo Vecchio, and Antonio Greco. 2023. "Epigenetic Mechanisms of Aging and Aging-Associated Diseases." *Cells* 12 (8). https://doi.org/10.3390/cells12081163.

Trepanowski, John F., Robert E. Canale, Kate E. Marshall, Mohammad M. Kabir, and Richard J. Bloomer. 2011. "Impact of Caloric and Dietary Restriction Regimens on Markers of Health and Longevity in Humans and Animals: A Summary of Available Findings." *Nutrition Journal* 10 (1). https://doi.org/10.1186/1475-2891-10-107.

Tumurkhuu, Munkhtuya, Shahnawaz Ali Bhat, Md Zakir Hossain et al. 2025. "Editorial: Cellular and Molecular Mechanisms in Metabolic Disorders: Role of Inflammation and Oxidative Stress." *Frontiers in Pharmacology* 16 (March). https://doi.org/10.3389/fphar.2025.1580553.

Uddin, Fathema, Charles M. Rudin, and Triparna Sen. 2020. "CRISPR Gene Therapy: Applications, Limitations, and Implications for the Future." *Frontiers in Oncology* 10 (August). https://doi.org/10.3389/fonc.2020.01387.

Vaiserman, Alexander, and Dmytro Krasnienkov. 2021. "Telomere Length as a Marker of Biological Age: State-of-the-Art, Open Issues, and Future Perspectives." *Frontiers in Genetics* 11 (January). https://doi.org/10.3389/fgene.2020.630186.

Wadden, Thomas A., Meghan L. Butryn, and Christopher Wilson. 2007. "Lifestyle Modification for the Management of Obesity." *Gastroenterology* 132 (6). https://doi.org/10.1053/j.gastro.2007.03.051.

Wang, Kang, Huicong Liu, Qinchao Hu et al. 2022. "Epigenetic Regulation of Aging: Implications for Interventions of Aging and Diseases." *Signal Transduction and Targeted Therapy* 7 (1). https://doi.org/10.1038/s41392-022-01211-8.

Xie, Sijia, Xinwei Luo, Feitong Hong et al. 2025. "Towards Precision Aging Biology: Single-Cell Multi-Omics and Advanced AI-Driven Strategies." *Aging and Disease*, March 17. https://doi.org/10.14336/AD.2025.0218.

Yapa, Nethmi M. B., Valerie Lisnyak, Boris Reljic, and Michael T. Ryan. 2021. "Mitochondrial Dynamics in Health and Disease." *FEBS Letters* 595 (8). https://doi.org/10.1002/1873-3468.14077.

Zhang, Lei, Louise E. Pitcher, Matthew J. Yousefzadeh, Laura J. Niedernhofer, Paul D. Robbins, and Yi Zhu. 2022. "Cellular Senescence: A Key Therapeutic Target in Aging and Diseases." *Journal of Clinical Investigation* 132 (15). https://doi.org/10.1172/JCI158450.

Zhang, Mingliang, Caiyuan Zhou, Xinguo Li et al. 2025. "Interactions between Gut Microbiota, Host Circadian Rhythms, and Metabolic Diseases." *Advances in Nutrition* 16 (6): 100416. https://doi.org/10.1016/j.advnut.2025.100416.

Ziada, Adam S., Marie-Soleil R. Smith, and Hélène C. F. Côté. 2020. "Updating the Free Radical Theory of Aging." *Frontiers in Cell and Developmental Biology* 8 (September). https://doi.org/10.3389/fcell.2020.575645.

7 Ageing, Neural Communication, and Neurodegeneration

Vijay Kumar

7.1 INTRODUCTION

Ageing is a natural and complex biological process involving physiological and functional decline in living organisms with time and is marked by progressive deterioration of cells, tissues, and organs, leading to reduced ability to repair, maintain homeostasis, and adapt to environmental stresses (Hou et al. 2019). At the physiological level, ageing is driven by different factors like cellular damage, genetic mutations, oxidative stress, telomere shortening, and reduced regenerative capacity. All these conditions lead to different neurological disorders, cardiological disorders, and malfunctioned immune system (Wyss-Coray 2016). As per United Nations reports (2015), in the next 35 years, the number of people aged 60 and older in the world will reach almost 2.1 billion people. Most of the contribution to this growth will come from the developing regions; however, people aged above 80 years will be from developed countries (Gaspar, Diogo, and Joana 2023). The disease-free brains in the elderly population are very rare and brain ageing might continue with neurodegeneration. The progression of ageing depends upon human genetics and different environmental factors. Despite the improvements in health care and life expectancy, the most observed neurological disorders in the older population are Alzheimer's disease (AD) and Parkinson's disease (PD), and progression of these disorders increases with age (Elobeid et al. 2016). The molecular studies showed the presence of abnormal deposits of aggregated proteins like amyloid-β (Aβ), α-synuclein, and hyperphosphorylated tau (p-tau) in the brain tissue of old-aged individuals. However, the rate of cognitive impairments depends upon the level of these deposits, which still remains unclear (Dean III et al. 2014). The factors that may also play a vital part in brain ageing and neurodegeneration are immune cells and secreted communication factors, which are essential for tissue homeostasis (Gaspar, Diogo, and Joana 2023). The studies on animal models have suggested that the signals from the circulatory system can increase or decrease the various aspects of ageing and cognition (Hou et al. 2019). The most common age-related neurological brain disorders and their estimated prevalence as reported in studies are shown in Table 7.1. The frequency of AD in people aged ≥95 in the United States is ~50% while PD prevalence increases with age (Hou et al. 2019). In this chapter, we will discuss the biological processes associated with ageing, neural communication, and neurodegeneration and also highlight the two most prevalent brain disorders.

7.2 THE BIOLOGY OF AGEING

Ageing as described is an inevitable permanent phenomenon normally associated with easily identifiable characteristic changes in physical functioning and overall appearance of living beings. The research on diverse mammalian species led to the identification of various molecular and cellular processes and biomarkers linked with ageing. There are nine commonly known "hallmarks of ageing" identified by Lopez-Otin et al. and further classified into primary, antagonistic, and integrative hallmarks (López-Otín et al. 2013). The primary hallmarks include genome instability, telomere

DOI: 10.1201/9781003610274-9

TABLE 7.1
Age Associated Neurodegenerative Diseases

Disease	Prevalence	Main Symptoms	Risk Factors	Neurological Hallmarks
Alzheimer's disease (AD)	Crossed ~50 million over the world till 2023 (Gustavsson et al. 2023)	Defects in learning and memory, cognitive impairment, speech difficulties	Age, family history of head trauma, female gender, genetics, environmental factors	Aβ plaques, neurofibrillary tangles, neuronal loss, neuroinflammation
Parkinson's disease (PD)	2%–3% of global population aged over 65	Muscle stiffness, tremors, changes in speech and gait	Environmental factors, male gender, genetics, ethnicity, age	α-Synuclein containing Lewy bodies, grey matter atrophy, loss of dopaminergic neurons
Amyotrophic lateral sclerosis (ALS)	0.6–3.8 per 100 000 population in 2019 (Longinetti and Fang 2019), in Japan number exceeds 23.46 per 100,000 persons (Wolfson et al. 2023).	Progressive motor defects with muscle weakness, spasms, and atrophy	Environmental and risk exposure like pesticides, smoking, head injury, genetics	TAR-DNA binding protein 43 aggregation
Huntington's disease (HD)	2.7 per 100,000 persons globally (Medina et al. 2022)	Cognitive loss, chorea, dystonia, loss of coordination, behavioural difficulties	Genetic mutation in HTT, inheritance	Neuronal loss, psychiatric symptoms, striatal atrophy
Cockayne syndrome	2–3 per million globally (Karikkineth et al. 2017)	Growth failures, neurological disorders, premature ageing, photosensitivity, eye disorders	Genetics (mutation in CSA or CSB genes)	Growth retardation, neurodegeneration

attrition, dysfunction of proteostasis, and epigenetic modifications. Among these, genomic instability is widely recognized as a major driving force of ageing and remains a primary central focus of the established theories of ageing. The mitochondrial dysfunction, uncontrolled nutrient sensing, and cellular senescence are the antagonistic hallmarks while stem cells depletion and altered intercellular communication come under integrative hallmarks, which are mainly outcomes of the primary and antagonistic hallmarks. The brain tissue known as neural tissue primarily consists of postmitotic cells like neurons and oligodendrocytes, which are highly sensitive to the ageing effects because postmitotic cells are more exposed to DNA damage compared to dividing cells (López-Otín et al. 2013).

7.2.1 The Primary Hallmarks of Ageing

Genomic instability and DNA damage: Genome integrity and stability are ubiquitously thriving under vulnerability to various external, chemical, physical, and biological agents, and through indigenous challenges like infidelity in DNA replication, malfunctioning during chromosomal segregation, oxidative stresses, and continuous hydrolytic reactions. Different types of DNA damage caused by external or internal agents include point mutations, translocations, DNA single and double-strand breaks, insertions, deletions, telomere shortening, and gene disruption caused by viruses, base mismatches, etc. (Hou et al. 2019; López-Otín, Blasco et al. 2023). All these factors contribute to both normal and disease-associated ageing. DNA damage caused by oxidative stress via reactive oxygen species (ROS) leads to increased inflammation, accelerates ageing, and overall increases the susceptibility to neurodegeneration. The major DNA repair pathways in human are double-strand break repair, base excision repair (BER), mismatch repair, nucleotide excision repair (NER), and direct reversal. As the age progresses, the efficiency of all these repair pathways is compromised and leads to the accumulation of a damaged genomic DNA pool in the cytosol. Mutations in the genes of these pathways lead to the mutated phenotype of genes and overall increase the risk of neurodegeneration and ageing (Miller et al. 2022). All these alterations in the genomic DNA may affect the vital genes and essential transcriptional processes, resulting in damaging the cells and overall leading to the compromised organismal homeostasis. This becomes more important when the DNA alterations causes impact on stem cells and hamper their functions of tissue regeneration, and lead to their exhaustion which further promotes ageing and makes people more prone to age-related disorders (Miller et al. 2022; López-Otín, Blasco et al. 2023; Figure 7.1).

Besides alterations in genomic DNA, factors affecting the mitochondrion DNA (mtDNA) also contribute to ageing and age-related pathological conditions. Numerous studies showed that mtDNA is strongly affected by age-linked mutations, DNA modifications due to high replicative index, and a less-efficient DNA repair mechanism. Also, deep sequencing data analysis of aged cells showed that mtDNA mutational load may increase significantly through clonal expansion, and increased expansion of mtDNA mutations with age is also observed in the primary oocytes, somatic tissues, and more importantly in the lymphoblasts of patients with neurological disorders (Greaves et al. 2014).

Telomere attrition: Telomeres are the genomic portions composed of DNA and protein at the ends of the chromosomes. The ageing and age-related disorders are also associated with DNA damage at the end of linear chromosomes (telomeric regions). DNA polymerase during replication could not complete the copy of telomeric regions of eukaryotic DNA. Subsequently, after several rounds of cell division, telomeric regions at the ends of the chromosomes undergo significant shortening in length, which leads to genomic instability and induces cells to go finally for apoptosis or cell

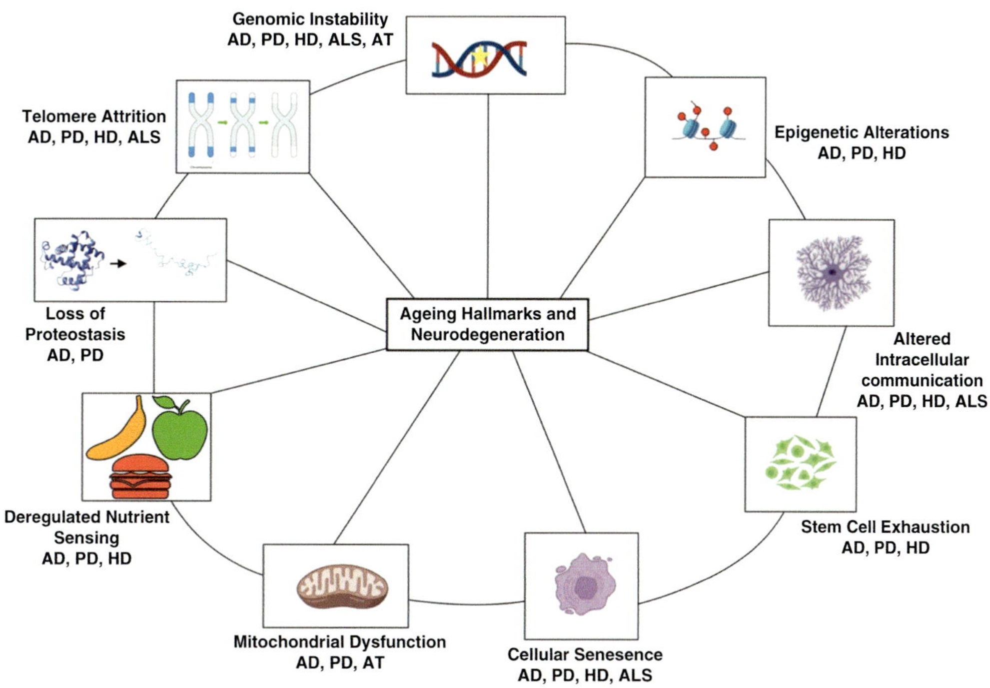

FIGURE 7.1 Hallmarks of ageing (nine characteristic hallmarks of ageing seen in neurodegenerative diseases—AD, Alzheimer's disease; PD, Parkinson's disease; HD, Huntington's disease; ALS, amyotrophic lateral sclerosis; AT, ataxia telangiectasia).

senescence. The telomerase enzyme with its reverse transcriptase activity prevents the cells from these detrimental effects of telomere shortening and elongates the telomere ends. However, expression of telomerase enzyme varies and mostly depends upon the cell types like most of somatic cells in mammals do not express telomerase enzyme, which leads to the cumulative shortening of chromosome ends with telomeric regions throughout life (López-Otín, Pietrocola et al. 2023). Also, there are several reports where telomere deletions in malignant cells reduce carcinogenesis by limiting the life span of corresponding cells. The genomic instability unequivocally favours the oncogenesis while telomeric attrition in contrast antagonizes malignancy and this is the reason why telomeric shortening considered a hallmark of ageing separate from genomic instability (López-Otín, Pietrocola et al. 2023). Studies on the animal models with genetic modifications have shown the linkage of telomeric shortening with cell senescence and organisms' ageing. Also, early ageing in the telomerase-deficient mice can be revert back if there is genetically activation of the telomerase gene (Tomás-Loba et al. 2008; Figure 7.2).

DNA damage caused by endogenous or exogenous agents contributes to both normal and pathological ageing. A decrease in the efficiency of DNA repair, excessive DNA damage, changes in nuclear architecture and telomere shortening favour the ageing process.

Epigenetic modifications: The epigenetics modifications that are associated with ageing include DNA methylation patterns, irregular post-translational modification of histones, abnormalities in chromatin remodelling, and non-regulated functions of non-coding RNAs. Altogether, these

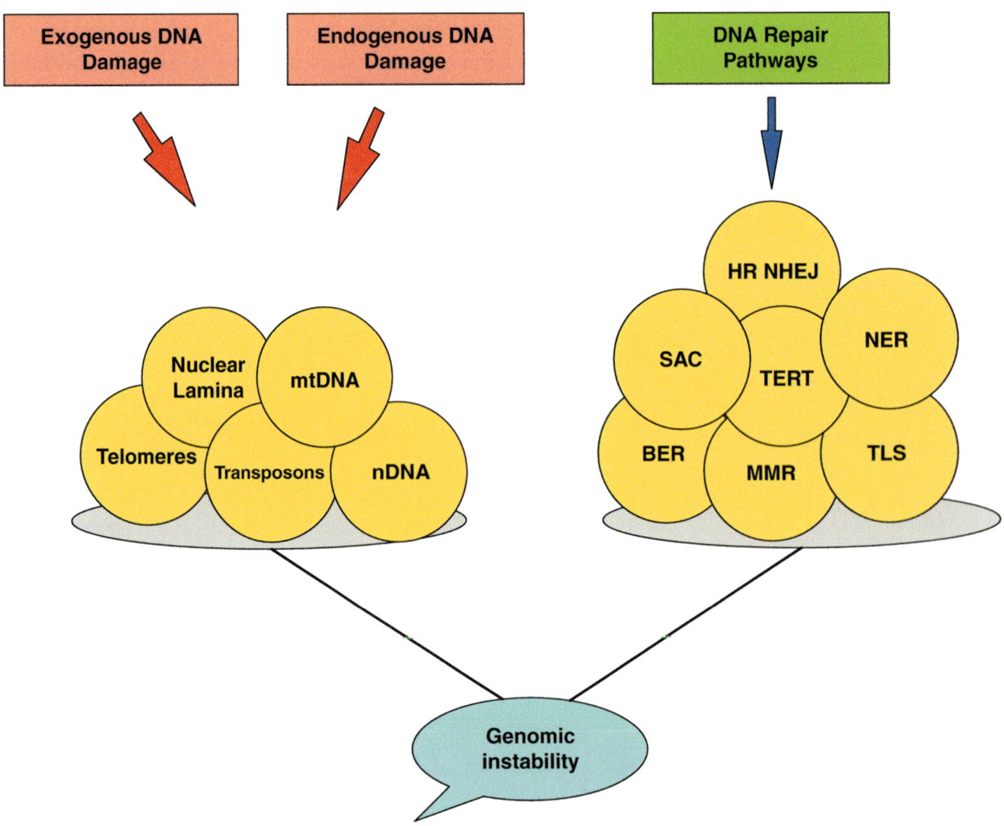

FIGURE 7.2 Loss of cellular integrity owing to genomic instability, telomere attrition, and epigenetic alterations. BER, base excision repair; NER, nucleotide excision repair; HR, homologous recombination; NHEJ, non-homologous end joining; MMR, mismatch repair; SAC, spindle assembly checkpoint; TERT, telomerase reverse transcriptase; TLS, *trans*lesion synthesis.

epigenetic regulatory alterations produce a significant effect on gene expression and other molecular and cellular processes lead to the progression of several age-dependent human disorders like cancer, neurological diseases, bone diseases, and metabolic disorders. A wide range of enzymatic systems contribute to the formation and maintenance of epigenetic functions, and these are DNA methyltransferases, methylases, histone acetylases, demethylases, and protein assemblies involved in the remodelling of chromatins or in the synthesis and growth of non-coding RNAs (López-Otín, Blasco et al. 2023). The changes in the DNA methylation patterns are interrelated with ageing; however, there is no definitive proof that shows they are actually involved in ageing. Among other epigenetic regulatory mechanisms, widespread loss of histones and tissue-dependent variations in their posttranslational modifications are also closely associated with ageing and neurodegeneration. Studies showing that an increase in the histone expression led to extended life span in *Drosophila* while elevated histone H4K16 acetylation or H3K4 trimethylation is present in fibroblasts of aged people and progeroid patients (E. S. Oh and Petronis 2021).

Loss of proteostasis: Proteostasis, a protein homeostasis, is the delicate balance of protein synthesis, folding, trafficking, and degradation in cells. It ensures that proteins maintain their correct structure and function while preventing the accumulation of misfolded or damaged proteins. As the organisms' age, this balance becomes disrupted, leading to a condition known as loss of proteostasis. This phenomenon is linked to numerous age-related diseases, including neurodegenerative disorders, cardiovascular diseases, and certain types of cancer. The onus of maintaining the proteostasis in the mammalian cells relies on the good regulation of proteosome and on the process of autophagy, as well as machinery involved in the ubiquitination and lysosomal system (López-Otín et al. 2013). Ageing and age-related disorders such as PD, AD, ALS, and epilepsy are linked with damaged or dysfunctional protein homeostasis, which leads to accretion of unfolded, misfolded, glycated, oxidized, or ubiquitinylated proteins, which form the insoluble aggregates as internal inclusion bodies or extracellular amyloid plaques (Martinez-Miguel et al. 2021).

7.2.2 Antagonistic Hallmarks of Ageing

Mitochondrial dysfunction and mitophagy: With ageing, the efficiency of our body system to neutralize the ROS and oxidative stress decreases and leads to cumulative accumulation of oxidative damage which is associated with deteriorating mitochondrial function. Other intertwined processes linked to mitochondrial dysfunction include mutations in mtDNA, destabilization of respiratory complexes due to deficient proteostasis, reduction in mitochondrial biogenesis, protein oxidation, and overall changes in the mitochondrial dynamics (Amorim et al. 2022). This situation hampers the functioning of mitochondria to cellular bioenergetics, increases the production of ROS, and triggers the accidental membrane permeabilization to apoptosis and cell death. Rationally with the beginning, the function of mitochondria is to maintain health and their progressive deterioration contribute to ageing phenotype. Also, neurons are highly metabolically active cells, require a high amount of energy to perform its functional activities, and changes in the functioning of mitochondria with ageing affects the overall neuronal functions and in some cases leads to neurological disorders (Keogh and Chinnery 2015). The quality mechanisms which pioneered in maintaining the mitochondrial in optimal condition are proteosome degradation, removal of damaged proteins, fusion and fission reactions in mitochondria, use of proteases to get rid of damaged proteins, and mitophagy. Among these, mitophagy has received considerable attention in recent years. Mitophagy is one of the kinds of autophagy where it specifically degrades the dysfunctional mitochondria and plays a vital role in preventing the age-related health disorders especially neurological diseases. The underlying mechanisms of mitophagy include receptor-mediated and PINK1/Parkin-dependent mitophagy (Hou et al. 2019).

Cellular senescence: Cellular senescence, first identified by Hayflick in 1965 in fibroblast cells, is a stress-induced state in the cell where it goes for stable cell-cycle arrest and senescence-related phenotype like SASP (senescence-associated secretary phenotype) expressed in the cells with age

(Hayflick 1965; Kültz 2005). Cellular senescence is a natural response of the cells to maintain a healthy survival cells pool while clearing the damaged cells. Rate of cell senescence varies in cells of different tissues and about 2–20× fold variation in the rates among individuals aged <35 years old to >65 years old and affects certain cell types like fibroblast, immune cells, and endothelial cells; however, all cells undergo senescence with age (López-Otín, Blasco et al. 2023). Cells having mitochondria dysfunction associated senescence have a low level of NAD^+:NADH ratio, and this leads to growth arrest and stops cells from SASP formation through IL-1. The cells having a higher amount of accumulated DNA damage become prone to senescence and marked for growth arrest. For example, neurons with their high energy requirements and increased amount of ROS in neurons lead to the accumulation of a high amount of DNA damage in brain tissue, and it increases with age and linked with neurodegeneration (Nacarelli et al. 2019).

Deregulated nutrient sensing and altered metabolism: Nutrient-sensing networks are mostly conserved throughout the evolution and consist of extracellular ligands like insulin, insulin growth factors (IGFs), tyrosine kinase receptor, and intracellular signalling cascades like PI3K-AKT and Ras-MEK-ERK pathways. It also includes transcription factors like E26 and FOXOs, which are involved in different cellular processes (Slack et al. 2015). It has been reported that genetically reducing the activity of nutrient sensing networks can increase the life and health span in a wide variety of animal models. There is also a correlation of epigenetic age with nutrient sensing in human cells (López-Otín, Blasco et al. 2023). Nutrient-sensing pathways like mTOR, AMPK, and Sirtuins are potential therapeutic targets for neurological disorders and are being explored (López-Otín et al. 2013). The studies showed that overnutrition increases the progression of ageing through multiple ways like triggering the internal nutrient sensors like MTORC1 and acetyltransferase EP300, and inhibition of nutrient scarcity detecting sensing elements like AMPK, SIRT1, and SIRT3. While on the opposite side fasting and dieting constraints inhibit MTORC1 and EP300 and activate AMPK, SIRT1, and SIRT3 and overall increase the longevity in diverse animal models including primates (López-Otín, Blasco et al. 2023).

7.2.3 Integrative Hallmarks of Ageing

Stem cell exhaustion: Every individual needs functional stem cells for maintaining optimal health in the later part of life. However, like other cell types, the functional and proliferative power of stem cells declines with organism's age. This reduced functional power of stem cells can be caused due to DNA damage, malfunctional proteostasis, epigenetic dysregulation, inefficient DNA repair, telomerase inactivation, and cell senescence (Oh, Lee, and Wagers 2014). Numerous studies have shown that stem cell function can be improved by increasing the expression of certain genes like HSP90 or FOXO4 transcription factor and rapamycin treatment. Also, in aged animals, exposure to young blood by joining the circulatory system of old and young, surgical procedure called heterochronic parabiosis helps in improving the stem cell function in liver, spinal cord, brain, and muscle and overall reduces the ageing and restores the cognitive processes. Therefore, targeting stem cell regeneration might be a potential therapeutic strategy to attenuate age-related neurodegenerative disorders (Villeda et al. 2014).

Altered intercellular communication and immune function: Ageing is a natural inevitable process linked to gradual alterations in intercellular communication resulting in the accumulation of noise, leading to compromised hermetic and homeostatic regulation. It leads to variations in the level of different hormones like insulin, leptin, adiponectin, and growth factors like IGF1 and further regulates the neuronal damage and neurodegeneration (Amorim et al. 2022). The immune system plays an essential role in shaping the brain during development, and both nervous and immune systems undergo systematic changes with age, leading to further loss of regulation among these systems, which further contribute to neurodegeneration. Recent studies showed that the pathological mechanism underlying neurodegeneration might involve the interactions of microglial cells (innate resident immune cells in the central nervous system) with astrocytes and oligodendrocytes. The coss talks among mast cells, microglia, T cells, and oligodendrocytes are vital in the functioning of the immune system and also help in preventing the progression of neurodegeneration. Inflammation is a primary protective event that occurs in response to any abuse in the cells and becomes upregulated with the advancement of age and classified into different regimes like low grade, asymptomatic, chromic, systematic, and controlled. Low levels of inflammation might have some good effects, but uncontrolled or sustained inflammation is highly risky and can lead to age-linked neurodegenerative disorders like AD and PD (Cao and Zheng 2018).

7.3 NEURAL COMMUNICATION IN AGEING

Neural communication occurs primarily through synapses, where electrical signals (action potentials) trigger the release of neurotransmitters from a presynaptic neuron to a postsynaptic neuron involving ion channels, receptors, and secondary messengers and further supported by glial cells like astrocytes that regulate the synaptic environment (Verkhratsky and Nedergaard 2017). Efficient communication depends on the structural integrity of neurons (axons, dendrites) and the precision of synaptic transmission, and finally the brain's ability to adapt through plasticity, which is its power to form or strengthen connections. With age, synaptic efficacy declines due to reduced neurotransmitter release, receptor density, and postsynaptic responsiveness. Studies showed there is a decrease in synaptic density in the prefrontal cortex and hippocampus regions critical for memory and executive function (Morrison and Baxter 2012). For example, the number of dendritic spines, the sites of excitatory synapses, diminishes with age and leads to changes in signal transmission. This is in part linked to altered calcium signalling, where dysregulation in aged neurons leads to weaker synaptic plasticity, like long-term potentiation, a key mechanism for learning (Kumar et al. 2007).

Aging affects neurotransmitter systems variably including dopamine levels, which drop significantly in the basal ganglia, contributing to motor slowing and PD risk, with losses of 5%–10% per decade after age 40 (Bäckman et al. 2010). Acetylcholine is vital for memory but declines in the cholinergic system notably in the basal forebrain, correlating with AD susceptibility (Schliebs and Arendt 2011). Glutamate and GABA, the brain's primary excitatory and inhibitory transmitters, also show imbalances disrupting the excitatory-inhibitory equilibrium and network stability. All this contributes to slower processing speeds and reduces inter-regional communication evident in the tasks requiring coordination across brain areas (Peters 2006). Ageing increases oxidative stress and inflammation, damaging the neuronal membranes and DNA. ROS accumulate due to declining mitochondrial efficiency and impairs the energy supply for synaptic activity (Mattson and Arumugam 2018). Chronic low-grade inflammation, termed "inflammaging," involves microglia activation, releasing cytokines that disrupt synaptic pruning and plasticity (Franceschi et al. 2018). Moreover, weakened plasticity heightens vulnerability to neurodegenerative diseases like Alzheimer's, where early synaptic failure precedes neuron death (Saxena and Caroni 2011).

7.4 NEURODEGENERATION AND NEUROLOGICAL DISORDERS

Neurodegeneration involves the neuronal loss marked by progressive death of neuronal cells and is characterized by cognitive and motor dysfunction, depending on the brain area where loss of neurons occurs (Miller et al. 2022). The changes linked with ageing affect the brain and lead to neurodegeneration and are further associated with different neurodegenerative disorders. We will discuss the most prevalent neurodegenerative diseases AD and PD in this section and shed some lights on other important neurological disorders.

7.4.1 Alzheimer's Disease (AD)

AD is one of the common neurological disorders and primarily affects the brain, leading to a gradual loss of memory, cognitive abilities, and changes in behavioural patterns. In senior individuals, it is the prime cause of dementia and becoming one of the most lethal, expensive, and burdening diseases of this century. The characteristic clinical feature of AD includes learning defects, late memory loss, mood swings, behavioural issues, and disorientation. It was first reported in 1906 by Dr Alois Alzheimer and characterized by deposition of abnormal protein aggregates in the brain. These deposits hamper the transmission of nerve impulses between the neurons and lead to cell death and overall deterioration of brain tissue (Gadhave et al. 2024; Scheltens et al. 2021). It has been reported that the defects in at least three proteins, namely, presenilin 1, presenilin 2, and amyloid precursor proteins (APPs) are associated with risk factors in AD. The most prominent molecular features in AD are Aβ plaques and p-tau neurofibrillary tangles in the brain. A lot of studies suggested that both these Aβ and tau proteins misfold, self-assemble, and propagate through an internal mechanism that closely matches the seeded propagation and aggregation of prion protein in Creutzfeldt–Jakob disease (Jucker and Walker 2023). The sequential cleavage of APP by β-secretase and γ-secretase led to the formation of a short abnormal peptide containing 36–43 amino acids called Aβ. The exact mechanism that causes AD is still a matter of controversy and requires intensive research; however, factors that might lead to AD include mitochondrial dysfunction, DNA damage, Aβ plaques, tau aggregation and deposition, oxidative stress, and inflammation. At present, there is no clear remedy for the AD, FDA-approved acetylcholinesterase inhibitors like donepezil and galantamine might help in alleviating some symptoms but cannot halt the progression of the disease (Lane, Hardy, and Schott 2018). As per the Aβ hypothesis, deposition of Aβ and toxicity is the starting factor in AD and is thought to progress in two stages—early crucial impetus of AD is Aβ proteopathy where Aβ forms oligomers and fibrils that lead to the deposition of Aβ plaques in the brain, resulting in the loss of neurons. The second stage involves the widespread tauopathy and secondary changes like inflammation, vascular abnormalities, tangles, and neurodegeneration (Jucker and Walker 2023). To find the therapeutic drugs, targeting tauopathy is the potential target for immunization-based strategies while Aβ proteopathy becomes the attractive target for early prevention or measures.

Other important risk factors that might be involved in AD are neuroinflammatory responses, mitochondrial dysfunction, and DNA damage. Neuroinflammatory response is a brain response to internal or external stress, reactive oxidative stress, and accumulation of p-tau or Aβ plaques and all these events lead to the release of pro-inflammatory cytokines owing to abnormal activation of microglial and astrocytes (Hardy 2009). There is sufficient evidence which suggests that DNA damage and inefficient DNA repair machinery exacerbates the progression of AD. Genes involved in DNA damage repair like BER gene expression vary based on the tissue type, cell type, and age and might decrease with the advancement of neurodegeneration. The studies showed there is a decrease in the levels of certain enzymes like Uracil DNA glycosylase, β-OGG1 glycosylase, and Polβ when compared between the post-mortem brain of AD patient of the same age as a control without AD (Weissman et al. 2007). Also, malfunctioning of certain solute transporters is associated with the progression of AD and glutamate transporters are one of them (Scott et al. 2011). Recent studies showed the role of glutamate transporters in neurodegeneration and modulating the activity of glutamate transporters using PAMs (positive allosteric modulators) might serve as a potential therapeutic strategy (van Veggel et al. 2024; 2025). The ageing hallmarks that we discussed in the previous section are directly or indirectly associated with the progression of AD and other neurodegenerative diseases. For example, telomere shortening or instability is associated with ageing and neurodegeneration, and defects in the telomere maintenance are linked with AD patients. Mitophagy plays an essential function as it reduces inflammation and thereby protects cells from neurodegeneration. However, it has been reported that the patients with AD have defective mitophagy based upon the cross-species studies on mice and *C. elegans* animal models and on human iPSC-derived AD neurons (Fang et al. 2019). Altogether, to find a potential therapeutic measure for AD, targeting one or two ageing hallmarks might serve as important therapeutic strategies in future.

7.4.2 Parkinson's Disease

PD is a progressive neurodegenerative disorder that primarily affects movement, though it can also involve non-motor symptoms. It is caused by the gradual loss of dopamine-producing neurons in a region of the brain called the substantia nigra, which plays a key role in coordinating movement. Dopamine is a neurotransmitter that facilitates smooth, purposeful muscle activity, and its depletion leads to the characteristic symptoms of PD. The disease was first described by James Parkinson in 1817 in his work "An Essay on the Shaking Palsy," and it remains the second most common neurological disorder globally, affecting millions of people. The exact cause associated with PD is not fully understood but it is believed that genetic and environmental factors play an important role in combination (Poewe et al. 2017). Approximately 15% of PD reports are linked with family history with mutation in certain genes like LRRK2, PARK7, and SNCA (Farrer 2006). Environmental factors including exposure to pesticides and heavy metals have been reported to fasten the progression of PD (Ascherio and Schwarzschild 2016). Another important consideration is ageing, and the aged population becomes more prone to neurodegeneration including PD; however, early onset can also affect the younger individual. The prominent diagnostics include neuromuscular dysfunction, which affects the motor abilities and speed, rigidity, or rest tremor. At the molecular level, different pathways and mechanisms are involved in PD including α-synuclein proteostasis, mitochondrial dysfunction, oxidative stress, calcium imbalance, and neuroinflammation (Hou et al. 2019).

The characteristic pathological feature of PD is the presence of Lewy bodies, which are abnormal protein aggregates primarily made up of α-synuclein proteins in the brain. These aggregates disrupt the neuronal function and lead to cell death; however, a precise underlying mechanism is still not fully understood. Aggregation of α-synuclein and defects in mitochondrial function are the prominent synergetic hallmarks of PD. It has been reported that the mitochondrial dysfunction promotes aggregation of α-synuclein in degenerating neurons in PD patients (Melki 2015). There is no cure for PD and most of the treatments include only managing the symptoms and improving the quality of life. The cornerstone of therapy is levodopa, which is a precursor of dopamine and crosses the blood–brain barrier and raises the dopamine levels. It is normally combined with carbidopa to reduce the risk of side effects (Olanow, Stern, and Sethi 2009).

7.4.3 Other Neurological Disorders

We highlighted in detail the two most prominent neurological diseases AD and PD while other neurological disorders associated with ageing are ALS, HD, Cockayne syndrome, and Taxia telangiectasia. We briefly discuss some of these in this section.

Amyotrophic lateral sclerosis (ALS) is a devastating neurodegenerative disease that affects nerve cells in the brain and spinal cord, leading to progressive muscle weakness and eventually leading to paralysis. ALS is characterized by the degeneration of motor neurons, the nerve cells responsible for controlling voluntary muscle movements such as walking, speaking, and breathing. The term "amyotrophic" refers to muscle atrophy due to the lack of nourishment, while "lateral sclerosis" describes the hardening of the spinal cord's lateral columns as motor neurons die. The exact cause of ALS remains elusive in most cases and there is no cure for ALS and treatments mainly aim to slow the progression and manage symptoms. Riluzole, approved in 1995, modestly extends survival by reducing glutamate excitotoxicity (Dharmadasa and Kiernan 2018).

Huntington's disease (HD) is a rare, inherited neurodegenerative disorder that progressively destroys nerve cells in the brain, leading to movement disorders, cognitive decline, and psychiatric symptoms. HD primarily affects the basal ganglia that regulate movement, emotion, and cognition. The disease is driven by the degeneration of neurons, especially in the striatum, a key component of the basal ganglia (Gadhave et al. 2024). It is mainly caused by a single mutation in the HTT gene on chromosome 4 and especially the expansion of CAG trinucleotide repeats. These expanded

CAG repeats produce abnormal Huntingtin protein, which aggregates in the neurons and disrupts the cellular processes like protein clearance, mitochondrial function, and leads to cell death (Bates et al. 2015). At present, there is no cure for HD, and treatments focus on symptom management. Tetrabenazine and deutetrabenazine, approved by the FDA, reduce chorea by modulating dopamine levels, while antipsychotics or antidepressants address psychiatric symptoms (McGarry et al. 2020).

7.5 THERAPEUTIC APPROACHES IN NEURODEGENERATIVE DISORDERS

The potential therapeutic approaches to find treatment for neurological disorders associated with ageing are NAD^+ supplementation, mitophagy induction, inhibiting cellular senescence, targeting proteins aggregation, neuroinflammation, and nutrient-sensing metabolism artefacts.

7.5.1 NAD^+ (Nicotinamide Adenine Dinucleotide) Supplementation

NAD^+ supplementation has gained a lot of attention as a potential therapeutic strategy for neurodegenerative diseases due to its critical role in cellular metabolism, energy production, and maintenance of neuronal health. NAD^+ is a coenzyme involved in redox reactions and serves as a substrate for enzymes like sirtuins, PARPs (poly-ADP-ribose polymerases), and CD38, which regulate DNA repair, gene expression, and mitochondrial functions, and all the processes implicated in neurodegeneration. NAD^+ levels naturally decline with age, and this depletion exacerbates these pathological processes. Studies in mouse models showed that NAD^+ precursors like nicotinamide mononucleotide (NMN) or nicotinamide riboside improve cognitive function, reduce amyloid plaques, and enhance synaptic plasticity. NAD^+ precursors have been shown to protect dopaminergic neurons in models of Parkinson's by improving mitochondrial dynamics and reducing oxidative stress (Hou et al. 2019).

7.5.2 Mitophagy Induction

Mitophagy is the selective autophagic process that targets and removes damaged or dysfunctional mitochondria maintaining cellular homeostasis and is particularly relevant in neurodegeneration, where mitochondrial dysfunction is a hallmark of neurodegenerative diseases. Several compounds and approaches have been explored to enhance mitophagy, with evidence from preclinical and some clinical studies. Boosting NAD^+ levels activate sirtuins (e.g., SIRT1), which deacetylate PGC-1α and upregulate mitophagy genes, or enhance PINK1-Parkin signalling indirectly via mitochondrial health. Research showed that NMN restore mitophagy in AD mouse models by enhancing NAD^+-SIRT1 signalling, reducing Aβ burden (Fang et al. 2019).

7.5.3 Inhibition of Cellular Senescence

Cellular senescence is a state of irreversible cell-cycle arrest triggered by stressors like DNA damage, oxidative stress, or telomere shortening. While it serves as a protective mechanism against cancer, the accumulation of senescent cells contributes to ageing and age-related diseases, including neurodegeneration. Inhibiting cellular senescence or selectively clearing senescent cells known as senolysis has emerged as a therapeutic strategy against neurodegeneration. A combination of Dasatinib and Quercetin proves to reduce the burden of senescent cells, tau pathology, and cognitive deficits (Musi et al. 2018). SASP inhibitors block inflammatory signalling (e.g., NF-κB and JAK/STAT) to reduce SASP without killing senescent cells.

7.5.4 Targeting Protein Aggregation

Protein aggregation is a central pathological feature in neurodegenerative diseases. Misfolded proteins including Aβ and tau in AD, α-synuclein in PD, huntingtin in HD, and TDP-43 in ALS form aggregates that disrupt cellular function, trigger inflammation, and cause neuronal death. Targeting protein aggregation involves preventing misfolding, enhancing clearance, or mitigating toxicity. Small-molecule chaperons stabilize the native protein conformation and prevent it from misfolding. For example, Tafamidis stabilizes transthyretin to prevent amyloidosis and protect toxicity (Eisele et al. 2015).

7.5.5 Targeting Metabolism and Inflammation

Targeting metabolism and inflammation in neurodegeneration addresses two interconnected pathological motors including metabolic dysfunction, which impairs neuronal energy supply and homeostasis, and chronic inflammation, which exacerbates neuronal damage. These processes are hallmarks of neurological diseases. Boosting mitochondrial function by increasing the NAD^+ availability via supplementation of NAD^+ precursors showed a good response in AD models (Fang et al. 2019). The use of inflammation inhibitors like MCC950, which is an NLRP3 inhibitor, reduces the inflammation and Aβ pathology in the AD mouse model (Coll et al. 2015).

7.6 CONCLUSION

Ageing is the most significant risk factor for neurodegenerative diseases like Alzheimer's, Parkinson's, and Huntington's. As we age, the brain undergoes natural changes, accumulating oxidative stress, DNA damage, and impaired cellular repair processes, which increase the vulnerability to neurodegeneration. These processes can lead to the progressive loss of neurons' structure and function and might further lead to brain disorders. While ageing itself doesn't guarantee neurodegeneration, the cumulative wear and tear on biological systems creates a fertile ground for these disorders to emerge, especially when combined with genetic or environmental factors. There have been consistent efforts to find evidence-based treatments for neurodegenerative diseases but till now neither effective treatments nor potent protective strategies are identified. In future, more studies are required to understand the connection between ageing hallmarks and neurodegeneration, including DNA damage and mitochondrial dysfunction which are central, and research investigating the connection between the nucleus and mitochondria will help in understanding the connection between ageing and neurodegenerative diseases. Neurodegenerative disorders are highly complicated in nature with strong influence of genetic and environmental factors toward disease progression, and therefore, a single drug or single pathway targeted treatment strategies might not be sufficient to combat these diseases. Essentially, more adequate holistic approaches or combinational therapies might serve as potential therapeutic measures including multitargeted pharmaceutical evidence-based strategies combined with non-pharma approaches or changes in lifestyle might serve as potential ways to combat or slow down the progression of neurological disorders.

REFERENCES

Amorim, João A, Giuseppe Coppotelli, Anabela P Rolo, Carlos M Palmeira, Jaime M Ross, and David A Sinclair. 2022. "Mitochondrial and Metabolic Dysfunction in Ageing and Age-Related Diseases." *Nature Reviews Endocrinology* 18 (4): 243–58. https://doi.org/10.1038/s41574-021-00626-7.

Ascherio, Alberto, and Michael A Schwarzschild. 2016. "The Epidemiology of Parkinson's Disease: Risk Factors and Prevention." *The Lancet Neurology* 15 (12): 1257–72. https://doi.org/10.1016/S1474-4422(16)30230-7.

Bäckman, Lars, Ulman Lindenberger, Shu-Chen Li, and Lars Nyberg. 2010. "Linking Cognitive Aging to Alterations in Dopamine Neurotransmitter Functioning: Recent Data and Future Avenues." *Neuroscience & Biobehavioral Reviews* 34 (5): 670–77. https://doi.org/10.1016/j.neubiorev.2009.12.008.

Bates, Gillian P, Ray Dorsey, James F Gusella, Michael R Hayden, Chris Kay, Blair R Leavitt, Martha Nance et al. 2015. "Huntington Disease." *Nature Reviews Disease Primers* 1 (1): 15005. https://doi.org/10.1038/nrdp.2015.5.

Cao, Wei, and Hui Zheng. 2018. "Peripheral Immune System in Aging and Alzheimer's Disease." *Molecular Neurodegeneration* 13 (1): 51. https://doi.org/10.1186/s13024-018-0284-2.

Coll, Rebecca C, Avril A B Robertson, Jae Jin Chae, Sarah C Higgins, Raúl Muñoz-Planillo, Marco C Inserra, Irina Vetter et al. 2015. "A Small-Molecule Inhibitor of the NLRP3 Inflammasome for the Treatment of Inflammatory Diseases." *Nature Medicine* 21 (3): 248–55. https://doi.org/10.1038/nm.3806.

Dean III, Douglas C, Beth A Jerskey, Kewei Chen, Hillary Protas, Pradeep Thiyyagura, Auttawat Roontiva, Jonathan O'Muircheartaigh et al. 2014. "Brain Differences in Infants at Differential Genetic Risk for Late-Onset Alzheimer Disease: A Cross-Sectional Imaging Study." *JAMA Neurology* 71 (1): 11–22. https://doi.org/10.1001/jamaneurol.2013.4544.

Dharmadasa, Thanuja, and Matthew C Kiernan. 2018. "Riluzole, Disease Stage and Survival in ALS." *The Lancet Neurology* 17 (5): 385–86. https://doi.org/10.1016/S1474-4422(18)30091-7.

Eisele, Yvonne S, Cecilia Monteiro, Colleen Fearns, Sandra E Encalada, R Luke Wiseman, Evan T Powers, and Jeffery W Kelly. 2015. "Targeting Protein Aggregation for the Treatment of Degenerative Diseases." *Nature Reviews Drug Discovery* 14 (11): 759–80. https://doi.org/10.1038/nrd4593.

Elobeid, Adila, Sylwia Libard, Marina Leino, Svetlana N Popova, and Irina Alafuzoff. 2016. "Altered Proteins in the Aging Brain." *Journal of Neuropathology & Experimental Neurology* 75 (4): 316–25. https://doi.org/10.1093/jnen/nlw002.

Fang, Evandro F, Yujun Hou, Konstantinos Palikaras, Bryan A Adriaanse, Jesse S Kerr, Beimeng Yang, Sofie Lautrup et al. 2019. "Mitophagy Inhibits Amyloid-β and Tau Pathology and Reverses Cognitive Deficits in Models of Alzheimer's Disease." *Nature Neuroscience* 22 (3): 401–12. https://doi.org/10.1038/s41593-018-0332-9.

Farrer, Matthew James. 2006. "Genetics of Parkinson Disease: Paradigm Shifts and Future Prospects." *Nature Reviews Genetics* 7 (4): 306–18. https://doi.org/10.1038/nrg1831.

Franceschi, Claudio, Paolo Garagnani, Paolo Parini, Cristina Giuliani, and Aurelia Santoro. 2018. "Inflammaging: A New Immune–Metabolic Viewpoint for Age-Related Diseases." *Nature Reviews Endocrinology* 14 (10): 576–90. https://doi.org/10.1038/s41574-018-0059-4.

Gadhave, Dnyandev G, Vrashabh V Sugandhi, Saurav Kumar Jha, Sopan N Nangare, Gaurav Gupta, Sachin Kumar Singh, Kamal Dua, Hyunah Cho, Philip M Hansbro, and Keshav Raj Paudel. 2024. "Neurodegenerative Disorders: Mechanisms of Degeneration and Therapeutic Approaches with Their Clinical Relevance." *Ageing Research Reviews* 99:102357. https://doi.org/10.1016/j.arr.2024.102357.

Gaspar, Filipa, Silva Diogo, and Trigo Joana. 2023. "Ageing in the Brain : Mechanisms and Rejuvenating Strategies." *Cellular and Molecular Life Sciences* 80 (7): 1–21. https://doi.org/10.1007/s00018-023-04832-6.

Greaves, Laura C, Marco Nooteboom, Joanna L Elson, Helen A L Tuppen, Geoffrey A Taylor, Daniel M Commane, Ramesh P Arasaradnam et al. 2014. "Clonal Expansion of Early to Mid-Life Mitochondrial DNA Point Mutations Drives Mitochondrial Dysfunction during Human Ageing." *PLOS Genetics* 10 (9): e1004620.

Gustavsson, Anders, Nicholas Norton, Thomas Fast, Lutz Frölich, Jean Georges, Drew Holzapfel, Tunahan Kirabali et al. 2023. "Global Estimates on the Number of Persons across the Alzheimer's Disease Continuum." *Alzheimer's & Dementia* 19 (2): 658–70. https://doi.org/10.1002/alz.12694.

Hardy, John. 2009. "The Amyloid Hypothesis for Alzheimer's Disease: A Critical Reappraisal." *Journal of Neurochemistry* 110 (4): 1129–34. https://doi.org/10.1111/j.1471-4159.2009.06181.x.

Hayflick, L. 1965. "The Limited in Vitro Lifetime of Human Diploid Cell Strains." *Experimental Cell Research* 37 (3): 614–36. https://doi.org/10.1016/0014-4827(65)90211-9.

Herrmann, Markus, Irene Pusceddu, Winfried März, and Wolfgang Herrmann. 2018. "Telomere Biology and Age-Related Diseases," Clinical Chemistry and Laboratory Medicine (CCLM), 56 (8): 1210–22. https://doi.org/doi:10.1515/cclm-2017-0870.

Hou, Yujun, Xiuli Dan, Mansi Babbar, Yong Wei, Steen G Hasselbalch, Deborah L Croteau, and Vilhelm A Bohr. 2019. "Ageing as a Risk Factor for Neurodegenerative Disease." *Nature Reviews Neurology* 15 (10): 565–81. https://doi.org/10.1038/s41582-019-0244-7.

Jucker, Mathias, and Lary C Walker. 2023. "Alzheimer's Disease: From Immunotherapy to Immunoprevention." *Cell* 186 (20): 4260–70. https://doi.org/10.1016/j.cell.2023.08.021.

Karikkineth, Ajoy C, Morten Scheibye-Knudsen, Elayne Fivenson, Deborah L Croteau, and Vilhelm A Bohr. 2017. "Cockayne Syndrome: Clinical Features, Model Systems and Pathways." *Ageing Research Reviews* 33:3–17. https://doi.org/10.1016/j.arr.2016.08.002.

Keogh, Michael J, and Patrick F Chinnery. 2015. "Mitochondrial DNA Mutations in Neurodegeneration." *Biochimica et Biophysica Acta (BBA) - Bioenergetics* 1847 (11): 1401–11. https://doi.org/10.1016/j.bbabio.2015.05.015.

Kültz, Dietmar. 2005. "Molecular and Evolutionary Basis of the Cellular Stress Response." *Annual Review of Physiology* 67 (Volume 67, 2005): 225–57. https://doi.org/10.1146/annurev.physiol.67.040403.103635.

Kumar, Ashok, Jeffrey S Thinschmidt, Thomas C Foster, and Michael A King. 2007. "Aging Effects on the Limits and Stability of Long-Term Synaptic Potentiation and Depression in Rat Hippocampal Area CA1." *Journal of Neurophysiology* 98 (2): 594–601. https://doi.org/10.1152/jn.00249.2007.

Lane, Christopher A, John Hardy, and Jonathan M. Schott. 2018. "Alzheimer's Disease." *European Journal of Neurology* 25 (1): 59–70. https://doi.org/10.1111/ene.13439.

Longinetti, Elisa, and Fang Fang. 2019. "Epidemiology of Amyotrophic Lateral Sclerosis: An Update of Recent Literature." *Current Opinion in Neurology* 32 (5).

López-Otín, Carlos, Maria A Blasco, Linda Partridge, Manuel Serrano, and Guido Kroemer. 2013. "The Hallmarks of Aging." *Cell* 153 (6): 1194–1217. https://doi.org/10.1016/j.cell.2013.05.039.

López-Otín, Carlos, Maria A Blasco, Linda Partridge, Manuel Serrano, and Guido Kroemer. 2023. "Hallmarks of Aging: An Expanding Universe." *Cell* 186 (2): 243–78. https://doi.org/10.1016/j.cell.2022.11.001.

López-Otín, Carlos, Federico Pietrocola, David Roiz-Valle, Lorenzo Galluzzi, and Guido Kroemer. 2023. "Meta-Hallmarks of Aging and Cancer." *Cell Metabolism* 35 (1): 12–35. https://doi.org/10.1016/j.cmet.2022.11.001.

Martinez-Miguel, Victoria Eugenia, Celia Lujan, Tristan Espie--Caullet, Daniel Martinez-Martinez, Saul Moore, Cassandra Backes, Suam Gonzalez et al. 2021. "Increased Fidelity of Protein Synthesis Extends Lifespan." *Cell Metabolism* 33 (11): 2288–2300.e12. https://doi.org/10.1016/j.cmet.2021.08.017.

Mattson, Mark P, and Thiruma V Arumugam. 2018. "Hallmarks of Brain Aging: Adaptive and Pathological Modification by Metabolic States." *Cell Metabolism* 27 (6): 1176–99. https://doi.org/10.1016/j.cmet.2018.05.011.

McGarry, Andrew, Peggy Auinger, Karl Kieburtz, Michal Geva, Munish Mehra, Victor Abler, Igor D Grachev et al. 2020. "Additional Safety and Exploratory Efficacy Data at 48 and 60 Months from Open-HART, an Open-Label Extension Study of Pridopidine in Huntington Disease." *Journal of Huntington's Disease* 9 (2): 173–84. https://doi.org/10.3233/JHD-190393.

Medina, Alex, Yasamin Mahjoub, Larry Shaver, and Tamara Pringsheim. 2022. "Prevalence and Incidence of Huntington's Disease: An Updated Systematic Review and Meta-Analysis." *Movement Disorders* 37 (12): 2327–35. https://doi.org/10.1002/mds.29228.

Melki, Ronald. 2015. "Role of Different Alpha-Synuclein Strains in Synucleinopathies, Similarities with Other Neurodegenerative Diseases." *Journal of Parkinson's Disease* 5 (2): 217–27. https://doi.org/10.3233/JPD-150543.

Miller, Michael B, August Yue Huang, Junho Kim, Zinan Zhou, Samantha L Kirkham, Eduardo A Maury, Jennifer S Ziegenfuss et al. 2022. "Somatic Genomic Changes in Single Alzheimer's Disease Neurons." *Nature* 604 (7907): 714–22. https://doi.org/10.1038/s41586-022-04640-1.

Morrison, John H., and Mark G. Baxter. 2012. "The Ageing Cortical Synapse: Hallmarks and Implications for Cognitive Decline." *Nature Review Neuroscience* 13: 240–50. https://doi.org/10.1038/nrn3200

Musi, Nicolas, Joseph M Valentine, Kathryn R Sickora, Eric Baeuerle, Cody S Thompson, Qiang Shen, and Miranda E Orr. 2018. "Tau Protein Aggregation Is Associated with Cellular Senescence in the Brain." *Aging Cell* 17 (6): e12840. https://doi.org/10.1111/acel.12840.

Nacarelli, Timothy, Lena Lau, Takeshi Fukumoto, Joseph Zundell, Nail Fatkhutdinov, Shuai Wu, Katherine M Aird et al. 2019. "NAD+ Metabolism Governs the Proinflammatory Senescence-Associated Secretome." *Nature Cell Biology* 21 (3): 397–407. https://doi.org/10.1038/s41556-019-0287-4.

Oh, Edward Saehong, and Art Petronis. 2021. "Origins of Human Disease: The Chrono-Epigenetic Perspective." *Nature Reviews Genetics* 22 (8): 533–46. https://doi.org/10.1038/s41576-021-00348-6.

Oh, Juhyun, Yang David Lee, and Amy J Wagers. 2014. "Stem Cell Aging: Mechanisms, Regulators and Therapeutic Opportunities." *Nature Medicine* 20 (8): 870–80. https://doi.org/10.1038/nm.3651.

Olanow, C Warren, Matthew B Stern, and Kapil Sethi. 2009. "The Scientific and Clinical Basis for the Treatment of Parkinson Disease (2009)." *Neurology* 72 (21_supplement_4): S1-136. https://doi.org/10.1212/WNL.0b013e3181a1d44c.

Peters, Ruth. 2006. "Ageing and the Brain: This Article Is Part of a Series on Ageing Edited by Professor Chris Bulpitt." *Postgraduate Medical Journal* 82 (964): 84–8. https://doi.org/10.1136/pgmj.2005.036665.

Poewe, Werner, Klaus Seppi, Caroline M Tanner, Glenda M Halliday, Patrik Brundin, Jens Volkmann, Anette-Eleonore Schrag, and Anthony E Lang. 2017. "Parkinson Disease." *Nature Reviews Disease Primers* 3 (1): 17013. https://doi.org/10.1038/nrdp.2017.13.

Saxena, Smita, and Pico Caroni. 2011. "Selective Neuronal Vulnerability in Neurodegenerative Diseases: From Stressor Thresholds to Degeneration." *Neuron* 71 (1): 35–48. https://doi.org/10.1016/j.neuron.2011.06.031.

Scheltens, Philip, Bart De Strooper, Miia Kivipelto, Henne Holstege, Gael Chételat, Charlotte E Teunissen, Jeffrey Cummings, and Wiesje M van der Flier. 2021. "Alzheimer's Disease." *The Lancet* 397 (10284): 1577–90. https://doi.org/10.1016/S0140-6736(20)32205-4 (https://www.sciencedirect.com/science/article/abs/pii/S0140673620322054?via%3Dihub).

Schliebs, Reinhard, and Thomas Arendt. 2011. "The Cholinergic System in Aging and Neuronal Degeneration." *Behavioural Brain Research* 221 (2): 555–63. https://doi.org/10.1016/j.bbr.2010.11.058.

Scott, Heather A, Florian M Gebhardt, Ann D Mitrovic, Robert J Vandenberg, and Peter R Dodd. 2011. "Glutamate Transporter Variants Reduce Glutamate Uptake in Alzheimer's Disease." *Neurobiology of Aging* 32 (3): 553.e1. https://doi.org/10.1016/j.neurobiolaging.2010.03.008.

Slack, Cathy, Nazif Alic, Andrea Foley, Melissa Cabecinha, Matthew P. Hoddinott, and Linda Partridge. 2015. "The Ras-Erk-ETS-Signaling Pathway Is a Drug Target for Longevity." *Cell* 162 (1): 72–83. https://doi.org/10.1016/j.cell.2015.06.023.

Tomás-Loba, Antonia, Ignacio Flores, Pablo J Fernández-Marcos, María L Cayuela, Antonio Maraver, Agueda Tejera, Consuelo Borrás et al. 2008. "Telomerase Reverse Transcriptase Delays Aging in Cancer-Resistant Mice." *Cell* 135 (4): 609–22. https://doi.org/10.1016/j.cell.2008.09.034.

Veggel, Lieve van, Tamara A M Mocking, Hubert J Sijben, Rongfang Liu, Marina Gorostiola González, Majlen A Dilweg, Jeroen Royakkers et al. 2024. "Still in Search for an EAAT Activator: GT949 Does Not Activate EAAT2, nor EAAT3 in Impedance and Radioligand Uptake Assays." *ACS Chemical Neuroscience* 15 (7): 1424–31. https://doi.org/10.1021/acschemneuro.3c00731.

Veggel, Lieve van, Melissa Schepers, Assia Tiane, Vijay Kumar, Emily Willems, Ben Rombaut, Jurrie Noordijk et al. 2025. "EAAT3 Modulation: A Potential Novel Avenue towards Remyelination in Multiple Sclerosis." *Biomedicine & Pharmacotherapy* 186:117960. https://doi.org/10.1016/j.biopha.2025.117960.

Verkhratsky, Alexei, and Maiken Nedergaard. 2017. "Physiology of Astroglia." *Physiological Reviews* 98:239–89. https://doi.org/10.1152/physrev.00042.2016.

Villeda, Saul A, Kristopher E Plambeck, Jinte Middeldorp, Joseph M Castellano, Kira I Mosher, Jian Luo, Lucas K Smith et al. 2014. "Young Blood Reverses Age-Related Impairments in Cognitive Function and Synaptic Plasticity in Mice." *Nature Medicine* 20 (6): 659–63. https://doi.org/10.1038/nm.3569.

Weissman, Lior, Dong-Gyu Jo, Martin M Sørensen, Nadja C de Souza-Pinto, William R Markesbery, Mark P Mattson, and Vilhelm A Bohr. 2007. "Defective DNA Base Excision Repair in Brain from Individuals with Alzheimer's Disease and Amnestic Mild Cognitive Impairment." *Nucleic Acids Research* 35 (16): 5545–55. https://doi.org/10.1093/nar/gkm605.

Wolfson, Christina, Danielle E Gauvin, Foluso Ishola, and Maryam Oskoui. 2023. "Global Prevalence and Incidence of Amyotrophic Lateral Sclerosis." *Neurology* 101 (6): e613–23. https://doi.org/10.1212/WNL.0000000000207474.

Wyss-Coray, Tony. 2016. "Ageing, Neurodegeneration and Brain Rejuvenation." *Nature* 539 (7628): 180–86. https://doi.org/10.1038/nature20411.

8 Gut Microbiota, Composition and Relation with Age-Related Neurological Disorders

Yashik Bansal, Jillwin Joseph, and Nidhi Singla

8.1 INTRODUCTION

The human gut is a complex ecosystem comprising a diverse range of microorganisms that include bacteria, viruses (including phages), parasites and fungi (Thursby and Juge 2017; Backhed 2005; Neish 2009). These microorganisms, commonly known by different names such as resident microbes, microbiota, resident flora or commensal flora, outnumber human cells by approximately 10 times (Backhed 2005; Gill et al. 2006), but another estimate suggests a more humble proportion of almost 1:1 (Sender et al. 2016). Beyond the gut, the resident flora or microbiota are also commonly present in the oral cavity, respiratory tract, genital tract in females and the skin. Gut microbiota contribute to a large gene pool, called the microbiome, which is 100 times larger than the human genes in the body (encoding roughly 3 million genes compared to the human microbiome comprising approx. 23,000 genes) (Valdes 2018). The species composition varies both within an individual (known as alpha diversity) and between individuals (known as beta diversity) (Walters and Martiny 2020).

Gut microbiota exert a mutually beneficial relationship with the host, providing metabolic and immunological functions and mediating the regulation of the digestion through the extraction and absorption of various macro and micronutrients in the food, including fats, vitamins, amino acids and short-chain fatty acids (SCFAs) (Figure 8.1; Thursby and Juge 2017). They also aid in defence

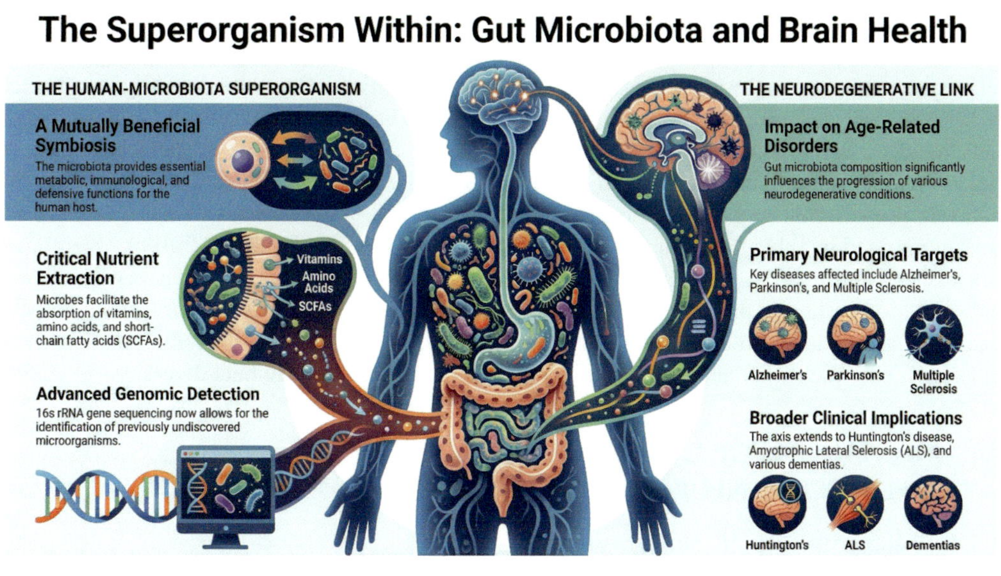

FIGURE 8.1 Gut microbiome and neurological disorders.

DOI: 10.1201/9781003610274-10

against invading pathogenic microorganisms. Therefore, this collective coexistence of the human host with its gut microbiota is often referred to as a superorganism (Gill et al. 2006; Luckey 1972). Characterizing the gut microbiota in the past was mainly done using conventional culture-based methods but the advent of newer sequencing-based methods such as whole genome sequencing has opened Pandora's box with the discovery of several new members that were missed by the culture-based methods. The newer, nucleic acid detection-based methods utilize the sequencing of the 16s rRNA gene (Ames et al. 2017; Johnson et al. 2019). This chapter delves briefly into the relationship between gut microbiota composition and its influence on the development of age-related neurodegenerative disorders such as Alzheimer's disease (AD), Parkinson's disease (PD), multiple sclerosis, Huntington's disease, Amyotrophic Lateral Sclerosis and dementias.

8.2 GUT MICROBIOTA COMPOSITION IN HEALTH

More than 2,000 species were reported from the Human Microbiome Project, which could be grouped under 12 broad phyla (Hugon et al. 2015; Li et al. 2014). However, the majority belong to the Bacteroidetes (Gram-negative bacteria) and Firmicutes (Gram-positive bacteria) and to the Actinobacteria, Proteobacteria, Fusobacteria and Verrucomicrobia (Hugon et al. 2015; Li et al. 2014; Arumugam et al. 2011). Bacteroidetes include species such as *Bacteroides fragilis*, *B. vulgatus*, *B. uniformis*, *Prevotella* spp. and *Alistipes finegoldii* whereas Firmicutes consists of *Faecalibacterium prausnitzii*, *Clostridium* spp., *Lactobacillus reuteri*, *Staphylococcus* spp., *Enterococcus faecium*, *Ruminococcus faecis* and *Roseburia intestinalis* (Rinninella et al. 2019a). Others include *Bifidobacterium* spp. (actinobacteria); *Escherichia coli*, *Helicobacter pylori* and *Shigella flexneri* (Proteobacteria); *Fusobacterium nucleatum* (Fusobacteria) and *Akkermansia muciniphila* (Verrucomicrobia). The composition is influenced by various factors, both intrinsic (i.e., host genetics) and extrinsic such as environmental factors including residence, diet, antibiotic consumption, mode of delivery, parasitic infections, body mass index (BMI), exercise and age. The influence of these extrinsic factors on the gut microbiota composition is discussed in the upcoming section.

8.3 CHANGES IN GUT MICROBIOTA—INFLUENCE OF EXTRINSIC FACTORS

As previously mentioned above, gut microbiota are susceptible to changes that could be observed within an individual (alpha diversity) or between individuals (beta diversity). Alpha diversity is observed as the individual ages and is influenced by factors such as the gestational age, mode of delivery, type of feed at the beginning of life, age and antibiotic intake. Beta diversity is influenced by a broader environmental factor having an interplay with each other such as the type of residence (developed vs. developing country), exercise levels, BMI, climate and type of diet . Beta diversity increases with a high fibre diet, with an increase in exercise and higher sports activities while it is inversely proportional to a diet rich in fats, lean proteins and with extremes of BMI. The changes are summarized in Table 8.1 (Thursby and Juge 2017; Rinninella et al. 2019a, 2019b). Overall, factors positively correlating with alpha diversity include full-term birth, vaginal delivery, breast milk and solid foods, whereas preterm birth, caesarian section delivery and artificial or formula milk are negatively associated with alpha diversity. The factors that increase beta diversity include exercise and a high fibre diet, whereas extremes of BMI and fat-rich diets reduce beta diversity.

8.4 MECHANISMS OF GUT MICROBIOTA INFLUENCING NEURODEGENERATION

8.4.1 Microglia and Neuroinflammation

Gut microbiota are known to influence the brain health through the gut–microbiota–brain axis. This involves the metabolites secreted by the microbiota that regulate the local intestinal immune system

TABLE 8.1
Factors Influencing Gut Microbiota Composition

Factor	Variable	Increased	Decreased
		Alpha Diversity	
Gestational age	<37 weeks	*Lactobacillus, Ruminococcus* spp., Lachnospiraceae, Peptostreptococcaceae, Clostridiaceae, Enterobacterales, *Enterococcus*	*Bifidobacterium* spp., *Atopobium* spp, *Bacteroides, Firmicutes*
	>37 weeks	*Bifidobacterium, Bacteroidetes, Ruminococcus* spp., Lachnospiraceae, Peptostreptococcaceae, Clostridiaceae, Enterobacterales	–
Mode of delivery	Vaginal birth	*Bifidobacterium* spp., *Bifidobacterium longum, Bifidobacterium catenulatum, Prevotella, Bacteroides fragilis, Lactobacillus, Staphylococcus, Streptococcus, Escherichia*	–
	Caesarean section	*Corynebacterium, Propionibacterium, Staphylococcus*	*Bacteroides, Escherchia, Shigella*
Feeding practices	Mothers milk	*Bifidobacterium, Lactobacillus, Staphylococcus, Enterococcus*	–
	Formula milk	*Bifidobacterium, Bacteroides, Clostridium, Clostridium difficile, Lactobacillus, Escherichia*	–
Human age	Adults	Bifidobacteriaceae, Coriobacteriaceae, Bacteroidaceae, Prevotellaceae, Rikenellaceae, Lachnospiraceae, Ruminococcaceae, *Proteobacteria, Fusobacteria, Akkermansia muciniphila*	–
	Elderly (>70 years)	*Proteobacteria*	Bifidobacteriaceae, *Clostridium*
Antibiotic intake	Vancomycin	–	*Lactobacillus, Clostridium*
	Fluoroquinolones	*Bacteroides*	*Bifidobacterium, Faecalibacterium, Oscillospira, Ruminococcus, Dialister, Alistipes*
	Clindamycin	–	Bifidobacteriaceae, *Lactobacillus*
		Beta Diversity	
BMI	Overweight/ obese	*Bifidobacterium, Lactobacillus, Firmicutes, Proteobacteria*	*Bacteroidetes, Bifidobacterium vulgatus*
	Underweight	Enterobacterales	*Roseburia, Ruminococcus, Clostridium* spp.
Exercise	Daily exercise	Clostridiales, Roseburia, Lachnospiraceae, Erysipelotrichaceae	Bacteroides, Lactobacillus
	Sedentary	*Firmicutes, Phascolarctobacterium, Roseburia, Intestinibacter, and Subdoligranulum*	Actinobacteria, Proteobacteria

(*Continued*)

TABLE 8.1 (*Continued*)
Factors Influencing Gut Microbiota Composition

Factor	Variable	Increased	Decreased
Diet	Vegan/ vegetarian	*Bacteroides, Prevotella, Bacteroidetes, Clostridium clostridioforme, Firmicutes prausnitzii, Klebsiella pneumoniae*	*Bifidobacteria, Bilophila, Clostridium* cluster XIV
	Gluten free	*Bifidobacteria, Clostridiaceae, Escherichia coli* (Enterobacteria)	Veillonellaceae, *Ruminococcus, Roseburia, Lactobacillus, Firmicutes prausnitzii, Clostridium lituseburense*
	Ketogenic	Enterobacteria, Desulfovibrio, Bacteroidetes, Parabacteroidetes, Akkemansia	*Bifidobacteria, Eubacterium rectale, Dialister, Firmicutes*
	Low fodmap diet (IBS diet)	–	*Bifidobacteria, Ruminococcus, Clostridium, Firmicutes prausnitzii*
	Western diet	*Ruminococcus torques, Enterobacteria, Bilophila, Alistipes, Bacteroides, Akkemansia*	*Bifidobacteria, Roseburia, Eubacterium rectale, Ruminococcus bromii, Lactobacillus, Prevotella*
	Mediterranean diet	*Bifidobacteria, Lactobacillus, Lachnospiraceae, Bacteroidetes*	*Clostridium, Enterobacteria*

to establish homeostasis (Figure 8.2). One of the important receptors is aryl hydrocarbon (AH) receptor, which responds to microbial and dietary (tryptophan) metabolites, and maintains the structural and functional integrity of the intestinal lining by facilitating tight junctions and synthesis of antimicrobial peptides by the local immune cells. Similarly, AH receptors are present in the microglial and astrocyte populations in the brain, which are very critical in regulating immune mechanisms and blood–brain barrier integrity and preventing inflammation of neurons. Impaired functions of intestinal AH receptor have been known to affect the integrity of the intestinal lining, coupled with the production of pro-inflammatory cytokines, which in turn can directly affect the neurons in the brain through the gut–brain axis (Wekerle 2018). One of the important pathways responsible for metabolite-induced neuroinflammation is the structural integrity at two levels: (1) the epithelium lining the intestine and (2) the endothelium lining the blood–brain barrier. Chronic inflammatory responses in the gut compromise the blood–brain barrier by molecules through systemic circulation. Gut microbial dysbiosis impairs the AH receptor, through metabolites such as indole, kynurenine and serotonin, which increases pro-inflammatory cytokines. These cytokines traverse the compromised blood–brain barrier and induce neuroinflammation and progressive neurodegeneration (Rothhammer et al. 2018; Coretti et al. 2024). Localized microglial synthesis of pro-inflammatory cytokines in the brain also contributes to neuroinflammation.

AD is the most common neurodegenerative disorder in the elderly. It is characterized by impairment in cognition with or without behavioural changes and socially, a reduced quality of life. The pathophysiology of AD is characterized by the formation in the cerebral neurons, coupled with impaired clearance of a protein called as beta amyloid (Aβ). The beta amyloid oligomerizes and forms insoluble senile plaques in the affected neurons. Apart from beta amyloid deposition, neurofibrillary tangles form from the hyperphosphorylation of a protein called Tau, leading to neuronal damage and cognitive impairment (Grammas 2011).

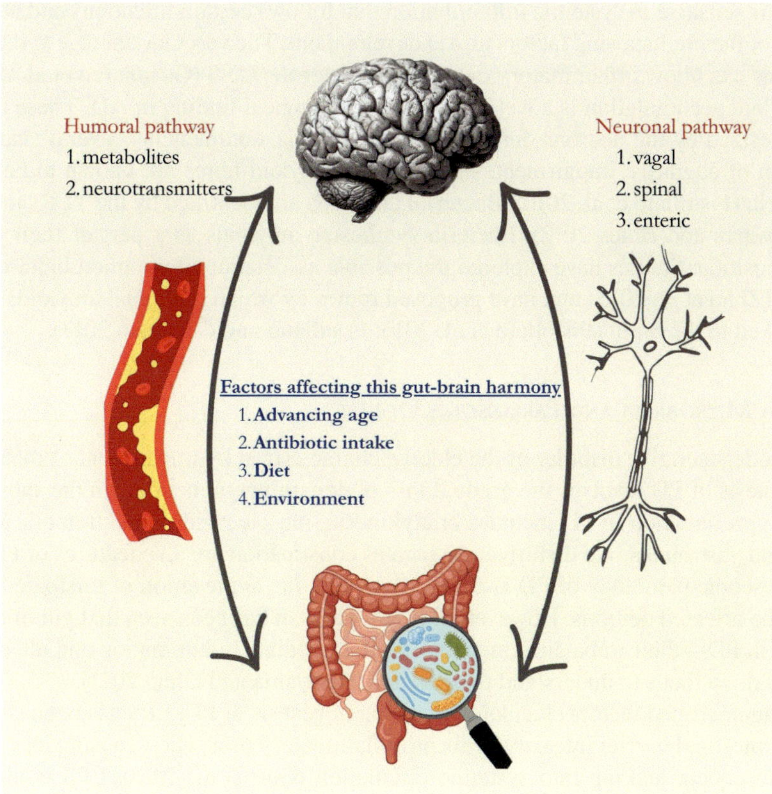

FIGURE 8.2 Communication between gut microbiome and gut microbiota and activities of brain neuroanatomical as well as humoral pathways.

8.4.2 Microbiota-Derived Neurotransmitters in Alzheimer's Disease

Gut microbiota have been identified to secrete certain neurotransmitters such as gamma-aminobutyric acid (GABA), glutamate, acetylcholine and dopamine, among a few others, that impact physiological and immune functions in the brain. They can also modulate the secretion of these chemicals by cerebral neurons, affecting synaptic functions related to cognition in AD and related disorders (Chen and Xu 2021). Microbe-derived GABA negatively affects the gut barrier integrity and inhibits norepinephrine in the CNS, which in turn leads to activation of microglial cells, causing neuroinflammation in AD (Conn et al. 2024).

Another important product of intestinal microbial metabolism is SCFAs. Anaerobic bacteria in the colon synthesize SCFAs such as butyrate, propionate and related molecules from dietary fibres (Louis and Flint 2017). SCFAs are known to have positive effects on gut and brain health by maintaining the structural integrity of the intestinal lining and BBB. However, microbiota dysbiosis can lead to a decrease in the availability of SCFAs, which in turn can predispose to the pathogenesis of AD (Marizzoni et al. 2020). Bacteroidetes is a predominant taxon in the human colon, even more abundant than *E. coli*. It has been identified that the Bacteroidetes population is enriched in the gut flora of AD. Evidence attributing the harmful role of Bacteroides in AD neurodegeneration describes toxin-mediated pathogenesis along with lipopolysaccharide (LPS) (Zhao and Lukiw 2018).

The BBB is sensitive to systemic inflammation that follows certain infections and has been identified as one of the predisposing factors to AD development. The vascular damage at the endothelial tight junctions can allow inflammatory cytokines to enter the CNS (Giridharan et al. 2019).

Beta amyloid accumulation is a well-described pathological finding in AD. These are predominantly synthesized by the neurons during AD pathogenesis, commencing several years before the manifestation of cognitive impairment. Interestingly, amyloid fibres are known to be synthesized by gut bacteria (Pistollato et al. 2016). Bacterial amyloids are identified by the TLRs in the immune system (Schwartz and Boles 2013). Bacteria synthesize amyloids as a part of their extracellular matrix composition. Studies have explored the possible association of gut microbiota-derived amyloids in AD (Zhao et al. 2015) and have proposed routes by which microbial amyloids are disseminated/trafficked to the brain (Pistollato et al. 2016; Friedland and Chapman 2017).

8.4.3 Gut Microbiota and Parkinson's Disease

PD is a neurodegenerative disorder of the elderly, characterized by impairment of motor functions. The pathogenesis of PD involves the gradual loss of dopaminergic neurons in the substantia nigra. The classical presentation in PD includes bradykinesia, muscle rigidity and tremors. Speech incoordination and gastrointestinal disturbances such as constipation are also features of PD (Han et al. 2022). Microscopic pathology of PD is characterized by the aggregation of misfolded α-synuclein proteins in the affected neurons. However, in recent years, it has been seen that gut microbiota play a major role in PD—their imbalance, metabolites that mediate inflammation and alterations in gut permeability draw focus to understand them better (Hirayama and Ohno 2021).

Among the microbial factors elucidated in the pathogenesis of PD, LPS has been shown to compromise the intestinal barrier integrity by its proinflammatory properties, causing loss of tight junctions and subsequent leaking into systemic circulation (Guo et al. 2013). LPS is also known to induce inflammatory responses at the blood–brain barrier, making it susceptible to harmful metabolites to reach the cerebral neurons. Elevated levels of LPS seen in PD patients probably account for the neuroinflammation leading to motor impairment (Brown et al. 2023). SCFAs, which are synthesized by gut bacteria, have been shown to be reduced in PD patients. However, in some studies, microbiota-derived SCFAs have been shown to worsen the neuroinflammation, suggesting their dual role in neurodegenerative diseases (Zhang et al. 2023).

One of the characteristic microscopic pathologies in PD is the accumulation of misfolded oligomers of a protein called α-synuclein (Stefanis 2012). This protein can spread in a prion-like manner, but can elicit inflammation in neurons. α-synuclein is expressed in monomeric forms in the neurons of the enteric nervous system and enteroendocrine cells. The expression of α-synuclein may be affected by microbiota dysbiosis and their metabolites. In PD patients, the expression of α-synuclein is higher than normal. It has been shown that the α-synuclein accumulation in the submucosal and myoenteric neurons precedes their appearance in the dopaminergic neurons in substantia nigra. This brings in the potential source of α-synuclein as derived from the gut environment. The distribution of α-synuclein from the gut to the brain is largely identified to be via the fibres of the vagus nerve that innervates the gut. Truncal vagotomy has been shown to reduce the risk of developing PD (Liddle 2018).

8.5 GUT MICROBIOTA IN NEURODEGENERATIVE DISORDERS VS. HEALTHY INDIVIDUALS

8.5.1 Microbiota Dysbiosis in Alzheimer's Disease

Studies comparing the intestinal microbiome diversity of AD patients with demographic-matched healthy controls have provided insights into the raised and reduced numbers of certain bacterial families and genera in AD patients. The majority of the studies have described microbial diversity specifically to members of Bacteroidetes, Firmicutes and Proteobacteria (Rinninella et al. 2019b). The microbiome diversity reported in individual studies shows variability in the bacterial communities with respect to their relative abundance in AD patients (Li et al. 2024). Certain bacterial phyla are reported with a higher relative abundance in some studies while the same phyla are seen in reduced numbers in other studies.

This variability may be attributed to the regional, dietary factors, and methodologies used for analyses. The bacterial population recognized in these studies is distributed at multiple levels of taxonomic hierarchy, showing significant variability. However, meta-analyses relevant to microbiome diversity in AD versus healthy controls have the advantage of larger participant numbers pooled from several studies, with multiple variables and rigorous statistical analyses. Results from meta-analyses have at least been able to identify Bacteroidetes and Akkermansia as commonly prevalent in AD (Jimenez Garcia et al. 2024; Li et al. 2024).

Variations in the microbiota in AD patients from human studies indicate a complex relationship between microbiome diversity and the clinical and pathological features such as cognitive deterioration, neuroinflammation, beta amyloid deposition and tauopathy or both.

8.5.2 Microbiota Dysbiosis in Parkinson's Disease

Several case–control studies have explored the association of gut microbial dysbiosis in PD, with most studies highlighting the differences in microbial taxa in PD versus age-matched healthy controls. The hypervariable region of the 16S rRNA gene of bacteria, especially the V4 region, has been widely used as the target for sequencing in most of the studies (Pérez-Bustamante et al. 2024), while relatively few studies have employed metagenomic shotgun sequencing for diversity analysis. The microbial dysbiosis was identified at several taxonomic levels from phyla up to the species level in these studies. With individual studies reporting minor inconsistencies in the distribution of certain taxa, the better approach for a more solid understanding is to identify the microbial dysbiosis from pooled data in meta-analyses. Meta-analyses of microbial dysbiosis in PD have shown consistently higher relative abundance of Akkermansiaceae, Bifidobacteriaceae, Verrumicrobiaceae, Ruminococcaceae and Lactobacillaceae, while bacteria belonging to Lachnospiraceae, Fecalibacterium, Prevotellaceae and Roseburia (Romano et al. 2021; Shen et al. 2021; Toh et al. 2022).

8.5.3 Microbiota Dysbiosis in Other Diseases

AD, being the commonest neurodegenerative disorder in humans, has, along with PD, attracted a significant focus from the scientific community with respect to understanding its relation to the microbiota. Other neurodegenerative disorders include but are not limited to amyotrophic lateral sclerosis (ALS), Lewy body disease (dementia), multiple sclerosis (MS), Huntington's disease, Multiple system atrophy (MSA), Friedreich ataxia and Spinal muscular atrophy. Gut microbiota dysbiosis has been studied in these different neurodegenerative disorders and a few studies documenting the changes in microbiota among patients suffering from such disorders are summarized in Table 8.2.

8.6 ANIMAL STUDIES ON GUT MICROBIOTA IN NEURODEGENERATIVE DISORDERS

8.6.1 Animal Studies on Gut Microbiota and Alzheimer's Disease

Several studies have shown the association of gut microbiota with neurological functions in standardized transgenic mouse models of AD. Fecal microbial diversity and the levels of SCFAs were observed to be relatively lower in the AD mice compared to wild-type mice. Metabolic studies in

TABLE 8.2
Gut Microbiota Dysbiosis in Other Neurodegenerative Diseases

Disorder	Increased	Decreased	Identification Method	References
Dementia with Lewy bodies	*Ruminococcus torques, Collinsella, Eggerthella*	*Agathobacter, Lachnospiraceae, Butyricicoccus, Coprococcus, Faecalibacterium*	*16SrRNA sequencing*	Nishiwaki et al. (2022)
Amyotrophic lateral sclerosis (ALS)	*Cyanobacteria, Lactobacillus, Bifidobacterium, Odoribacter, Akkermansia, Citrobacter*	*Veilloneaceae, Ruminococcus, Eubacterium,*	16SrRNA qPCR, NGS	Di Gioia et al. (2020)
Fronto temporal dementia	*Rhodospirillaceae, Melainabacteria*	–	–	Ji et al. (2024)
Multiple system atrophy (MSA)	*Ralstonia, Bacteroides*	*Ruminococcus, Lachnospiraceae, Firmicutes, Blautia, Coprococcus, Faecalibacterium, Roseburia*	16SrRNA, NGS	Engen et al. (2017)
Huntington's disease	*Actinobacteria, Desulfovibrionaceae, Intestinimonas, Bilophila, Lactobacillus, Oscillibacter, Gemmiger*	–	*16SrRNA sequencing*	Du et al. (2021)
Multiple sclerosis (MS)	*Methanobrevibacter, Akkermansia muciniphila*	*Prevotella, Faecalibacterium prausnitzii, Bacteroides coprophilus, Bacteroides fragilis*	Multiple	Mirza et al. (2020)

AD mouse models have shown several pathways associated with amyloid deposition and cognitive impairment (Zhang et al. 2017). The dysbiotic changes in gut microbial flora between the AD model compared to the wild-type litter showed a shifting predominance of inflammation-associated bacterial genus such as *Escherichia, Desulfovibrio, Akkermansia, Shigella* and *Blautia* in the former group. These mice were followed up for histopathological changes in the cerebral cortex (using immunohistochemistry). The AD group showed progressive accumulation of beta amyloid plaques with time, evidenced by increasing intensity of amyloid staining, whereas no plaques were demonstrated in the wild-type controls (Chen et al. 2020).

The relative predominance of Bacteroidetes in AD has been demonstrated in the mouse model, which supports the findings in human studies (Vogt et al. 2017). Also, dysbiotic changes in gut microbiome in AD mice have shown abnormally increased expression of NLRP3 inflammasome pathway and IL1β production in the brain, mediating neuroinflammation (Shukla et al. 2021). The role of LPS (endotoxin) in the accumulation of beta amyloid plaques has been demonstrated. Doses of LPS in the study group mice showed triplefold increased deposition of intracellular beta amyloid in the neocortex, the area for memory versus the phosphate-buffered saline-treated control group (Sheng et al. 2003).

Gut microbiota from AD mice have been shown to impair neurogenesis and decrease the production of brain-derived neurotropic factor, resulting in cognitive impairment in healthy mice (Kim et al. 2021). Transmission of gut microbiota from AD mice to co-housed healthy controls has been observed. Similar findings are reported in the household contacts of AD patients (Zhang et al. 2023). AD mice models have also shown improvement in cognitive functions, amyloid pathology and neurofibrillary tauopathy in response to microbiota from the gut of healthy mice (Kim et al. 2020).

8.6.2 Animal Studies on Gut Microbiota and Parkinson's Disease

A standardized mouse model of PD showed a decrease in the alpha diversity of intestinal microbiota with ageing, compared to the wild-type group. Parkinsonian motor deficits, gastrointestinal dysfunction and anxiety depression-like behavioural changes were also observed in this group (Radisavljevic et al. 2022). Microbiota from PD patients were shown to induce neuroinflammation and α-synuclein pathology in mice, compared to microbiota from healthy human controls. Also, microbial metabolites cause neuroinflammatory effects in germ-free mice (Sampson et al. 2016).

LPS from the dominance of pro-inflammatory Gammaproteobacteria mediated inflammatory damage to gut barrier integrity and development of motor symptoms in a PD mouse model (Gorecki et al. 2019). However, intestinal microbiota from healthy mice were able to improve the dysbiosis and microglial functions in PD mice, mainly by suppressing LPS-TLR4-mediated neuroinflammation (Sun et al. 2018; Zhao et al. 2021). SCFAs have also been shown to mitigate the neuroinflammation in PD models (Hou et al. 2021). However, the role of SCFAs in inflammation has to be explored further, as some studies show them as contributors to neuroinflammation (Sun et al. 2018).

Overall, from animal models of these two well-recognized neurodegenerative diseases, we could broadly understand the role of gut microbial diversity and that their dysbiosis and metabolites are responsible for the neuroinflammation, microscopic pathologies in brain, and cognitive and motor symptoms. Overall, proteobacterial members are identified in most of the studies as influencers of the pathogenetic mechanisms in these disorders. Although Bacteroidetes, Lachnospiraceae members and *Akkermansia* are identified as the significant taxa in most of these studies, there have been variations among them and other taxa when it comes to the relative abundance in the gut of mice used in these studies. This could be attributed to the differences in the various transgenic mouse models used in individual studies; each model can have its own unique pathways in the gut microbiota–brain axis communication. Fecal microbiota from healthy mice have shown improved outcomes in both diseases in mouse models.

8.7 FUTURE PERSPECTIVES—RESEARCH IN PIPELINE

The interplay between gut microbiota and neurological disorders has garnered significant interest in recent times. The clinicaltrials.gov database of the National Center for Biotechnology Information was searched to find the clinical trials registered on the website by using the keywords "neurological disorders" under condition/disease and "gut microbiota" under intervention/treatment (Clinicaltrials 2025a). A total of 199 clinical trials were found on the website with an equal number of observational ($n = 99$) and interventional studies ($n = 100$). Similar searches were performed to find studies specific to multiple sclerosis (Clinicaltrials 2025b), AD (Clinicaltrials 2025c) and PD (Clinicaltrials 2025d) that are summarized in Table 8.3. Broadly, these studies are exploring the interplay between gut microbiota and neurodegeneration, the therapeutic potential of gut microbiota modulation by supplements such as prebiotics, probiotics and faecal microbiota transplantation.

The treatment paradigms could witness a revolutionary change as more research provides evidence-based data, leading to a shift in the role of gut microbiota composition from association to causation. Targeted probiotic therapy, microbes that are engineered to suit the cessation of neurodegenerative diseases coupled with the use of dietary modification approaches, could become an integral part of disease management in the future.

TABLE 8.3

Clinical Trials Exploring the Interplay between Gut Microbiota and Neurological Disorders that Are Registered on the Clinicaltrials.gov

NCT Number	Concise Title	Type	Sample	Phase	Present Status
		Multiple Sclerosis ($n=19$)			
NCT03262870	"GM and MS"	O	40	–	UNK
NCT03183869	"Fecal microbial transplantation in relapsing MS patients"	I	14	II	T
NCT02736279	"Impact of tecfidera on GM"	O	25	–	UNK
NCT02580435	"Deciphering the role of the GM in MS"	O	520	–	UNK
NCT06220409	"GM composition, cognitive function, and physical fitness in MS patients"	I	29	NA	C
NCT02471560	"Tecfidera and the GM"	I	36	IV	C
NCT04148313	"A pilot study to explore the role of gut flora in MS"	O	100	–	UNK
NCT04929145	"The gut microbiome in adult MS"	O	152	–	C
NCT06746896	"Retrospective study on the role of SCFA modulation in MS"	O	40	–	ANR
NCT05539729	"Vancomycin study in MS (MS)"	I	12	I	R
NCT05808101	"Gut microbiome and depression"	O	51	–	T
NCT05844826	"Evaluation of the microbiome in MS"	I	60	NA	R
NCT01353547	"Genes and environment in MS"	O	5000	–	R
NCT03594487	"FMT of FMP30 in relapsing-remitting MS"	I	30	I	ANR
NCT05934188[a]	"Exploring the gut–brain axis in ageing and neurodegeneration"	O	200	–	R
NCT04038541	"Prebiotic vs probiotic in MS"	I	37	NA	C
NCT05779449	"Targeting the gut dysbiosis to treat inflammation-driven synaptopathy in MS"	I	100	NA	R
NCT04014413	"Safety and efficacy of fecal microbiota transplantation"	I	450	NA	R
NCT06596746[a]	"Neurodegenerative diseases progression markers (MARKERS-NDD)"	O	600	–	R
		Alzheimer's disease ($n=13$)			
NCT03593941	"Ageing gut brain interactions"	O	60	–	C
NCT03827733	"GM and Ads"	O	200	–	UNK
NCT06039267	"Brain health & the microbiome"	O	90	–	ANR
NCT04100889	"A non-interventional pilot study to explore the role of gut flora in AD"	O	100	–	UNK

(Continued)

TABLE 8.3 (*Continued*)
Clinical Trials Exploring the Interplay between Gut Microbiota and Neurological Disorders that Are Registered on the Clinicaltrials.gov

NCT Number	Concise Title	Type	Sample	Phase	Present Status
NCT06199193	"Design of personalised supplements based on the GM through artificial intelligence for Alzheimer's patients"	I	120	NA	R
NCT05145881	"Effect of Probiotics in AD"	I	40	NA	ANR
NCT04094129	"Evaluation of the effect of wismemo on Alzheimer's dementia patients"	I	0	NA	W
NCT03487380	"Taxonomic and functional composition of the intestinal microbiome: a predictor of rapid cognitive decline in patients with AD"	I	86	NA	T
NCT05943925@	"Dementias and microbiota composition: is possible to revert the dementia symptoms reverting the microbiota composition?"	O	240	–	R
NCT05601856	"Gut microbiome and blood indices in patients with AD and their spousal caregivers"	O	104	–	R
NCT06094452	"Effects of 24-week computerized cognitive training in patients with MCI and AD"	I	84	NA	C
NCT06083350	"Effects of yeast beta-glucan on cognitive function in patients with mild cognitive impairment"	I	144	NA	ANR
NCT06507254	"Polyphenols and cognitive decline"	I	300	NA	R
	Parkinson's disease ($n = 27$)				
NCT03710668	"GM and PD"	O	50		UNK
NCT06197698	"A proof-of-concept study evaluating the microbiota–gut–brain axis"	I	100	NA	R
NCT04730245	"Microbiota modification for the treatment of motor complication of Parkinson´s disease"	I	14	I	C
NCT05787756	"Constipation and changes in the gut flora in PD"	O	80	–	R
NCT03808389	"FMT for PD"	I	49	NA	C
NCT03876327	"FMT as a potential treatment for PD"	I	10	II/III	C
NCT04148326	"A pilot study to explore the role of gut flora in PD"	O	100	–	UNK

(*Continued*)

TABLE 8.3 (*Continued*)

Clinical Trials Exploring the Interplay between Gut Microbiota and Neurological Disorders that Are Registered on the Clinicaltrials.gov

NCT Number	Concise Title	Type	Sample	Phase	Present Status
NCT03958708	"Modulation of GM by rifaximin in PD patients"	I	20	I/ II	UNK
NCT04159727	"GM dysbiosis and translocation during IBD and Parkinson"	I	15	NA	C
NCT06762028	"Exploring the relationship between L-dopa responsiveness and small intestinal microbiome in PD"	O	100	–	NR
NCT02784145	"Effects of resistant starch on bowel habits, fecal short chain fatty acids and GM in Parkinson disease"	I	95	NA	C
NCT01536769	"Intestinal and nasal microbiota of patients with idiopathic PD"	O	150	–	C
NCT05469997	"Ketogenic diet interventions in PD: safeguarding the gut microbiome"	I	50	NA	ANR
NCT04652843	"Single-center pathophysiological study of the role of inflammation, changes in the intestinal epithelial barrier and the intestinal microbiota in PD"	I	77	NA	T
NCT04854291	"A trial of FMT in PD patients"	I	51	NA	C
NCT04956939	"Levodopa response and gut microbiome in patients with PD"	O	38	–	C
NCT03671785	"Study of the fecal microbiome in patients with PD"	I	15	I	C
NCT04119596	"Study of the genome, gut metagenome and diet of patients with incident PD"	O	138	–	T
NCT05558787	"AADC/ TDC in advanced PD"	O	48	–	C
NCT03026231	"Characterization of fecal microbiome changes after administration of PRIM-DJ2727 in PD patients"	I	0	I/ II	W
NCT04180865	"The Dutch Parkinson and cognition study"	O	150	–	UNK
NCT04847011	"Deciphering the impact of exposures from the gut microbiome-derived molecular complex in human health and disease"	I	183	NA	C
NCT05573542	"Early signs of Parkinsons disease in IBS"	O	120	–	EI

(*Continued*)

TABLE 8.3 (*Continued*)
Clinical Trials Exploring the Interplay between Gut Microbiota and Neurological Disorders that Are Registered on the Clinicaltrials.gov

NCT Number	Concise Title	Type	Sample	Phase	Present Status
NCT03645226	"Gut microbiota across early stages of synucleinopathy: from high-risk relatives, REM sleep behavior disorder to early Parkinson's disease"	O	441	–	C
NCT04855344	"Deep brain stimulation therapy and intestinal microbiota"	O	30	–	UNK
NCT03667404	"Resistant maltodextrin for gut microbiome in Parkinson's disease: safety and tolerability study"	I	30	II	ANR
NCT05353868	"Gut microbiota in the progression of alpha-synucleinopathies"	O	490	–	R

[a] Study exploring MS, AD and PD, @study common to both AD and PD.
NA, not available; R, recruiting; EI, enrolment by invitation; T, terminated; NR, not yet recruiting; C, completed; ANR, active not recruiting; NLA, no longer available; UNK, unknown status; W, withdrawn; I, interventional study; O, observational study; GM , gut microbiota; MS, multiple sclerosis; PD, Parkinson's disease; AD, Alzheimer's disease; IBS, irritable bowel syndrome; REM, rapid eye movement; IBD, inflammatory bowel disease; FMT, fecal microbiota transplant; SCFA, short chain fatty acids; MCI, mild cognitive impairment; AADC, aromatic L-amino acid decarboxylase; TDC, tyrosine decarboxylase.

8.8 CONCLUSIONS

Gut microbiota play an indispensable role in the homeostasis of the central nervous system, through the gut–brain axis. Gut microbiota could undergo alterations in response to various stimuli, such as diet, antibiotic intake, environmental factors and aging, resulting in a decrease in beneficial species and a concomitant rise of pro-inflammatory species. This causes a shift in the gut–brain axis, thereby contributing to neurodegeneration and subsequent onset of age-related neurodegenerative disorders including but not limited to dementias, PD, AD and multiple sclerosis. This is chiefly mediated through the influence of metabolites such as SCFAs and LPS on the permeability of the blood–brain barrier, the activation of the microglia and neuronal function. Recent studies have shown promising therapeutic potentials of microbiota modulation via diet, pre and probiotic supplementation. Numerous clinical trials are ongoing to better understand the interplay between gut microbiota and age-related neurodegeneration.

REFERENCES

Ames, N. J., Ranucci, A., Moriyama, B., and Wallen, G. R. 2017. The human microbiome and understanding the 16S rRNA gene in translational nursing science. *Nursing Research* 66, 184–97. https://doi.org/10.1097/NNR.0000000000000212.

Arumugam, M., Raes, J., Pelletier, E. et al. 2011. Enterotypes of the human gut microbiome. *Nature* 473, 174–80. https://doi.org/10.1038/nature09944.

Backhed, F. 2005. Host-bacterial mutualism in the human intestine. *Science* 307, 1915–20. https://doi.org/10.1126/science.1104816.

Brown, G. C., Camacho, M., and Williams-Gray, C. H. 2023. The endotoxin hypothesis of Parkinson's disease. *Movement Disorders* 38, 1143–55. https://doi.org/10.1002/mds.29432.

Chen, Y., Fang, L., Chen, S. et al. 2020. Gut microbiome alterations precede cerebral amyloidosis and microglial pathology in a mouse model of Alzheimer's disease. *BioMed Research International* 8456596. https://doi.org/10.1155/2020/8456596.

Chen, Y. and Xu, J. 2021. Regulation of neurotransmitters by the gut microbiota and effects on cognition in neurological disorders. *Nutrients* 13, 2099. https://doi.org/10.3390/nu13062099.

Clinicaltrials Website. National Center for Biotechnology Information, National Institute of Health, U.S. Department of Health and Human Services. 2025a. https://clinicaltrials.gov/search?cond=neurological%20disorders&intr=Gut%20microbiota (accessed January 17, 2025).

Clinicaltrials Website. National Center for Biotechnology Information, National Institute of Health, U.S. Department of Health and Human Services. 2025b. https://clinicaltrials.gov/search?cond=Multiple%20Sclerosis&intr=Gut%20microbiota&limit=50 (accessed January 17, 2025).

Clinicaltrials Website. National Center for Biotechnology Information, National Institute of Health, U.S. Department of Health and Human Services. 2025c. https://clinicaltrials.gov/search?cond=Alzheimer%20Disease&intr=Gut%20microbiota&limit=50 (accessed January 17, 2025).

Clinicaltrials Website. National Center for Biotechnology Information, National Institute of Health, U.S. Department of Health and Human Services. 2025d. https://clinicaltrials.gov/search?cond=Parkinson%20Disease&intr=Gut%20microbiota&limit=50&page=1 (accessed January 17, 2025).

Conn, K. A., Borsom, E. M., and Cope, E. K. 2024. Implications of microbe-derived γ-aminobutyric acid (GABA) in gut and brain barrier integrity and GABAergic signaling in Alzheimer's disease. *Gut Microbes* 16, 2371950. https://doi.org/10.1080/19490976.2024.2371950.

Coretti, L., Buommino, E., and Lembo, F. 2024. The aryl hydrocarbon receptor pathway: a linking bridge between the gut microbiome and neurodegenerative diseases. *Frontiers in Cellular Neuroscience* 18, 1433747. https://doi.org/10.3389/fncel.2024.1433747.

Di Gioia, D., Bozzi Cionci, N., Baffoni, L. et al. 2020. A prospective longitudinal study on the microbiota composition in amyotrophic lateral sclerosis. *BMC Medicine* 18, 153. https://doi.org/10.1186/s12916-020-01607-9.

Du, G., Dong, W., Yang, Q. et al. 2021. Altered gut microbiota related to inflammatory responses in patients with Huntington's disease. *Frontiers in Immunology* 11, 603594. https://doi.org/10.3389/fimmu.2020.603594.

Engen, P. A., Dodiya, H. B., Naqib, A. et al. 2017. The potential role of gut-derived inflammation in multiple system atrophy. *Journal of Parkinson's Disease* 7, 331–46. https://doi.org/10.3233/JPD-160991.

Friedland, R. P., and Chapman, M. R. 2017. The role of microbial amyloid in neurodegeneration. *PLoS Pathogens* 13, e1006654. https://doi.org/10.1371/journal.ppat.1006654.

Gill, S. R., Pop, M., DeBoy, R. T. et al. 2006. Metagenomic analysis of the human distal gut microbiome. *Science* 312, 1355–9. https://doi.org/10.1126/science.1124234.

Giridharan, V. V., Masud, F., Petronilho, F., Dal-Pizzol, F., and Barichello, T. 2019. Infection-induced systemic inflammation is a potential driver of Alzheimer's disease progression. *Frontiers in Aging Neuroscience* 11, 122. https://doi.org/10.3389/fnagi.2019.00122.

Gorecki, A. M., Preskey, L., Bakeberg, M. C. et al. 2019. Altered gut microbiome in Parkinson's disease and the influence of lipopolysaccharide in a human α-synuclein over-expressing mouse model. *Frontiers in Neuroscience* 13, 839. https://doi.org/10.3389/fnins.2019.00839.

Grammas, P. 2011. Neurovascular dysfunction, inflammation and endothelial activation: implications for the pathogenesis of Alzheimer's disease. *Journal of Neuroinflammation* 8, 26. https://doi.org/10.1186/1742-2094-8-26.

Guo, S., Al-Sadi, R., Said, H. M., and Ma, T. Y. 2013. Lipopolysaccharide causes an increase in intestinal tight junction permeability in vitro and in vivo by inducing enterocyte membrane expression and localization of TLR-4 and CD14. *American Journal of Pathology* 182, 375–87. https://doi.org/10.1016/j.ajpath.2012.10.014.

Han, M. N., Finkelstein, D. I., McQuade, R. M., and Diwakarla, S. 2022. Gastrointestinal dysfunction in Parkinson's disease: current and potential therapeutics. *Journal of Personalized Medicine* 12, 144. https://doi.org/10.3390/jpm12020144.

Hirayama, M., and Ohno, K. 2021. Parkinson's disease and gut microbiota. *Annals of Nutrition & Metabolism* 77 (Suppl 2), 28–35. https://doi.org/10.1159/000518147.

Hou, Y., Li, X., Liu, C. et al. 2021. Neuroprotective effects of short-chain fatty acids in MPTP induced mice model of Parkinson's disease. *Experimental Gerontology* 150, 111376. https://doi.org/10.1016/j.exger.2021.111376.

Hugon, P., Dufour, J.-C., Colson, P., Fournier, P.-E., Sallah, K., and Raoult, D. 2015. A comprehensive repertoire of prokaryotic species identified in human beings. *Lancet Infectious Diseases* 15, 1211–19. https://doi.org/10.1016/S1473-3099(15)00293-5.

Ji, D., Chen, W. Z., Zhang, L., Zhang, Z. H., and Chen, L. J. 2024. Gut microbiota, circulating cytokines and dementia: a Mendelian randomization study. *Journal of Neuroinflammation* 21, 2. https://doi.org/10.1186/s12974-023-02999-0.

Jimenez-García, A. M., Villarino, M., and Arias, N. 2024. A systematic review and meta-analysis of basal microbiota and cognitive function in Alzheimer's disease: a potential target for treatment or a contributor to disease progression? *Alzheimer's & Dementia (Amsterdam, Netherlands)* 16, e70057. https://doi.org/10.1002/dad2.70057.

Johnson, J. S., Spakowicz, D. J., Hong, B. Y. et al. 2019. Evaluation of 16S rRNA gene sequencing for species and strain-level microbiome analysis. *Nature Communications* 10, 5029. https://doi.org/10.1038/s41467-019-13036-1.

Kim, M. S., Kim, Y., Choi, H. et al. 2020. Transfer of a healthy microbiota reduces amyloid and tau pathology in an Alzheimer's disease animal model. *Gut* 69, 283–94. https://doi.org/10.1136/gutjnl-2018-317431.

Kim, N., Jeon, S. H., Ju, I. G. et al. 2021. Transplantation of gut microbiota derived from Alzheimer's disease mouse model impairs memory function and neurogenesis in C57BL/6 mice. *Brain, Behavior, and Immunity* 98, 357–65. https://doi.org/10.1016/j.bbi.2021.09.002.

Li, H., Cui, X., Lin, Y., Huang, F., Tian, A., and Zhang, R. 2024. Gut microbiota changes in patients with Alzheimer's disease spectrum based on 16S rRNA sequencing: a systematic review and meta-analysis. *Frontiers in Aging Neuroscience* 16, 1422350. https://doi.org/10.3389/fnagi.2024.1422350.

Li, J., Jia, H., Cai, X. et al. 2014. An integrated catalog of reference genes in the human gut microbiome. *Nature Biotechnology* 32, 834–41. https://doi.org/10.1038/nbt.2942.

Liddle, R. A. 2018. Parkinson's disease from the gut. *Brain Research* 1693 (Pt B), 201–6. https://doi.org/10.1016/j.brainres.2018.01.010.

Louis, P., and Flint, H. J. 2017. Formation of propionate and butyrate by the human colonic microbiota. *Environmental Microbiology* 19, 29–41. https://doi.org/10.1111/1462-2920.13589.

Luckey, T. D. 1972. Introduction to intestinal microecology. *American Journal of Clinical Nutrition* 25, 1292–4.

Marizzoni, M., Cattaneo, A., Mirabelli, P. et al. 2020. Short-chain fatty acids and lipopolysaccharide as mediators between gut dysbiosis and amyloid pathology in Alzheimer's disease. *Journal of Alzheimer's Disease* 78, 683–97. https://doi.org/10.3233/JAD-200306.

Mirza, A., Forbes, J. D., Zhu, F. et al. 2020. The multiple sclerosis gut microbiota: a systematic review. *Multiple Sclerosis and Related Disorders* 37, 101427. https://doi.org/10.1016/j.msard.2019.101427.

Nadiia, R., Lydia, R., Kress, M. 2024. The gut microbiome and the brain. *Current Opinion in Supportive and Palliative Care*. https://doi.org/10.1097/spc.0000000000000717.

Neish, A. S. 2009. Microbes in gastrointestinal health and disease. *Gastroenterology* 136, 65–80. https://doi.org/10.1053/j.gastro.2008.10.080.

Nishiwaki, H., Ueyama, J., Kashihara, K. et al. 2022. Gut microbiota in dementia with Lewy bodies. *NPJ Parkinson's Disease* 81, 169. https://doi.org/10.1038/s41531-022-00428-2.

Pérez-Bustamante, I. S., Cruz-Flores, R., López-Carvallo, J. A., and Sánchez-Serrano, S. 2024. Effect of the 16S rRNA gene hypervariable region on the microbiome taxonomic profile and diversity in the endangered fish Totoaba macdonaldi. *Microorganisms* 12, 2119. https://doi.org/10.3390/microorganisms12112119.

Pistollato, F., Sumalla Cano, S., Elio, I., Masias Vergara, M., Giampieri, F., and Battino, M. 2016. Role of gut microbiota and nutrients in amyloid formation and pathogenesis of Alzheimer disease. *Nutrition Reviews* 74, 624–34. https://doi.org/10.1093/nutrit/nuw023.

Radisavljevic, N., Cirstea, M., Bauer, K. et al. 2022. Effects of gut microbiota alterations on motor, gastrointestinal, and behavioral phenotype in a mouse model of Parkinson's disease. *Journal of Parkinson's Disease* 12, 1479–95. https://doi.org/10.3233/JPD-223165.

Rinninella, E., Cintoni, M., Raoul, P. et al. 2019. Food components and dietary habits: keys for a healthy gut microbiota composition. *Nutrients* 11, 2393. https://doi.org/10.3390/nu11102393.

Rinninella, E., Raoul, P., Cintoni, M. et al. 2019. What is the healthy gut microbiota composition? A changing ecosystem across age, environment, diet, and diseases. *Microorganisms* 7, 14. https://doi.org/10.3390/microorganisms7010014.

Romano, S., Savva, G. M., Bedarf, J. R., Charles, I. G., Hildebrand, F., and Narbad, A. 2021. Meta-analysis of the Parkinson's disease gut microbiome suggests alterations linked to intestinal inflammation. *NPJ Parkinson's Disease* 7, 27. https://doi.org/10.1038/s41531-021-00156-z.

Rothhammer, V., Borucki, D. M., Tjon, E. C. et al. 2018. Microglial control of astrocytes in response to microbial metabolites. *Nature* 557 (7707), 724–8. https://doi.org/10.1038/s41586-018-0119-x.

Sampson, T. R., Debelius, J. W., Thron, T. et al. 2016. Gut microbiota regulate motor deficits and neuroinflammation in a model of Parkinson's disease. *Cell* 167, 1469–80.e12. https://doi.org/10.1016/j.cell.2016.11.018.

Schwartz, K., and Boles, B. R. 2013. Microbial amyloids--functions and interactions within the host. *Current Opinion in Microbiology* 16, 93–9. https://doi.org/10.1016/j.mib.2012.12.001.

Sender, R., Fuchs, S. and Milo, R. 2016. Revised estimates for the number of human and bacteria cells in the body. *PLoS Biology* 14, e1002533. https://doi.org/10.1371/journal.pbio.1002533.

Shen, T., Yue, Y., He, T. et al. 2021. The association between the gut microbiota and Parkinson's disease, a meta-analysis. *Frontiers in Aging Neuroscience* 13, 636545. https://doi.org/10.3389/fnagi.2021.636545.

Sheng, J. G., Bora, S. H., Xu, G., Borchelt, D. R., Price, D. L., and Koliatsos, V. E. 2003. Lipopolysaccharide-induced-neuroinflammation increases intracellular accumulation of amyloid precursor protein and amyloid beta peptide in APPswe transgenic mice. *Neurobiology of Disease* 14, 133–45. https://doi.org/10.1016/s0969-9961(03)00069-x.

Shukla, P. K., Delotterie, D. F., Xiao, J. et al. 2021. Alterations in the gut-microbial-inflammasome-brain axis in a mouse model of Alzheimer's disease. *Cells* 10, 779. https://doi.org/10.3390/cells10040779.

Stefanis L. 2012. α-Synuclein in Parkinson's disease. *Cold Spring Harbor Perspectives in Medicine* 2, a009399. https://doi.org/10.1101/cshperspect.a009399.

Sun, M. F., Zhu, Y. L., Zhou, Z. L. et al. 2018. Neuroprotective effects of fecal microbiota transplantation on MPTP-induced Parkinson's disease mice: gut microbiota, glial reaction and TLR4/TNF-α signaling pathway. *Brain, Behavior, and Immunity* 70, 48–60. https://doi.org/10.1016/j.bbi.2018.02.005.

Thursby, E., and Juge, N. 2017. Introduction to the human gut microbiota. *The Biochemical Journal* 474, 1823–36. https://doi.org/10.1042/BCJ20160510.

Toh, T. S., Chong, C. W., Lim, S. Y. et al. 2022. Gut microbiome in Parkinson's disease: New insights from meta-analysis. *Parkinsonism & Related Disorders* 94, 1–9. https://doi.org/10.1016/j.parkreldis.2021.11.017.

Valdes, A. M., Walter, J., Segal, E., Spector, T. D. 2018. Role of the gut microbiota in nutrition and health. *BMJ* 361, 36–44. https://doi.org/10.1136/bmj.k2179.

Vogt, N. M., Kerby, R. L., Dill-McFarland, K. A. et al. 2017. Gut microbiome alterations in Alzheimer's disease. *Scientific Reports* 7, 13537. https://doi.org/10.1038/s41598-017-13601-y.

Walters, K. E., and Martiny, J. B. H. 2020. Alpha-, beta-, and gamma-diversity of bacteria varies across habitats. *PloS One* 15, e0233872. https://doi.org/10.1371/journal.pone.0233872.

Wekerle H. 2018. Brain inflammatory cascade controlled by gut-derived molecules. *Nature* 557 (7707), 642–3. https://doi.org/10.1038/d41586-018-05113-0.

Xie, L., Li, P., Baiyun, L., Hongwei, C., Xiang, M. 2023. Research progress on the association between trimethylamine/trimethylamine-N-oxide and neurological disorders. *Postgraduate Medical Journal* 100 (1183), 283–88. https://doi.org/10.1093/postmj/qgad133.

Zhang, L., Wang, Y., Xiayu, X. et al. 2017. Altered gut microbiota in a mouse model of Alzheimer's disease. *Journal of Alzheimer's Disease* 60, 1241–57. https://doi.org/10.3233/JAD-170020.

Zhang, X., Tang, B., and Guo, J. 2023. Parkinson's disease and gut microbiota: from clinical to mechanistic and therapeutic studies. *Translational Neurodegeneration* 12, 59. https://doi.org/10.1186/s40035-023-00392-8.

Zhang, Y., Shen, Y., Liufu, N. et al. 2023. Transmission of Alzheimer's disease-associated microbiota dysbiosis and its impact on cognitive function: evidence from mice and patients. *Molecular Psychiatry* 28, 4421–37. https://doi.org/10.1038/s41380-023-02216-7.

Zhao, Y., and Lukiw, W. J. 2018. Bacteroidetes neurotoxins and inflammatory neurodegeneration. *Molecular Neurobiology* 55, 9100–7. https://doi.org/10.1007/s12035-018-1015-y.

Zhao, Y., Dua, P., and Lukiw, W. J. 2015. Microbial sources of amyloid and relevance to amyloidogenesis and Alzheimer's disease (AD). *Journal of Alzheimer's Disease & Parkinsonism* 5, 177. https://doi.org/10.4172/2161-0460.1000177.

Zhao, Z., Ning, J., Bao, X. Q. et al. 2021. Fecal microbiota transplantation protects rotenone-induced Parkinson's disease mice via suppressing inflammation mediated by the lipopolysaccharide-TLR4 signaling pathway through the microbiota-gut-brain axis. *Microbiome* 9, 226. https://doi.org/10.1186/s40168-021-01107-9.

9 Clinical Management of Geriatric and Psychiatric Cases

Implications of Surgery and Hospitalization on Gut–Brain Axis

Monica Saini, Kanwarbir Singh Gill, Harmail Singh, Mantaz Kaur Dhillon, Harnoor Kaur Sibia, and Simran Kaur

9.1 INTRODUCTION

9.1.1 Overview of Geriatric Psychiatry

Geriatric psychiatry is a domain that emphasizes on the mental well-being of older adults, incorporating expertise in psychiatry, neurology and geriatrics. As the global population endure to age, geriatric psychiatry has turn out to be an essential aspect of clinical practice and research. In the year 2024, about 830 million people were aged 65 or older globally, and this figure is projected to reach 1.7 billion by the year 2054. A significant growth will be seen in Asia as its elderly inhabitants will become more than doubles, chiefly due to population rise, better health care and longer life expectation (Van Teutem 2024; Table 9.1).

Elder adults often undergo psychiatric disorders, for instance, depression, anxiety, dementia and delirium, frequently combined with chronic medical conditions, social seclusion and functional impairments. The health management of older adults involves expert knowledge because of the multifaceted interplay of psychiatric indications, medical comorbidities and psychosocial problems. Studies show that psychiatric conditions are underdiagnosed in senior citizens. For instance, it was stated that psychiatric conditions alike depression in older adults are connected with complex illness and mortality because of their interface with chronic diseases, especially cardiovascular disease (Katon 2003).

9.1.2 Ageing and Mental Health Disorders: Common Comorbidities

Mental health conditions in older adults are often comorbid with physical illnesses, which aggravate both illnesses. Depression, dementia and anxiety disorders are among the most predominant psychiatric

TABLE 9.1

Predicted Number of People Aged 65 Years and Older Worldwide and by Region, in Years 2024 and 2054

Year	World 65+ (Millions)	Asia 65+ (Millions, Approx.)	Europe 65+ (Millions, Approx.)	Africa 65+ (Millions, Approx.)
2024	830	~500	~150	~50
2054	1,700	~1,000	~200	~120

UN World Population Prospects 2024; Our World in Data, Teutem 2024.

DOI: 10.1201/9781003610274-11

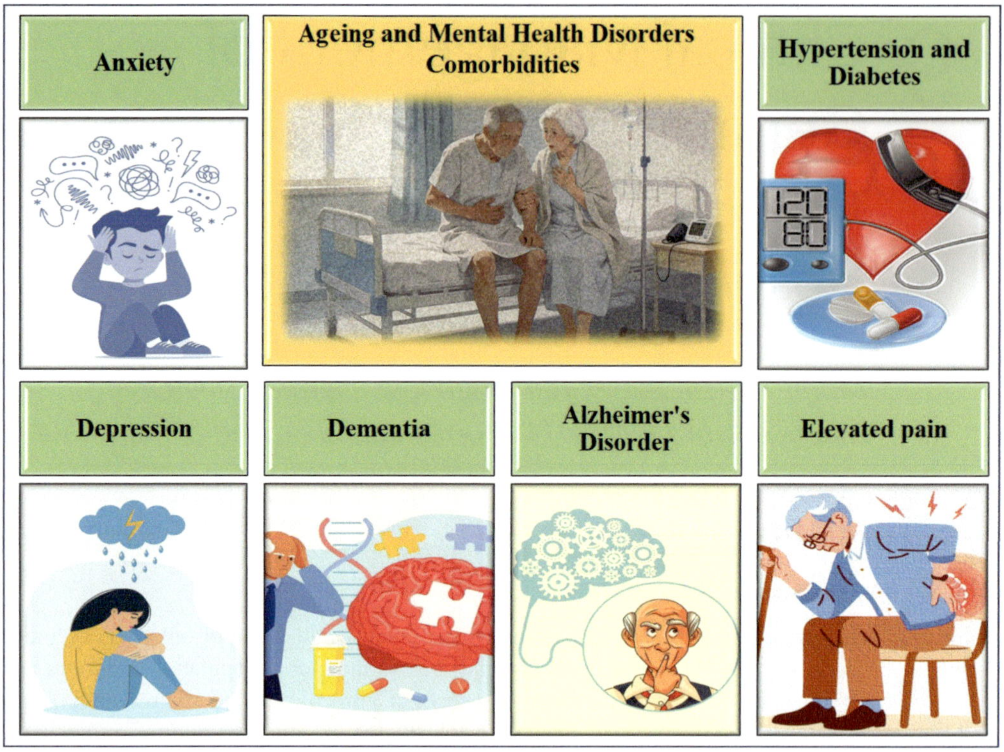

FIGURE 9.1 Age-related mental and physical disorders.

TABLE 9.2
The Burden of Disease in Older People

Disease	Impact to Total Burden (%)
Cardiovascular diseases	30.3
Malignant neoplasms	15.1
Chronic respiratory diseases	9.5
Musculoskeletal illnesses	7.5
Neurological and mental conditions	6.6

Prince et al. (2014).

disorders in geriatric inhabitants (Figure 9.1). Particularly, depression is connected with amplified risks for cardiac incidents in individuals including type 2 diabetes complications and decreased functional capacity (Inoue et al. 2020). The incidence of depression amongst Whites and Hispanics was approximately threefold higher than that of African Americans. Dementia diagnosis and elevated pain severity were associated with increased depression levels (Steffens et al. 2009; Table 9.2).

At present, an approximately 6.7 million Americans ageing 65 and above are surviving with Alzheimer's dementia. Without medical developments to prevent, reduce or treat the disease, this figure could upsurge to 13.8 million by 2060. In 2019, official death reports stated 121,499 deaths from Alzheimer's, making it the sixth-leading cause of death in the United States. Though, in 2020 and 2021, when COVID-19 turned out to be a major cause of death, Alzheimer's ranked seventh.

Among adults aged 65 and older, it persists the fifth-leading cause of death. Amid 2000 and 2019, mortalities from stroke, heart disease and HIV dropped, whereas deaths from Alzheimer's amplified by over 145% (Alzheimer's Association 2023). Comorbidities such as diabetes, high blood pressure and stroke additionally worsen male dementia care, requiring a consolidative approach. Anxiety conditions also remain extremely prevalent, often worsening medical illnesses such as chronic obstructive pulmonary disease (COPD) and heart disease, to increased healthcare applications. In a systematic review and meta-analysis, the pooled prevalence of anxiety disorders in older COPD patients was 36% (Sulaimanova and Kyzy 2025).

9.1.3 Clinical Challenges in Caring Older Adults

The medical management of the geriatric population represents several challenges because of the connection of ageing, mental well-being and physical comorbidities. One of the most substantial problems is polypharmacy, which signifies to the usage of various medicines to manage multiple conditions. The occurrence of multimorbidity was 34.70% amongst the elderly patients with chronic illness. More than half (60.59%) of the patients with multiple morbidity had two co-occurring chronic diseases. "Degenerative/digestive diseases," "metabolic diseases" and "cardiovascular diseases" were three exact multimorbidity outlines (Lu et al. 2021). Elder adults are predominantly prone to adverse drug outcomes, drug–drug interactions and reactions and medicine non-compliance, all of which can aggravate both physical and mental health circumstances (Mangin, Heath, and Jamoulle 2012). In a research study by Christopher et al. (2022), the main issues of medication use recognized were: nonadherence, unfavourable drug events, availability, polypharmacy, inappropriate medications, belief regarding medications, insufficiency of knowledge and awareness and lack of deprescribing. Another task is cognitive deterioration, which complicates the diagnosis and dealing of psychiatric disorders. Cognitive diminishing can mask signs of depression or anxiety, indicating the underdiagnosis or misdiagnosis of diseases. Besides, older adults with psychiatric disorders often encounter social isolation, financial stress and deficit of family support, furthermore complicating their supervision and management (Prince et al. 2014; Figure 9.2).

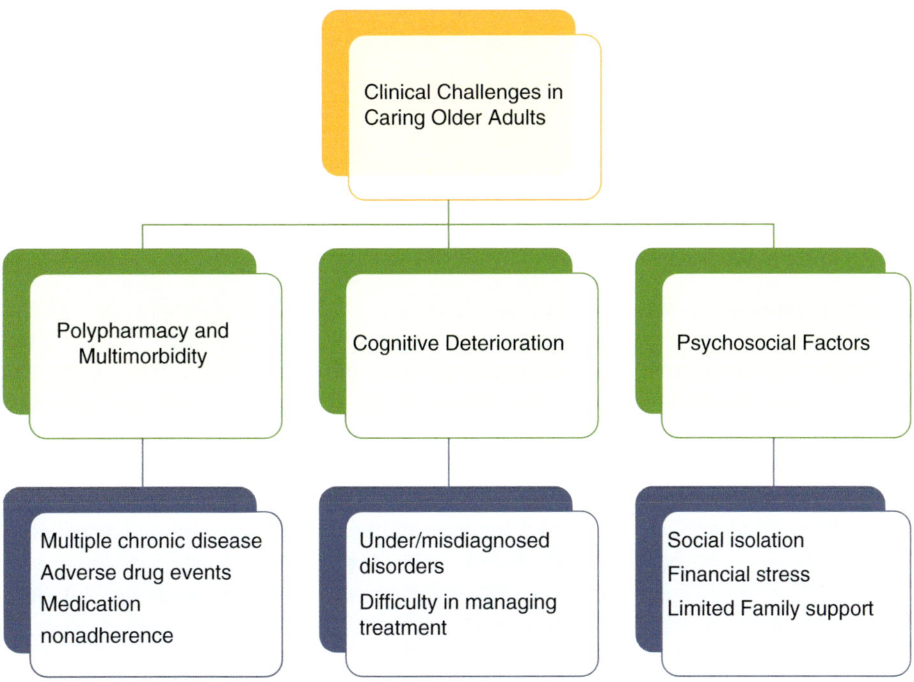

FIGURE 9.2 Challenges in clinical care of older adults.

9.1.4 The Role of Multidisciplinary Approach in Geriatric Care

Geriatric psychiatric attention often needs a multidisciplinary team work, including professionals from several fields, including psychiatry, primary care, neurology, nursing, social work and occupational therapy. Multidisciplinary teams are crucial for complete care, referring both medical and psychosocial needs. In a research study by Gaugler et al. (2005), multidisciplinary interference for dementia patients, which comprises medical management, caregiver support and social services, was associated with enhanced patient results and reduced caregiver burden. Multidisciplinary units help make certain that multifaceted issues such as medication management, cognitive decline and psychosocial aspects are dealt with in a coordinated manner. Interaction amongst healthcare workers is vital to avert fragmented care, which can lead to adverse consequences such as expanded hospitalizations and excessive healthcare expenses.

9.1.5 Ethical Concerns in Geriatric Psychiatry

Ethical issues are a central matter of concern in geriatric psychiatry, predominantly when dealing with consent and decision-making in older adults, cognitive loss and end-of-life care.

- **Consent and decision-making in elder adults**: As cognitive decline advances in older adults, the capability to make well-versed decisions becomes progressively compromised. The valuation of decision-making ability is vital in ensuring that patients are included in their care to the greatest amount possible. As reported by Appelbaum (2007), decision-making capability should be evaluated across four spheres: considerate, appreciation, intellectual and stating a choice. When this capacity is weakened, family members or legal councils may be important to step in, managed by the ethics of beneficence and non-maleficence.
- **Ethical difficulties in end-of-life care for psychiatric patients**: End-of-life maintenance for psychiatric patients, predominantly individuals with incurable illnesses like progressive dementia, gives away exceptional ethical dilemmas. Choices concerning the withdrawal of treatment, soothing care and do-not-resuscitate instructions should be completed with careful consideration of the patient's desires, worth of life and probable suffering. It has been found in the studies that elder adults with psychiatric disorders and conditions are less probable to get palliative care facilities, weighing the necessity for better consideration of ethical matters in this extent (Sampson et al. 2006).

9.1.6 Gut–Brain Axis: Key Concepts

The GBA is a significantly evolving area in geriatric psychiatry, emphasising the bidirectional interaction amongst the gastrointestinal system and the central nervous system. This axis includes neural, hormonal and immunologic pathways and is affected by the gut microbiota. Contemporary research proposes that dysbiosis or a disproportion in the gut microbiota might be a factor in psychiatric disorders, for instance, depression, anxiety and even cognitive deterioration (Cryan et al. 2019).

The gut–brain association may have specific significance in the elderly population due to variations in the microbiome associated with age, diet and medication use. For instance, findings from a study have shown that changes in the microbiome can impact inflammation and neuroinflammation, possibly adding to neurodegenerative diseases (Zhu et al. 2020). Hence, considering the GBA may deliver new therapeutic goals for managing psychiatric situations in older adults, with interventions extending from dietary alterations to probiotics.

The gut microbiome has been progressively known for its vital functioning in modifying not only gastrointestinal health but also psychological health and behaviour. Research studies have confirmed the influence of gut bacteria on temperament, cognition and neurodevelopment across numerous biochemical mechanisms, as well as the production of neurotransmitters like serotonin, dopamine and GABA, in addition to SCFAs and other microbial metabolites (Cryan and Dinan 2012).

9.2 PSYCHIATRIC DISORDERS AND GUT–BRAIN AXIS DYSREGULATION

Psychiatric disorders in late life more and more appear linked to dysregulation of the microbiome–GBA, pointing to gut dysbiosis' effects on depression, anxiety, psychosis and treatment responses in older adults. Amalgamating gut health with geriatric psychiatry provides a mechanistic way for refining pharmacological and non-pharmacological elderly care plans (Góralczyk-Bińkowska, Szmajda-Krygier, and Kozłowska 2022; Kumar et al. 2023; Patil and Mehdi 2025).

9.2.1 Depression, Anxiety, and Microbiome

Major alterations are seen in the gut microbiome of patients with major depressive disorders and anxiety disorders, such as (Cao et al. 2025; Kumar et al. 2023):

- ↑ pro-inflammatory taxa
- ↓ SCFA-producing bacteria.
- ↑ Parabacteroides, Oscillibacter, Streptococcus, and Eggerthella (depression)
- ↓ Prevotellaceae, Ruminococcaceae, and Faecalibacterium (depression)
- ↑ Enterobacteriaceae (anxiety).

Mechanistically, these alterations affect the GBA via several interconnected pathways supporting a causal link between microbial shifts, neuroinflammation and mood symptoms (Bibbò et al. 2022; Góralczyk-Bińkowska, Szmajda-Krygier, and Kozłowska 2022; Kumar et al. 2023):

- ↓ SCFAs, intestinal barrier integrity,
- ↑ lipopolysaccharide (LPS)-driven inflammation
- altered tryptophan–kynurenine metabolism, 5-HT and GABA signalling

9.2.2 Gut Dysbiosis in Schizophrenia and Bipolar Disorder

In schizophrenia and first-episode psychosis, many have reported decreased microbial diversity and characteristic shifts in important taxa, including variations in *Lactobacillus, Bifidobacterium, Escherichia* and *Clostridium* species. Further, antipsychotic treatment itself modifies the microbiota that leads to increased *Bifidobacterium* and *Escherichia coli* and decreased *Clostridium coccoides* and *Lactobacillus*—alterations that are associated with metabolic side effects like weight gain (Góralczyk-Bińkowska, Szmajda-Krygier, and Kozłowska 2022; Theleritis et al. 2024).

Bipolar disorder is also characterized by gut microbiota aberrations, although findings are heterogeneous; consistent signals include altered abundance of *Lactobacillus, Faecalibacterium* and *Ruminococcus* compared with healthy controls. Notably, adjunctive probiotic supplementation has been associated with reduced rehospitalization and improved depressive symptoms and cognition in bipolar patients, highlighting the therapeutic potential of microbiome-modulating approaches (Cocean and Vodnar 2024; Obi-Azuike et al. 2023; Ortega et al. 2023).

9.2.3 Psychopharmacology in Older Adults and the Gut–Brain Axis

Psychotropic medications commonly prescribed in late-life psychiatry—antidepressants, antipsychotics, mood stabilizers, benzodiazepines—exert bidirectional interactions with the gut microbiota that may be especially relevant in older adults with multimorbidity and polypharmacy. Many antidepressants and antipsychotics possess direct antimicrobial properties, selecting for specific taxa and potentially aggravating dysbiosis, while age-related changes in gastric acidity, motility, and immune function further modulate these effects (Góralczyk-Bińkowska, Szmajda-Krygier, and Kozłowska 2022; Liu et al. 2023; Ortega et al. 2023; Shaikh et al. 2025).

Second-generation antipsychotics, frequently used for psychosis and behavioural disturbances in dementia, are strongly linked to weight gain, insulin resistance, and metabolic syndrome, effects that are increasingly attributed to drug-induced shifts in gut microbial composition and SCFA profiles. For older adults, these microbiome-mediated metabolic effects amplify vascular and cognitive risk, making microbiota-sensitive drug selection, dose minimization and metabolic monitoring integral to geriatric psychopharmacology (Abidin et al. 2025; Ortega et al. 2023; Theleritis et al. 2024).

9.2.4 Non-Pharmacological Approaches in Late Life

Dietary modulation offers a foundational, low-risk strategy to influence the GBA in elderly patients with psychiatric disorders. Diets rich in fibre, polyphenols and omega-3 fatty acids—such as Mediterranean-style patterns—enhance SCFA-producing bacteria, improve barrier integrity, and attenuate systemic inflammation associated with depression, anxiety and cognitive decline (Bibbò et al. 2022; Kumar et al. 2023; Patil and Mehdi 2025).

Probiotics and prebiotics, often termed psychobiotics when targeted at mental health, have shown promising effects on mood, stress reactivity and cognitive performance, including older adults. Proposed mechanisms include modulation of HPA-axis activity, enhancement of GABA and serotonin production, reduction of neuroinflammation, and restoration of an eubiotic microbiome profile, though strain specificity, dose and inter-individual variability remain key limitations. Psychotherapy, while not directly altering microbes, can reduce chronic stress and HPA-axis activation, indirectly supporting microbiome stability and gut barrier function in late life (Cocean and Vodnar 2024; Hoffman et al. 2020; Kumar et al. 2024; Patil and Mehdi 2025; Shaikh et al. 2025).

9.2.5 Integrating Gut–Brain Health in Geriatric Care

Incorporating gut–brain considerations into psychiatric care plans for older adults requires a structured, multimodal approach that combines assessment, pharmacological optimization, and lifestyle interventions. Clinically, this may include screening for gastrointestinal symptoms, diet quality, recent antibiotic or proton-pump inhibitor (PPI) use and metabolic parameters alongside standard psychiatric evaluation, thereby identifying candidates for targeted microbiome-supportive strategies (Abidin et al. 2025; Shaikh et al. 2025).

Care plans can then integrate: cautious psychotropic prescribing with attention to microbiome liabilities; individualized dietary counselling to increase fibre and anti-inflammatory nutrients; time-limited, evidence-based psychobiotic supplementation; and psychotherapy aimed at stress reduction and behaviour change, ideally delivered within interdisciplinary geriatric mental-health teams. As the evidence base grows, microbiome-informed, personalized interventions are likely to become a core component of precision psychiatry in the elderly, though robust longitudinal trials in this age group are still urgently needed (Cocean and Vodnar 2024; Góralczyk-Bińkowska, Szmajda-Krygier, and Kozłowska 2022; Kumar et al. 2024; Ortega et al. 2023; Patil and Mehdi 2025).

9.3 SURGERY AND GUT–BRAIN AXIS IN GERIATRIC PATIENTS

Surgical procedures in elderly patients have turned out to be increasingly conventional as life expectancy growths, leading to an excessive number of geriatric populations requiring various surgical interventions. The most recurrent elective surgeries carried out on older adults include cataract and lens operations, spinal fusions, cardiac surgeries, laminectomies, total or partial hip and knee replacement procedures (Caring Senior Service 2025). These surgeries' objective is to enhance the quality of life by focusing on mobility issues, vision damage and chronic pain frequently confronted in older adults. Crucial surgeries in this group characteristically comprise thoracic, abdominal cancer resections and breast and prostate cancer operations along with cholecystectomies. Whereas emergency surgeries are mostly linked to traumatic incidents, for instance, hip fracture repairs, indicating the greater risk of falls and fractures amongst the elderly (Potestio and Carney 2024; Caring Senior Service 2025).

The supervision and results of these surgical interventions in the elderly are complex by the incidence of multiple comorbidities, reduced physiological sources and frailty. Such aspects increase the chances of postoperative difficulties and mortality, requiring comprehensive preoperative evaluations and personalized perioperative care approaches. For instance, spinal surgeries bring a higher possibility of infection and nerve impairment, necessitating vigilant postoperative physiotherapy to reinstate function. Orthopaedic surgical procedures like hip and knee replacements also comprise risks associated with immovability and systemic inflammation (Deiner, Westlake, and Dutton 2014; Masutani et al. 2020).

Developing research features the significant interaction between surgical involvements and GBA dysbiosis—a disturbance in the microbiota equilibrium that disturbs both gut and brain well-being. Surgical stress and pressure, anaesthesia and perioperative medicines like antibiotics and opioid use can disturb gut microbiota, directing to amplified neuroinflammation and cognitive dysfunction post-surgery. This association is particularly critical in neurosurgical measures, where gut dysbiosis may worsen neurological recovery. Management of gut microbiota earlier and after surgical procedure could significantly advance recovery results and decrease perioperative neurocognitive conditions in the elderly adults (Danehower 2021; Wang et al. 2021; Table 9.3).

9.3.1 Preoperative and Postoperative Management in the Elderly

Preoperative and postoperative care in aged patients is crucial to lessen risks and progress surgical results, given their augmented vulnerability because of physiological ageing, comorbidities and reduced practical reserve. A complete preoperative assessment, often comprising a Comprehensive Geriatric Assessment (CGA), assesses cognitive functioning, frailty, nutritional grade, polypharmacy and comorbid circumstances to guide personalized care plans (Schipa et al. 2023). Assessment tools like the Mini-Cog besides Timed Up-and-Go test recognize risks for delirium, restlessness, mobility problems and postoperative difficulties. Local anaesthesia and multimodal analgesia are

TABLE 9.3
Relationship Among Common Geriatric Procedures with Gut–Brain Axis Dysbiosis

Types of Surgery	Gut–Brain Axis Dysbiosis Mechanism	Impact on Recovery and Neurological Functioning
Neurosurgical (e.g., organic/brain injury surgeries)	Surgical pressure and anaesthesia upset gut microbiota; intensified blood–brain barrier (BBB) permeability	Neuroinflammation aggravated; decreased neurological recovery (Danehower 2021; Wang et al. 2021)
Major abdominal surgeries (e.g., prostate/cancer resections, cholecystectomy)	Perioperative antibiotics, opioids and surgical anxiety cause microbial disproportion	Postoperative cognitive dysfunction; systemic swelling affecting recovery (Danehower 2021; Wang et al. 2021)
Orthopaedic surgical procedures (hip/knee replacements)	Immovability, opioid use and antibiotics disturb gut flora composition	Systemic inflammation, late functional regaining, probable cognitive decline (Danehower 2021)
Cataract and lens operations	Anaesthesia and perioperative medicines may moderately disturb gut microbiota	Negligible gut–brain axis disturbance; minor or no cognitive impact (Danehower 2021)

TABLE 9.4
Relation of Pre/Postoperative Maintenance and Care in Geriatric Population with Gut–Brain Axis Dysbiosis

Care Period	Gut–Brain Axis Dysbiosis Mechanism	Effect on Recovery and Neurocognitive Functioning
Preoperative	Undernourishment and polypharmacy disrupt gut microbiota (Matsumoto et al. 2025)	Upsurges baseline neuroinflammation and susceptibility (Olotu et al. 2019)
Intraoperative	Anaesthesia, antibiotics and opioids change microbiota range (Lim and Lee 2019)	Might upsurge blood–brain barrier absorbency and inflammation (Wang et al. 2021)
Postoperative	Stress, hassle, opioid use and immovability led to microbiome disproportion (Olotu et al. 2019)	Causes delirium and postoperative cognitive dysfunction (Inouye, Westendorp, and Saczynski 2013)

chosen to lessen cognitive dysfunction and opioids' side effects (Lim and Lee 2019). Preserving intraoperative normothermia and examining anaesthetic depth also diminish the risk of delirium in patients (Evered et al. 2021).

Post-surgery, cautious monitoring for delirium by means of tools like the Confusion Assessment Method (CAM) is essential, as it impacts 30%–50% of geriatric postoperative patients and stretches hospitalisation (Inouye, Westendorp, and Saczynski 2013). Initial mobilization, hydration, pain regulation and keeping sensory aids encourage recovery. Postoperative cognitive dysfunction (POCD) is accomplished by minimalizing sedatives and strengthening adequate oxygenation (Kotekar, Shenkar, and Nagaraj 2018). GBA dysbiosis influences postoperative healing such as surgical tension and stress and antibiotics and opioids disturb gut microbiota, encouraging systemic tenderness and neuroinflammation and adding delirium and cognitive deterioration. Modifying gut microbiota via probiotics or nutritional interventions may enhance results in elderly surgical patients (Olotu et al. 2019; Table 9.4).

9.3.2 Ethical and Procedural Protocols for Surgery in Geriatric Psychiatry

- **Informed consent and capacity**: Screening with UBACC (The University of California, San Diego Brief Assessment of Capacity to Consent); use timing, simple language, helpful feedback; assign proxies if capacity reserved for surrogacy (Prusaczyk et al. 2016).
- **Risk versus benefit**: Preoperative psychiatric assessment for depression/anxiety; frailty evaluation directs elective surgery choices (Qaffas et al. 2023).
- **Pain and psychotropics**: Multimodal analgesia chosen; minimalize sedatives/opioids to diminish delirium risk (Qaffas et al. 2023).
- **Postoperative monitoring**: Multidisciplinary approach for delirium/post-traumatic stress disorder (PTSD) assessment; re-evaluate decision competence ethically (Prusaczyk et al. 2016; Qaffas et al. 2023)

9.3.3 Impact of Surgery on Gut Health: Nutritional and Microbiome Changes

Surgery intensely disturbs gut health via manifold perioperative aspects. Mechanical bowel formulation produces "bacterial genocide," dropping total bacteria 34.7-fold, declining helpful Bifidobacterium and Lactobacillus, whereas increasing infective Proteobacteria and Enterobacteriaceae for up to 30 days (Stavrou 2016). Antibiotics further reduce microbial variety by 25%–33%, destroying *Firmicutes* and raising *Bacteroidetes*, with healing taking months (Stavrou 2016; Zheng et al. 2023). Opioids and stress hormones alike norepinephrine increase bacterial virulency (*P. aeruginosa*, *E. faecalis*), leading to translocation and contamination (Stavrou 2016). Nutritional deficiency through fasting/parenteral nutrition assists *Proteobacteria* upon nutrient-dependent *Firmicutes*, damaging short-chain fatty acid (SCFA) production crucial for mucosal integrity (Zheng et al. 2023). These variations lead to dysbiosis, anastomotic leakages and sepsis. Probiotics moderately restore variety, dipping difficulties by 37% (Stavrou 2016)

9.3.4 Post-Surgical Cognitive Decline: The Role of the Gut–Brain Axis

Post-surgical cognitive decline, incorporating POCD and delirium, disturbs 25%–60% of geriatric patients, led by GBA dysbiosis instead of isolated brain pathology (Yang et al. 2024; Abdullah, Khan, and Hassan 2025). Surgical disturbance, anaesthesia (sevoflurane/isoflurane), antibiotics and opioids diminish microbial diversity, exhausting SCFA producers alike *Lachnospiraceae*, *Ruminococcaceae* and *Butyrivibrio*, whereas uplifting pro-inflammatory *Proteobacteria*, *Bacteroidetes* and *Mucispirillum* (Zhang et al. 2023; Yu, Zhu, and Tang 2024). This stimulates leaky gut, LPS endotoxemia and Th17/IL-17 surges, conceding blood–brain barrier (BBB) integrity, triggering NLRP3 inflammasomes/microglia and damaging hippocampal synaptic plasticity/brain-derived neurotrophic factor (BDNF) (Yang et al. 2024; Abdullah, Khan, and Hassan 2025). Metabolomic changes show ↓thiamine, spermidine, LCPUFAs and SCFAs (butyrate), intensifying neuroinflammation and depressive behaviours (Zhang et al. 2023). Faecal microbiota relocation from POCD donors' transmission deficits; perioperative probiotics/prebiotics reinstate diversity, lower IL-6/TNF-α, and reduce POCD by 30%–50% (Yu, Zhu, and Tang 2024; Yang et al. 2024).

- **Dysbiosis profile**: ↓*Lachnospiraceae/Blautia/Eubacterium/Butyrivibrio*, ↑*Bacteroidetes/Proteobacteria/Mucispirillum* post-anaesthesia/surgical procedure, continuing 7–30 days (Zhang et al. 2023; Yu, Zhu, and Tang 2024).
- **Leaky gut/endotoxemia**: Operation disturbs tight connections; LPS/IL-17 translocation through Th17 cells initiates systemic neuroinflammation and NLRP3 activation (Abdullah, Khan, and Hassan 2025; Yang et al. 2024).
- **Metabolite deficits**: ↓SCFAs (butyrate), thiamine, spermidine, LCPUFAs damage BBB/BDNF/synaptogenesis; ↑TMAO indorses oxidative stress (Zhang et al. 2023; Yu, Zhu, and Tang 2024).
- **Mechanisms**: VagHPA axis/HPA dysregulation; microglial M1 alteration, hippocampal PSD95↓, IL-6/1β↑ produce memory/attention shortfalls (Yang et al. 2024; Abdullah, Khan, and Hassan 2025).
- **Interventions**: Probiotics/faecal microbiota transplantation (FMT) reinstates SCFA producers, ↓POCD 30%–50%; microbiota-intended approaches for geriatric risk stratification (Yu, Zhu, and Tang 2024).

9.3.5 Managing Delirium and Post-Surgical Complications in Geriatric Psychiatry

Handling postoperative delirium in elderly patients needs multicomponent approaches aiming risk factors such as dementia, age >65, comorbidities and surgery procedure stress (Robinson 2008). Diagnosis depends on assessment tools like the Confusion Assessment Method for ICU (CAM-ICU) and the Mini-Mental State Examination to find out various delirium subtypes, comprising hypoactive (71%), hyperactive (rare) and mixed kinds (29%) (Robinson 2008; Grover and Avasthi 2018). Non-pharmacologic management—orientation, sleep routines, early mobilization, physical and sensory aids—decreases incidence 33% and complexities like falls and infections (Grover and Avasthi 2018). Haloperidol (0.5–2mg) is first-line pharmacotherapy, better than benzodiazepines, with QT intensive care; atypical (quetiapine) as substitutes (Robinson 2008). Gut–brain dysbiosis after taking antibiotics/opioids aggravates neuroinflammation/delirium; probiotic bacterium may alleviate via SCFA restoration; however, evidence is developing (indirect from prior analyses; Table 9.5).

TABLE 9.5
Management of Delirium/Post-Surgical Difficulties and Gut–Brain Axis Dysbiosis

Management Approach	Gut–Brain Dysbiosis	Impact on Recovery
Non-pharmacologic (mobilization)	Counter indicates opioid/ antibiotic-induced↓firmicutes/SCFAs (Stavrou 2016)	↓Systemic swelling, neuroprotection via motility
Medication review (↓opioids/ anticholinergics)	Prevents↑proteobacteria/LPS translocation, virulency (Yang et al. 2024)	↓Leaky gut → neuroinflammation/ delirium hazard
Haloperidol/quetiapine	Does not straightly reinstate microbiota; check for further dysbiosis	Symptom regulate; indirect through declined stress cytokines
Probiotics (emerging)	Reinstates SCFA producers, ↓IL-6/ TNF-α post-surgery (indirect)	Possible↓POCD/delirium by BBB/ microglia variation

9.4 HOSPITALIZATION AND ITS PSYCHOLOGICAL IMPACT ON THE ELDERLY

Hospitalization is often medically necessary in elderly people due to systemic ailments, but it poses significant psychological and physical challenges to the patient's recovery. The prolonged stay at the hospital or in an ICU setting can initiate a cycle of stress, functional decline and compromised well-being that extends beyond the initial medical issue and leads to a distortion from reality, leading to cognitive dysfunction. To prevent worsening of the symptoms, addressing these multifaceted issues requires a holistic, systematic and multidisciplinary approach.

9.4.1 Hospitalization Stress: Psychological and Physical Effects on Older Adults

Hospital stays can be a major source of stress for the elderly, leading to a decline in both mental and physical health (Reynolds et al. 2022). Prolonged stays in the intensive care unit lead to patients' withdrawal from their surroundings, and patients are unable to reorient to their surroundings later.

Psychological effects: Common psychological responses include anxiety, depression and decreased sleep, with the risk directly proportional to the length of the patient's hospital stay, particularly for those with chronic conditions (Reynolds et al. 2022). Delirium is an acute confusional state marked by a change in mental function, disorientation and agitation. It affects up to one-third of elderly people over 70 and can persist post-discharge. In some severe cases, the intense stress of the hospital environment may even lead to the development of PTSD (Reynolds et al. 2022).

Physical decline and loss of independence: The physical consequences of hospitalization are very closely linked to mental health. Research indicates that a significant percentage of older patients, especially those over 85, leave the hospital with a greater level of disability than when they arrived initially at the hospital. This functional decline, often exacerbated by prolonged bed rest and immobility, directly contributes to a loss of independence, leading to feelings of helplessness, and worthlessness, which fuel the impending depression (Reynolds et al. 2022). A lack of focus on psychological factors during the discharge process can hinder a successful transition from hospital to home, worsening the patient's health (Hussein, Edwards, and Patel 2024).

9.4.2 The Effects of Prolonged Bed Rest on Gut Function and Cognition

Immobility, a frequent component of hospital care, triggers a systemic decline in older adults, impacting various physiological systems, including the gastrointestinal and nervous systems.

Gastrointestinal effects: Prolonged bed rest and inactivity slow down the metabolic rate of the body. Immobility leads to constipation due to decreased colonic peristalsis, which leads to increased water reabsorption in the colon, making stool harder and more difficult to pass. Furthermore, the prolonged recumbent position makes swallowing more difficult, potentially leading to gastric reflux (Knight, Nigam, and Jones 2009).

Metabolic rate: Inactivity for a long time causes a rapid drop in the basal metabolic rate (e.g., a decline of nearly 7% in the first 10–24 hours). This decline reflects the loss of lean muscle mass, which contributes to metabolic dysregulation like insulin resistance and impaired glucose tolerance.

Cognitive and nervous system effects: Bed rest is associated with sensory deprivation, which can precipitate psychological symptoms like aggression, depression and hopelessness. The bed rest is also linked to structural and functional changes in brain tissue. Particularly, in the head-down tilt position used to simulate microgravity, causes a cephalad fluid shift (fluid moving toward the head). This can alter cerebral circulation and is associated with structural changes to brain tissues (Knight, Nigam, and Jones 2009). There is a decline in executive function and cognitive distortions and effect on working memory and an increase in the reaction time. Elderly people living in a community setting, even for relatively brief episodes of bed rest, can lead to adverse functional outcomes and disability due to their diminished physiological reserve (Gill, Allore, and Guo 2004; Wick 2011). This cascade of functional decline is collectively referred to as deconditioning (Wick 2011).

9.4.3 Hospital-Acquired Infections and Gut Health in the Elderly

Older adults face a disproportionately higher incidence and severity of healthcare-associated infections (HAIs) and hospital-acquired infections due to age-related changes like immunosenescence (weakening of the immune system) and inflammaging (chronic, low-grade inflammation).

Mechanism: With age, the thymus (where T cells mature) contracts, which leads to a diminished pool of naïve T cells (T cells ready to respond to a new pathogen). This impairs the elderly adult's ability to mount an effective defence against novel hospital-acquired pathogens, such as multidrug-resistant organisms.

Impact on HAIs: This systemic vulnerability further increases the risk and severity of HAIs, including pneumonia, urinary tract infections (UTIs) and gut infections (Moffa and Tana 2025; Taylor and Oppenheim 1998).

9.4.4 Age-Related Gut Dysbiosis

Gut microbiota dysbiosis: A key factor is gut microbiota dysbiosis, characterized by reduced microbial diversity and a loss of beneficial bacteria (e.g., *Faecalibacterium prausnitzii*) that produce SCFAs (Moffa and Tana 2025). SCFAs are vital for maintaining the integrity of the epithelial barrier and regulating immune function.

Opportunistic pathogens: Even before hospitalization, the gut microbiota of older adults tends to have lower diversity and an increased proportion of opportunistic pathogens (e.g., Proteobacteria), a condition known as age-related dysbiosis (Moffa and Tana 2025).

Increased infection risk: Hospital-related factors, including polypharmacy and the use of broad-spectrum antibiotics, worsen this dysbiotic state, creating an environment permissive for colonization of opportunistic pathogens (Moffa and Tana 2025). This significantly increases the risk for common HAIs like UTIs, hospital-acquired pneumonia, and particularly *Clostridioides difficile* infection (*C. diff.*), which is the leading cause of healthcare-associated infective diarrhoea (Taylor and Oppenheim 1998; Inivos 2021). The resultant SCFA deficiency further impairs the gut barrier, leading to a "leaky gut" phenotype that facilitates pathogen translocation, increasing inflammation and the risk of sepsis (Moffa and Tana 2025).

Dietary changes (nothing by mouth): Changes in dietary pattern, periods of prolonged fasting, or Ryle tube feeding alter the nutrient availability to the gut, which rapidly changes microbial composition of the GI system (Cresci 2025).

PPIs: Commonly prescribed to hospitalized patients, PPIs alter the stomach's PH, making it less acidic. This acid is a natural barrier to pathogen entry, and its reduction makes the stomach more susceptible to bacterial and spore infections (Freedberg, Lebwohl, and Abrams 2014).

9.4.5 Procedural Protocols for Hospitalized Elderly Patients

Specialized protocols are essential to mitigate the risks associated with hospitalization in older adults. A multicomponent intervention and systematic package is the cornerstone of delirium prevention and management (Knowles et al. 2016). Key actions include:

Environmental adjustments: Minimizing ward shifting, ensuring that patients are cared by familiar staff during their stay at the hospital and providing a suitable environment with appropriate lighting, visible clocks and calendars for reorientation (National Institute for Health and Care Excellence (NICE) 2023; Knowles et al. 2016).

Addressing reversible factors: Ensuring adequate hydration (to prevent dehydration and manage constipation) and mobility with regular visits by the physiotherapists are helpful (encouraging early mobilization, walking and active range-of-motion exercises) (National Institute for Health and Care Excellence (NICE) 2023; Knowles et al. 2016).

Sensory and pain management: Resolving reversible causes of impairment (e.g., impacted earwax) and ensuring proper use of hearing and visual aids to prevent the impairments caused by sensory deprivation. (National Institute for Health and Care Excellence (NICE) 2023; Knowles et al. 2016).

Behavioural interventions: Promoting good sleep hygiene and using verbal and non-verbal techniques to de-escalate distress before considering pharmacological options like short-term haloperidol (National Institute for Health and Care Excellence (NICE) 2023; Knowles et al. 2016).

Ethical protocols for involuntary hospitalization: Virtuous care for the elderly hinges on balancing the principles and acting in the patient's best interest (Kemp et al. 2021; Alodhialah, Almutairi, and Almutairi 2024).

- **Capacity assessment**: The patient's right to informed consent or refusal is central, requiring a determination of their decision-making capacity (Kemp et al. 2021; Marks, Predescu, and Dunn 2021).
- **Involuntary hold**: If a patient lacks the capacity to make safe decisions (e.g., refusing treatment and desiring to leave against medical advice) and is at risk of harm to self, then physician may choose a mechanism for an involuntary hold, which can be based on specific criteria for psychiatric illness then further serial assessments should be done for the patient and giving the complete information of the illness to the patient (Marks, Predescu, and Dunn 2021).
- **Surrogate decision-making**: A legal guardian or a family member is employed when the patient lacks capacity and their wishes are unknown (Kemp et al. 2021; Alodhialah, Almutairi, and Almutairi 2024).

Best practices for communication with family and caregivers in ethical decision-making: Effective communication with the guardians is vital for ethical decision-making and patient support (Schapira et al. 2008). Educate the Family (Clarity): Provide clear information about the diagnosis and prognosis. Use diagrams to explain complex medical facts so that patients and attendants are able to understand the situation better (Schapira et al. 2008; Balkac and Ergun 2018).

- **Shared understanding**: Using the "ask-tell-ask" technique (ask a question, provide information, then ask for understanding and more details) (Schapira et al. 2008).

- **Role and preference clarity**: The healthcare team should understand the caregiver's role and encourage the appointment of a single main family contact to streamline information flow (Sherman 2019; Alodhialah, Almutairi, and Almutairi 2024).
- **Address disagreement directly**: If there is a conflict between family members or between the family and the clinical team, identify the source of the conflict (e.g., misunderstanding of the prognosis, guilt, differing spiritual views) (Sherman 2019).
- **Reach an ethics consultation**: When an ethical conflict cannot be resolved, a formal consultation with the hospital's Ethics Committee should be requested.
- **Schedule regular updates**: Schedule specific times for meetings to discuss the plan, even if there are no major updates.

9.4.6 Role of Psychosocial Support in Improving Outcomes during Hospitalisation

Psychosocial support to address the emotional, psychological and social needs of hospitalized patients and their families helps in significantly improving coping and clinical outcomes of the patients (Pant, Vyas, and Papathanassoglou 2025).

Mental health improvement: Psychosocial interventions at an early stage, such as Cognitive-Behavioural Therapy (CBT), Problem-Solving Therapy, and Reminiscence Therapy, have been shown to significantly lower the risk of depression and anxiety (Klausner and Alexopoulos 1999).

Impact on physical recovery: Uncontrolled stress associated with a critical illness or hospital stay can lead to immune and metabolic dysfunction, increased pain and sleep disturbances, thus further complicating the physical recovery of the patient (Pant, Vyas, and Papathanassoglou 2025). Non-mental health professionals, like nurses and ward staff, can be trained to deliver brief, standardized psychosocial interventions to ensure wider access for frail elderly patients (Klausner and Alexopoulos 1999).

9.4.7 Rehabilitation and Discharge Planning: Addressing Cognitive and Gut-Health Recovery

A comprehensive, multidisciplinary discharge plan, ideally initiated at admission, is crucial for a smooth transition and preventing readmissions (Gane et al. 2022).

Cognitive recovery and rehabilitation: Rehabilitation for older adults is a goal-oriented process that begins as soon as the patient is medically stable. The rehabilitation team, which may include rehabilitation psychologists and neuropsychologists, addresses emotional issues like depression and identifies cognitive problems (Nas et al. 2015).

Gut-health recovery and discharge: This involves post-discharge planning that includes a dietician to guide the elderly on nutritional guidance to support the reestablishment of a healthy gut microbiota, potentially including probiotics or a fibre-rich diet to boost SCFA production (Moffa and Tana 2025; Nas et al. 2015). Continuous monitoring for symptoms of gut dysbiosis, such as persistent diarrhoea, is essential.

Engaging patients and families: Successful discharge planning requires actively engaging patients and their families as active team members (Gane et al. 2022).

9.5 SPECIAL CONSIDERATIONS IN THE CLINICAL MANAGEMENT OF GERIATRIC PATIENTS

9.5.1 Cognitive Decline, Dementia and the Gut–Brain Connection

Cognitive decline involves executive dysfunction and memory loss, which progresses over time often preceding dementia such as Alzheimer's disease, impacted by gut microbiota dysbiosis via the

TABLE 9.6
Gut–Brain Axis Changes, Microbial Alterations and Cognitive Effects in Ageing and Dementia

Aspect	Key Findings	Microbial Changes	Clinical Correlations
Cognitive decline	↓α-Diversity, low SCFAs impair plasticity (Jaqua et al. 2024; Tana et al. 2025; He et al. 2024)	↓*Ruminococcaceae Faecalibacterium*, (Alkasir et al. 2016; Tana et al. 2025)	Memory loss, hippocampal atrophy, memory loss (Alkasir et al. 2016; Tana et al. 2025;)
Dementia (e.g., Alzhiemer's)	Lipopolysacchride-driven inflammation promotes amyloid/tau (Alkasir et al. 2016; He et al. 2024; Wang et al. 2025; Escobar et al. 2022)	↑*Escherichia/Shigella*, Bacteroides (Alkasir et al. 2016; Tana et al. 2025)	Amyloid burden, executive dysfunction (Tana et al. 2025; He et al. 2024)
Gut-brain connection	Dysbiosis → barrier leak → microglial activation → neuroinflammation (Wang et al. 2025; He et al. 2024)	↓Lachnospiraceae, Akkermansia, (Escobar et al. 2022; Tana et al. 2025)	Neuroinflammation, ↓BDNF (Jaqua et al. 2024; Tana et al. 2025)

GBA (He et al. 2024; Tana et al. 2025). Mild cognitive impairment (MCI) correlates with atrophy of hippocampus and amyloid-beta accumulation. This communication pathway works through immune activation and microbial metabolites (for example, SCFAs) and vagus nerve signalling (Tana et al. 2025; Escobar et al. 2022; Alkasir et al. 2016). Gut barrier integrity is compromised by pro-inflammatory microbes such as *Bacteroides, Escherichia*, and *Shigella*. Pro-inflammatory microbes stimulate microglial TLR4/NF-κB pathways by permitting LPS translocation, further promoting neuroinflammation (He et al. 2024; Wang et al. 2025; Alkasir et al. 2016). Publications on microbiota–cognition interactions document about depleted SCFA-producing microbes, for example, *Faecalibacterium* and *Ruminococcaceae* being associated with diminished cognitive performance and dementia (He et al. 2024; Tana et al. 2025).

Wang et al. (2025) document the linkage of dysbiosis to endothelial dysfunction, thus exacerbating pathology in vascular dementia.

Administration of probiotics (for example, *Lactobacillus*) ameliorates symptoms of MCI through inflammation modulation (Tana et al. 2025). Transplantation of faecal microbiota depreciates amyloid deposition in preclinical models, further pointing towards therapeutic viability (Escobar et al. 2022; Alkasir et al. 2016).

Daily exercise and plant-based diets shift microbiota towards anti-inflammatory profiles, thus leading to elevation of BDNF (Escobar et al. 2022; Jaqua et al. 2024; Table 9.6).

9.5.2 Ethical and Procedural Protocols in Complex Geriatric Cases

Rigorous ethical protocols such as balancing autonomy, beneficence, non-maleficence and justice, alongside structured procedural frameworks like CGA, are required in geriatric cases (comprising multimorbidity) (Alodhialah, Almutairi, and Almutairi 2024; İlgili, Arda, and Munir 2014). Doctors and nurses face difficulties in obtaining informed consent from patients with cognitive impairments, honouring advance directives amid family conflicts, in management of risks of polypharmacy, etc. (Alodhialah, Almutairi, and Almutairi 2024; Janssens et al. 2018). Despite ageism barriers, procedural protocols underline vulnerability protections, including assent from impaired subjects, proxy consents and oversight from ethics committees to ensure scientifically sound research (İlgili, Arda, and Munir 2014; L. Yu et al. 2020).

TABLE 9.7
Significant Protocols and Ethical Considerations

Protocol Area	Ethical Considerations	Procedural Steps
Informed consent	Use proxies/assent for patients with cognitive impairments; assess capacity (Alodhialah, Almutairi, and Almutairi 2024; İlgili, Arda, and Munir 2014)	Family involvement, simplified forms, visual aids; re-evaluate as needed (Yu et al. 2020)
Management of polypharmacy	Non-maleficence via interaction checks; patient-focused deprescribing (Alodhialah, Almutairi, and Almutairi 2024; Janssens et al. 2018)	Review of complete history, STOP/START criteria, prioritizing patient preferences and goals (British Geriatrics Society 2019)
Advanced directives/ end-of-life	Honour wishes; resolve family conflicts ethically (Alodhialah, Almutairi, and Almutairi 2024)	Documentation of preferences; multi-modal palliative planning (Fahed, Barron, and Steffens 2020; Janssens et al. 2018)

In countries like China, research protocols for geriatric patients mandate risk-benefit assessments, geriatric expertise in ethics committees and tailored informed consent processes using visual aids, simplified language along with family involvement (L. Yu et al. 2020). In multimorbid frail elderly patients, decisions for interventions prioritize balanced risk-benefits, intervention efficacy, alignment with goals and patient preferences often favouring palliative care over curative interventions (Janssens et al. 2018). CGA toolkits guide primary care practitioners through psychological, functional, physical and environmental evaluations, thus creating individualized care plans to reduce frailty progression (British Geriatrics Society 2019).

Adherence to procedural protocols decreases adverse outcomes. For example, capacity assessments using scales like GP-Cog ensure valid consent, while periodic medication reviews prevent drug interactions (Fahed, Barron, and Steffens 2020). Interprofessional collaboration along with training in geriatric ethics continues to be essential for resource allocation and navigating end-of-life decisions (İlgili, Arda, and Munir 2014; Table 9.7).

9.5.3 Polypharmacy and Its Impact on Gut Health and Cognition

Microbiota disruption and prevalence: Polypharmacy is defined as ≥5 drugs, affects 40%–50% of elderly individuals and changes gut microbiota composition significantly, leading to a reduction in alpha-diversity and increasing pro-inflammatory microbial population such as *Escherichia coli*, *Rothia mucilaginosa* and *Enterococcus faecalis* (Larsson et al. 2024; Maher, Hanlon, and Hajjar 2013). Drugs like metformin and PPIs (antacids) associate with oral bacteria translocation (e.g., Veillonella spp. and Streptococcus spp.) into the gastrointestinal tract, while beta-blocker antidepressants deplete butyrate producers like *Romboutsia timonensis* (Gemikonakli et al. 2022; Larsson et al. 2024).

Mechanisms linking gut dysbiosis to polypharmacy: Drugs induce dysbiosis through impaired barrier function and altered production of metabolites; direct anti-biotic effects. High drug burden index polypharmacy overrules diversity loss related to ageing but increases *Firmicutes* and *Bacteroidetes* shifts which even persist post-deprescribing (Gemikonakli et al. 2022; Rojas-Solé et al. 2023). Polypharmacy correlates with reduced SCFAs and inflammation essential for epithelial integrity and also increases gut metabolic modules for nitrate reduction and amino acid degradation (Larsson et al. 2024).

Gut–brain axis and its cognitive impacts: Gastrointestinal tract dysbiosis from polypharmacy leads to neuroinflammation via microglial activation and LPS leakage subsequently accelerating amyloid-beta pathology in Alzheimer's disease models (Tamene et al. 2025; Maher, Hanlon, and

TABLE 9.8
Impact of Polypharmacy on Gut and Mental Health

Aspect	Gut Health Effects	Cognitive Effects
Microbiota shifts	↑*E. coli*, Rothia; ↓romboutsia, butyrate producers (Larsson et al. 2024; Gemikonakli et al. 2022)	Dysbiosis → LPS-driven neuroinflammation (Tamene et al. 2025)
Diversity/metabolites	↓Alpha-diversity; ↑nitrate/amino acid modules (Larsson et al. 2024; Rojas-Solé et al. 2023)	↓SCFAs impair BDNF, hippocampal function (Gemikonakli et al. 2022)
Medication-specific	PPIs↑oral taxa; antidepressants↓firmicutes (Larsson et al. 2024)	Polypharmacy accelerates decline, reversible partially (Maher, Hanlon, and Hajjar 2013)

Hajjar 2013). Reduced SCFAs impair hippocampal neuroplasticity and BDNF; however, deprescribing partially reverses these impairments through recovery of microbiota (Gemikonakli et al. 2022).

Therapeutic implications and deprescribing: Tapering high-drug burden index polypharmacy regimens leads to partial restoration of diversity, mitigating frailty-linked cognition via Bifidobacteriaceae recovery (Gemikonakli et al. 2022). Antioxidant supplementation, which targets oxidative stress from polypharmacy–gut interactions, offers adjunctive potential (Rojas-Solé et al. 2023; Table 9.8).

9.5.4 Nutritional Interventions for Enhancing Gut–Brain Health

High-fibre diets improve microbial diversity: High-fibre nourishments from fruits, vegetables, whole grains and legumes like lentils act as prebiotics, promoting Bifidobacterium and Lactobacillus to generate SCFAs as butyrate, acetate and propionate, firming gut barrier integrity and lessening leaky gut-caused neuroinflammation (Patil and Mehdi 2025; Berding et al. 2020).

Mediterranean diet encourages anti-inflammatory microbiota: Mediterranean diet adherence rises *Faecalibacterium prausnitzii*, *Roseburia*, and *Prevotella*, uplifting SCFA production whereas inhibiting pathogens, associating with improved cognitive resilience, mood adaptation and decreased depression/anxiety via vagus-HPA axis alteration (Patil and Mehdi 2025; Aziz et al. 2024).

Plant-based diets boost neurotransmitter production: Plant-based diets improve Bacteroidetes/Actinobacteria, enhancing serotonin (90% gut-derived) and GABA fusion, alleviating dysbiosis-linked cognitive deterioration and emotional dysregulation via polyphenol-mediated antioxidant outcomes (Berding et al. 2020; Liu et al. 2025).

Probiotic supplement: Probiotics (Lactobacillus, Bifidobacterium strains) lessen neuroinflammation, control HPA responses and develop stress resilience; combined with prebiotics, they reinstate diversity lost to western diets, increasing BDNF and brain elasticity (Patil and Mehdi 2025).

Omega-3s and polyphenols target gut-brain signalling: Omega-3 fatty acids and polyphenols derived from olive oil/berries assist Akkermansia and butyrate producers, reinforcing BBB integrity and serotonin paths, contributing adjunctive support for anxiety and depression in addition to MCI prevention (Aziz et al. 2024; Berding et al. 2020; Table 9.9).

9.5.5 The Role of Family and Caregivers in Managing Gut and Mental Well-Being in the Elderly

Vital concrete support is provided by the family caregivers by regulating medication regimens, monitoring symptoms and confirming treatment compliance for geriatric patients with mental

TABLE 9.9
Key Interventions

Intervention	Gut Health Benefits	Brain Health Outcomes
High-fibre/prebiotics	↑SCFAs, Bifidobacterium; barrier restoration (Patil and Mehdi 2025; Berding et al. 2020)	↓Neuroinflammation, HPA regulation (Liu et al. 2025)
Mediterranean diet	↑Faecalibacterium, variety (Patil and Mehdi 2025; Aziz et al. 2024)	Cognitive elasticity, mood enhancement (Berding et al. 2020)
Plant-based diets	↑Bacteroidetes, antioxidants, polyphenols (Berding et al. 2020)	↑Serotonin/GABA, emotional constancy (Patil and Mehdi 2025)
Probiotics	Reinstate diversity, ↓pathogens (Patil and Mehdi 2025)	↑BDNF, reduce stress (Liu et al. 2025)

TABLE 9.10
Key Roles

Role Area	Gut Health Support	Mental Health Support
Medication/nutrition	Fibre/probiotics vs. dysbiosis (Hu et al. 2024)	Adherence averts relapse (Chang, Chiou, and Chen 2009)
Emotional nurture	Stress buffers block (Chadda 2014)	Reduces loneliness (Kar 2021)
Lifestyle	Activity improves SCFAs (Schulz and Sherwood 2008)	Enhances cognition (Hu et al. 2024)

disorders like dementia and schizophrenia, dipping relapse chances and cognitive decline (Hu et al. 2024; Chang, Chiou, and Chen 2009). This caution extends to gut strength management, where caregivers track polypharmacy consequences that disturb microbiota balance, encouraging adherence to probiotics besides fibre-rich diets to alleviate dysbiosis-linked neuroinflammation (Schulz and Sherwood 2008).

Caregivers offer compassion, acceptance and company, buffering geriatric patients' stress and loneliness, which aggravate GBA dysregulation and mental health decline (Chadda 2014). Psychological support nurtures resilience, dropping cortisol levels that damage gut barrier integrity and cognitive functioning, while family interferences enhance the patient overall mood besides caregiver well-being (Hu et al. 2024; Kar 2021).

Caregivers accomplish daily nutrition, choosing anti-inflammatory diets (e.g., Mediterranean patterns) to boost SCFA production and gut diversity, contradicting age-related dysbiosis that adds to cognitive impairment (Chang, Chiou, and Chen 2009). They enable bodily activity and social meeting, modifying sedentary-induced microbiota changes and depression in multimorbid senior adults (Schulz and Sherwood 2008). Caregivers undergo a high burden from monetary strain, disturbed routines and health deterioration, yet psychoeducation diminishes this by teaching management strategies like problem-centred support, supporting their part in holistic gut-mental well-being maintenance (Chadda 2014; Hu et al. 2024; Table 9.10).

9.6 EMERGING RESEARCH AND FUTURE DIRECTIONS

Emerging work on the microbiome–gut–brain axis in ageing suggests that preserving gut health may become a central pillar of geriatric psychiatry, guiding new diagnostic, therapeutic and ethical frameworks. Future research will likely move from associative studies toward precision, microbiome-informed interventions tailored to older adults with mental disorders (Aznou et al. 2025; L. Cao et al. 2025; Kadyan et al. 2025; Khan et al. 2024).

9.6.1 Recent Advances in Gut–Brain Research and Ageing:

Ageing is characteristically associated with GBA remodelling, which includes decreased microbiome diversity, an increase in pro-inflammatory taxa and a loss of SCFA producers, along with frailty, cognitive decline and neurodegeneration. Current reviews and multi-omics studies have shown that gut-derived metabolites (like imidazole propionate, isoamylamine and altered GABA/SCFA profiles) are correlated with brain-glucose hypometabolism, cortical thinning and microglial dysfunction in AD and similar age-related disorders (Chen et al. 2025; Kanimozhi and Sukumar 2025; Loh et al. 2024; Wang et al. 2024).

Mechanistically, advances in animal and human work highlight how age-related gut dysbiosis promotes neuroinflammation via increased gut permeability, systemic immune activation and disruption of blood–brain barrier integrity. Simultaneously, dietary polyphenols, fibre and targeted probiotics seem to partially restore microbiota balance, promoting antioxidant defence, and providing support for healthier ageing of the brain via the GBA (Aldriwesh et al. 2025; Cao et al. 2025; Khan et al. 2024; Loh et al. 2024).

9.6.2 Surgical and Hospitalization Innovations for Gut Protection

Hospitalization and surgery, especially in older adults, can profoundly disturb the intestinal microbiome through fasting, anaesthesia, opioids, antibiotics and bowel preparation, thereby increasing risks of delirium, infection and prolonged recovery. Perioperative studies show that colorectal surgery and standard surgical bowel preparation can reduce SCFA levels, favour pathobiont overgrowth and delay microbiome recovery for months after discharge (Kohn et al. 2025; Ruiz et al. 2025).

Emerging protocols aim to protect microbial resilience by using judicious, narrow-spectrum antibiotics, minimizing unnecessary PPIs, promoting early enteral nutrition, and incorporating perioperative probiotics or synbiotics to reduce postoperative infections and ileus. In critical care and cardiothoracic settings, clinician guides now emphasize monitoring gut function and microbiome-friendly nutrition as part of organ support, a shift particularly relevant for cognitively vulnerable older adults (Ghosh et al. 2025; Ruiz et al. 2025; Tabari, Milani, and Aghasizadeh 2025).

9.6.3 Future Microbiome-Based Interventions in Geriatric Psychiatry

Microbiome-based therapeutics for ageing include psychobiotic probiotics, prebiotics, postbiotics and FMT, all under active investigation for depression, anxiety, cognitive impairment and neurodegenerative diseases. Psychobiotics targeting SCFA production, tryptophan metabolism and HPA-axis modulation have shown improvements in mood and neuropsychiatric symptoms, with some data suggesting enhanced cognition and reduced neuroinflammation in older populations (Ahmed et al. 2025; Kadyan et al. 2025; Mallick et al. 2025).

FMT, already effective for recurrent *Clostridioides difficile* infection, is being explored as a means to transfer "youthful" or "long-lived" microbiome signatures to older hosts to improve cognition, metabolic health and systemic inflammation. Future geriatric psychiatry may integrate microbiome sequencing, machine-learning risk stratification and microbiome-guided antidepressant or antipsychotic choices to optimize efficacy while minimizing metabolic and cognitive side effects (Aldriwesh et al. 2025; Araújo-Filho 2025; Aznou et al. 2025; Kadyan et al. 2025).

9.6.4 Ethical Considerations in Geriatric Mental-Health Care

Microbiome-focused interventions in older adults raise specific ethical challenges around consent, vulnerability and equitable access. Many geriatric patients have cognitive impairment, sensory deficits or fluctuating capacity, making transparent communication about uncertainties, long-term risks and donor-related issues in FMT or biobanked products essential (Kadyan et al. 2025; Li, Mou, and Yao 2025; Wang et al. 2024).

There is also a justice dimension: advanced microbiome sequencing and personalized interventions may initially be available only to affluent patients or academic centres, risking widening disparities in mental-health care for older adults. Ethical frameworks will need to address data privacy for microbiome profiles, commercialization of donor material and the potential stigmatization of "unhealthy" microbiomes in already marginalized elderly populations (Ahmed et al. 2025; Aldriwesh et al. 2025; Aznou et al. 2025; Wang et al. 2024).

9.7 CONCLUSION

The convergence of ageing research, gut–brain science and psychiatric practice supports an integrated care model in which microbiome health is monitored and managed alongside mood, cognition and physical comorbidities in older adults. Practically, this implies geriatric care pathways that combine microbiome-sensitive psychopharmacology, diet and lifestyle optimization, perioperative gut protection, and selective use of microbiome-based therapeutics within ethically robust, patient-centred frameworks. Interdisciplinary collaboration among psychiatrists, geriatricians, neurologists, surgeons, microbiologists and ethicists will be crucial to translate gut–brain science into safe, equitable and evidence-based interventions that enhance mental health and quality of life in ageing populations.

DECLARATION OF GENERATIVE AI AND AI-ASSISTED TECHNOLOGIES IN THE WRITING PROCESS

During the preparation of this work, the author(s) used AI assistance to support language refinement, structural editing, and generation of alternative phrasings for sections of the text. The tool was not used for data analysis, interpretation of results or the creation or verification of scientific claims or references. All AI-generated suggestions were critically reviewed, revised and verified by the author(s), who take full responsibility for the accuracy, integrity and originality of the content.

REFERENCES

Abdullah, Ismail A., Sariya Khan, and Fatma E. Hassan. 2025. "Gut–Brain Axis and Perioperative Gut Microbiome in Postoperative Cognitive Dysfunction: Implications for Neurosurgical Patients." *Medical Sciences* 13 (4): 236. doi:10.3390/medsci13040236.

Abidin, Zakirah Zainal, Zaw Myo Hein, Che Mohd Nasril Che Mohd Nassir, Norshafarina Shari, and Muhammad Danial Che Ramli. 2025. "Pharmacological Modulation of the Gut-brain Axis: Psychobiotics in Focus for Depression Therapy." *Frontiers in Pharmacology* 16 (September): 1665419. doi:10.3389/fphar.2025.1665419.

Ahmed, Nafees, Vishwas Gaur, Madhu Kamle, Abhishek Chauhan, Ritu Chauhan, Pradeep Kumar, and Namita Ashish Singh. 2025. "Microbiome-based Therapeutics for Metabolic Disorders: Harnessing Microbial Intrusions for Treatment." *Frontiers in Medical Technology* 7 (October): 1695329. doi:10.3389/fmedt.2025.1695329.

Aldriwesh, Marwh G., Raniah S. Alotibi, Nasser Alqurainy, Shatha Alrabiah, Assad M. Arafah, Majed F. Alghoribi, and Reham Ajina. 2025. "The Role of Gut Microbiome in Aging-associated Diseases: Where Do We Stand Now and How Technology Will Transform the Future." *Gut Microbes* 18 (1): 2607076. doi:10.1080/19490976.2025.2607076.

Alkasir, Rashad, Jing Li, Xudong Li, Miao Jin, and Baoli Zhu. 2016. "Human Gut Microbiota: The Links with Dementia Development." *Protein & Cell* 8 (2): 90–102. doi:10.1007/s13238-016-0338-6.

Alodhialah, Abdulaziz M., Ashwaq A. Almutairi, and Mohammed Almutairi. 2024. "Ethical and Legal Challenges in Caring for Older Adults with Multimorbidities: Best Practices for Nurses." *Healthcare* 12 (16): 1585. doi:10.3390/healthcare12161585.

Alzheimer Association. 2023 Alzheimer's Disease Facts and Figures. 2023. *Alzheimer S & Dementia* 19 (4): 1598–695. doi:10.1002/alz.13016.

Appelbaum, Paul S. 2007. "Assessment of Patients' Competence to Consent to Treatment." *New England Journal of Medicine* 357 (18): 1834–40. doi:10.1056/nejmcp074045.

Araújo-Filho, Irami. 2025. "Microbiome-Guided Antidepressant Therapy: The Future of Personalized Psychiatry." *Biomedical Journal of Scientific & Technical Research* 60 (4). doi:10.26717/bjstr.2025.60.009477.

Aziz, Tariq, Nageen Hussain, Zunaira Hameed, and Lin Lin. 2024. "Elucidating the Role of Diet in Maintaining Gut Health to Reduce the Risk of Obesity, Cardiovascular and Other Age-Related Inflammatory Diseases: Recent Challenges and Future Recommendations." *Gut Microbes* 16 (1): 2297864. doi:10.1080/19490976.2023.2297864

Aznou, Anouar, Joost P H Drenth, Max Nieuwdorp, and Abraham S Meijnikman. 2025. "Exploring the Role of the Gut Microbiome Across the Lifespan: Implications for Aging and Metabolic Disorders." *Journal of the Endocrine Society* 9 (10): bvaf130. doi:10.1210/jendso/bvaf130.

Balkac, Melisa, and Erdogan Ergun. 2018. "Role of Infographics in Healthcare." *Chinese Medical Journal* 131 (20): 2514–17. doi:10.4103/0366-6999.243569.

Berding, Kirsten, Klara Vlckova, Wolfgang Marx, Harriet Schellekens, Catherine Stanton, Gerard Clarke, Felice Jacka, Timothy G Dinan, and John F Cryan. 2020. "Diet and the Microbiota–Gut–Brain Axis: Sowing the Seeds of Good Mental Health." *Advances in Nutrition* 12 (4): 1239–85. doi:10.1093/advances/nmaa181.

Bibbò, Stefano, Salvatore Fusco, Gianluca Ianiro, Carlo Romano Settanni, Daniele Ferrarese, Claudio Grassi, Giovanni Cammarota, and Antonio Gasbarrini. 2022. "Gut Microbiota in Anxiety and Depression: Pathogenesis and Therapeutics." *Frontiers in Gastroenterology* 1 (October). doi:10.3389/fgstr.2022.1019578.

British Geriatrics Society. 2019. "Comprehensive Geriatric Assessment Toolkit for Primary Care Practitioners." https://www.bgs.org.uk/sites/default/files/content/resources/files/2019-03-12/CGA%20Toolkit%20for%20Primary%20Care%20Practitioners_0.pdf.

Cao, Luqi, Saurabh Kumar Jha, Neha Gupta, Xiang Chen, Renuka Soni, Luoxi Yuan, Rashi Srivastava, and Zhe-Sheng Chen. 2025. "The Aging Gut–Brain Axis: Effects of Dietary Polyphenols and Metal Exposure." *Chronic Diseases and Translational Medicine* 11 (4): 251–68. doi:10.1002/cdt3.70026.

Cao, YuanYuan, YiRan Cheng, WenChao Pan, JianWei Diao, LingZhi Sun, and MiaoMiao Meng. 2025. "Gut Microbiota Variations in Depression and Anxiety: A Systematic Review." *BMC Psychiatry* 25 (1): 443. doi:10.1186/s12888-025-06871-8.

Caring Senior Service. 2025. "7 Most Common Surgeries for Older Adults." https://caringseniorservice.com/blog/common-surgeries-for-older-adults/.

Chadda, Rakesh K. 2014. "Caring for the Family Caregivers of Persons with Mental Illness." *Indian Journal of Psychiatry* 56 (3): 221. doi:10.4103/0019-5545.140616.

Chang, Hsing-Yi, Chii-Jun Chiou, and Nain-Sen Chen. 2009. "Impact of Mental Health and Caregiver Burden on Family Caregivers' Physical Health." *Archives of Gerontology and Geriatrics* 50 (3): 267–71. doi:10.1016/j.archger.2009.04.006.

Chen, Ce, Guo-Qing Wang, Dai-Di Li, and Feng Zhang. 2025. "Microbiota–gut–brain Axis in Neurodegenerative Diseases: Molecular Mechanisms and Therapeutic Targets." *Molecular Biomedicine* 6 (1): 64. doi:10.1186/s43556-025-00307-1.

Christopher, Christina, Bhuvan Kc, Sunil Shrestha, Ali Qais Blebil, Deepa Alex, Mohamed Izham Mohamed Ibrahim, and Norhasimah Ismail. 2022. "Medication Use Problems among Older Adults at a Primary Care: A Narrative of Literature Review." *Aging Medicine* 5 (2): 126–37. doi:10.1002/agm2.12203.

Claesson, Marcus J., Ian B. Jeffery, Susana Conde, Susan E. Power, Eibhlís M. O'Connor, Siobhán Cusack, Hugh M. B. Harris, et al. 2012. "Gut Microbiota Composition Correlates with Diet and Health in the Elderly." *Nature* 488 (7410): 178–84. doi:10.1038/nature11319.

Cocean, Ana- Maria, and Dan Cristian Vodnar. 2024. "Exploring the Gut-brain Axis: Potential Therapeutic Impact of Psychobiotics on Mental Health." *Progress in Neuro-Psychopharmacology and Biological Psychiatry* 134 (June): 111073. doi:10.1016/j.pnpbp.2024.111073.

Cresci, Gail a. M. 2025. "Understanding How Foods and Enteral Feedings Influence the Gut Microbiome." *Nutrition in Clinical Practice* 40 (3): 555–74. doi:10.1002/ncp.11285.

Cryan, John F., and Timothy G. Dinan. 2012. "Mind-Altering Microorganisms: The Impact of the Gut Microbiota on Brain and Behaviour." *Nature Reviews. Neuroscience* 13 (10): 701–12. doi:10.1038/nrn3346.

Cryan, John F., Kenneth J. O'Riordan, Caitlin S. M. Cowan, Kiran V. Sandhu, Thomaz F. S. Bastiaanssen, Marcus Boehme, Martin G. Codagnone, et al. 2019. "The Microbiota-Gut-Brain Axis." *Physiological Reviews* 99 (4): 1877–2013. doi:10.1152/physrev.00018.2018.

Danehower, Sarah. 2021. "Targeting Gut Dysbiosis as a Means to Enhance Recovery from Surgical Brain Injury." *Surgical Neurology International* 12 (May): 210. doi:10.25259/sni_72_2021.

Deiner, Stacie, Benjamin Westlake, and Richard P. Dutton. 2014. "Patterns of Surgical Care and Complications in Elderly Adults." *Journal of the American Geriatrics Society* 62 (5): 829–35. doi:10.1111/jgs.12794.

Dey, A. B. 2017. "World Report on Ageing and Health." *The Indian Journal of Medical Research* 145 (1): 150–51. doi:10.4103/0971-5916.207249.

Escobar, Yael-Natalie H., Devin O'Piela, Loren E. Wold, Amy R. Mackos, Yael-Natalie H. Escobar, Devin O'Piela, Loren E. Wold, and Amy R. Mackos. 2022. "Influence of the Microbiota-Gut-Brain Axis on Cognition in Alzheimer's Disease." *Journal of Alzheimer S Disease* 87 (1): 17–31. doi:10.3233/jad-215290.

Evered, Lisbeth A., Matthew T.V. Chan, Ruquan Han, Mandy H.M. Chu, Benny P. Cheng, David A. Scott, Kane O. Pryor, et al. 2021. "Anaesthetic Depth and Delirium after Major Surgery: A Randomised Clinical Trial." *British Journal of Anaesthesia* 127 (5): 704–12. doi:10.1016/j.bja.2021.07.021.

Fahed, Mario, Gregory C. Barron, and David C. Steffens. 2020. "Ethical and Logistical Considerations of Caring for Older Adults on Inpatient Psychiatry during the COVID-19 Pandemic." *American Journal of Geriatric Psychiatry* 28 (8): 829–34. doi:10.1016/j.jagp.2020.04.027.

Foster, Jane A., and Karen-Anne McVey Neufeld. 2013. "Gut–Brain Axis: How the Microbiome Influences Anxiety and Depression." *Trends in Neurosciences* 36 (5): 305–12. doi:10.1016/j.tins.2013.01.005.

Freedberg, Daniel E., Benjamin Lebwohl, and Julian A. Abrams. 2014. "The Impact of Proton Pump Inhibitors on the Human Gastrointestinal Microbiome." *Clinics in Laboratory Medicine* 34 (4): 771–85. doi:10.1016/j.cll.2014.08.008.

Gane, Elise M., Veronika Schoeb, Petrea Cornwell, Cassandra Ranatunga Cooray, Brooke Cowie, and Tracy A. Comans. 2022. "Discharge Planning of Older Persons from Hospital: Comparison of Observed Practice to Recommended Best Practice." *Healthcare* 10 (2): 202. doi:10.3390/healthcare10020202.

Gaugler, J. E., R. L. Kane, R. A. Kane, and R. Newcomer. 2005. "Early Community-Based Service Utilization and Its Effects on Institutionalization in Dementia Caregiving." *The Gerontologist* 45 (2): 177–85. doi:10.1093/geront/45.2.177.

Gemikonakli, Gizem, John Mach, Fan Zhang, Martyn Bullock, Trang Tran, Emad El-Omar, and Sarah N Hilmer. 2022. "Polypharmacy with High Drug Burden Index (DBI) Alters the Gut Microbiome Overriding Aging Effects and Is Reversible with Deprescribing." *The Journals of Gerontology Series A* 78 (2): 213–22. doi:10.1093/gerona/glac195.

Ghosh, Angaj, Tess Evans, Calum J. Walsh, Matthew J. Maiden, Timothy P. Stinear, and Adam M. Deane. 2025. "The Gastrointestinal Microbiome in Critical Illness: A Clinician's Guide to Mechanisms, Emerging Tools, and Therapeutic Questions." *Critical Care and Resuscitation* 27 (4): 100132. doi:10.1016/j.ccrj.2025.100132.

Gill, T. M., H. Allore, and Z. Guo. 2004. "The Deleterious Effects of Bed Rest among Community-Living Older Persons." *The Journals of Gerontology Series A* 59 (7): M755–61. doi:10.1093/gerona/59.7.m755.

Góralczyk-Bińkowska, Aleksandra, Dagmara Szmajda-Krygier, and Elżbieta Kozłowska. 2022. "The Microbiota–Gut–Brain Axis in Psychiatric Disorders." *International Journal of Molecular Sciences* 23 (19): 11245. doi:10.3390/ijms231911245.

Grover, Sandeep, and Ajit Avasthi. 2018. "Clinical Practice Guidelines for Management of Delirium in Elderly." *Indian Journal of Psychiatry* 60 (7): 329. doi:10.4103/0019-5545.224473.

He, Jindi, Jiaxun Wu, Juan Liu, Hangcai Wu, and Heliu Hua. 2024. "Cognitive Impairment and the Gut-Brain Axis during 2014–2023: A Bibliometric Analysis." *Frontiers in Neurology* 15 (July): 1407956. doi:10.3389/fneur.2024.1407956.

He, Jindi, Jiaxun Wu, Juan Liu, Hangcai Wu, and Heliu Hua. 2024. "Cognitive Impairment and the Gut-brain Axis During 2014–2023: A Bibliometric Analysis." *Frontiers in Neurology* 15 (July): 1407956. doi:10.3389/fneur.2024.1407956.

Hoffman, Kevin W., Jakleen J. Lee, Cheryl M. Corcoran, David Kimhy, Thorsten M. Kranz, and Dolores Malaspina. 2020. "Considering the Microbiome in Stress-Related and Neurodevelopmental Trajectories to Schizophrenia." *Frontiers in Psychiatry* 11 (July): 629. doi:10.3389/fpsyt.2020.00629.

Hu, Mi, Xuping Li, Yu Zhu, Zhiyan Chen, Chong Lai, Ruijie Liu, Shuiyuan Xiao, and Michael R Phillips. 2024. "The Role of Family Caregiving in the Management of Individuals with Mental Illnesses and the Outcome of Family-Based Interventions for Mental Illnesses in China: A Scoping Review." *The Lancet Regional Health - Western Pacific* 56 (September): 101184. doi:10.1016/j.lanwpc.2024.101184.

Hussein, Yasmin, Sarah Edwards, and Harnish P Patel. 2024. "Psychological Impact of Hospital Discharge on the Older Person: A Systematic Review." *Geriatrics* 9 (6): 167. doi:10.3390/geriatrics9060167.

İlgili, Önder, Berna Arda, and Kerim Munir. 2014. "Ethics in Geriatric Medicine Research." *Turk Geriatri Derg* 17 (2): 188–95.

Inivos. 2021. "5 Most Common Hospital Acquired Infections (HAIs)." https://inivos.com/5-most-common-hospital-acquired-infections-hais/#:~:text=Methicillin%2Dresistant%20Staphylococcus%20aureus%20(MRSA,ranges%20from%2030%20to%2070%25.

Inoue, Kosuke, James Beekley, Atsushi Goto, Christie Y. Jeon, and Beate R. Ritz. 2020. "Depression and Cardiovascular Disease Events among Patients with Type 2 Diabetes: A Systematic Review and Meta-Analysis with Bias Analysis." *Journal of Diabetes and Its Complications* 34 (12): 107710. doi:10.1016/j.jdiacomp.2020.107710.

Inouye, Sharon K, Rudi Gj Westendorp, and Jane S Saczynski. 2014. "Delirium in Elderly People." *The Lancet* 383 (9920): 911–22. doi:10.1016/s0140-6736(13)60688-1.

Janssens, Jeroen F. a. M., Susanne J. De Kort, Wilco P. Achterberg, Susan Kurrle, Ngaire Kerse, Ian D. Cameron, and Dorothea P. Touwen. 2018. "Medical and Moral Considerations Regarding Complex Medical Decisions in Older Patients with Multimorbidity: A Compact Deliberation Framework." *BMC Geriatrics* 18 (1): 25. doi:10.1186/s12877-018-0707-5.

Jaqua, Ecler E., Mai-Linh N. Tran, Pedro Alvarez, Monica Gupta, and Jessica Yoong. 2024. "Dementia and Cognitive Decline: A HEALM Approach." *American Journal of Lifestyle Medicine* 19 (5): 779–94. doi:10.1177/15598276241291508.

Kadyan, Saurabh, Gwoncheol Park, Tejinder P Singh, Cole Patoine, Saiful Singar, Teresa Heise, Christin Domeier, et al. 2025. "Microbiome-based Therapeutics towards Healthier Aging and Longevity." *Genome Medicine* 17 (1): 75. doi:10.1186/s13073-025-01493-x.

Kanimozhi, N V, and M Sukumar. 2025. "Aging Through the Lens of the Gut Microbiome: Challenges and Therapeutic Opportunities." *Archives of Gerontology and Geriatrics Plus* 2 (2): 100142. doi:10.1016/j.aggp.2025.100142.

Kar, Nilamadhab. 2021. "Family Members as Caregivers of Older People with Mental Illness: Issues and Interventions." *Journal of Geriatric Care and Research* 8 (1): 21–8.

Katon, Wayne J. 2003. "Clinical and Health Services Relationships between Major Depression, Depressive Symptoms, and General Medical Illness." *Biological Psychiatry* 54 (3): 216–26. doi:10.1016/s0006-3223(03)00273-7.

Kemp, Candace L., Jason Lesandrini, Jennifer Craft Morgan, and Elisabeth O. Burgess. 2021. "The Ethics in Long-Term Care Model: Everyday Ethics and the Unseen Moral Landscape of Assisted Living." *Journal of Applied Gerontology* 41 (4): 1143–52. doi:10.1177/07334648211049806.

Khan, Romeesa, Claudia M. Di Gesù, Juneyoung Lee, and Louise D. McCullough. 2024. "The Contribution of Age-related Changes in the Gut-brain Axis to Neurological Disorders." *Gut Microbes* 16 (1): 2302801. doi:10.1080/19490976.2024.2302801.

Klausner, Ellen J., and George S. Alexopoulos. 1999. "The Future of Psychosocial Treatments for Elderly Patients." *Psychiatric Services* 50 (9): 1198–204. doi:10.1176/ps.50.9.1198.

Knight, John, Yamni Nigam, and Aled Jones. 2009. "Effects of Bedrest 2: Gastrointestinal, Endocrine, Renal, Reproductive and Nervous Systems." *Nurs Times* 105 (22): 24–7.

Knowles, R., V. Laxton, E. Caine, A. Verran, A. Uddin, R. Hartley, M. Wade, M. Galliver, and A. Rahman. 2016. "7delirium: Diagnosis, Prevention and Management: A Multidisciplinary Problem with a Multidisciplinary Solution." *Age and Ageing* 45 (suppl_1): i2. doi:10.1093/ageing/afw024.07.

Kohn, Julia, Alexander Troester, Zachary Ziegert, Julia Frebault, Sonja Boatman, Maria Martell, Harika Nalluri-Butz, et al. 2025. "The Role of Surgical and Perioperative Factors in Shaping Gut Microbiome Recovery After Colorectal Surgery." *Antibiotics* 14 (9): 881. doi:10.3390/antibiotics14090881.

Kotekar, Nalini, Anshul Shenkar, and Ravishankar Nagaraj. 2018. "Postoperative Cognitive Dysfunction & Ndash; Current Preventive Strategies." *Clinical Interventions in Aging* Volume 13 (November): 2267–73. doi:10.2147/cia.s133896.

Kumar, Akash, Bhagavathi Sundaram Sivamaruthi, Swarnima Dey, Yogesh Kumar, Rishabha Malviya, Bhupendra G. Prajapati, and Chaiyavat Chaiyasut. 2024. "Probiotics as Modulators of Gut-Brain Axis for Cognitive Development." *Frontiers in Pharmacology* 15 (February): 1348297. doi:10.3389/fphar.2024.1348297.

Kumar, Akash, Jhilam Pramanik, Nandani Goyal, Dimple Chauhan, Bhagavathi Sundaram Sivamaruthi, Bhupendra G. Prajapati, and Chaiyavat Chaiyasut. 2023. "Gut Microbiota in Anxiety and Depression: Unveiling the Relationships and Management Options." *Pharmaceuticals* 16 (4): 565. doi:10.3390/ph16040565.

Larsson, Anna, Ulrika Ericson, Daniel Jönsson, Mariam Miari, Paschalis Athanasiadis, Gabriel Baldanzi, Louise Brunkwall, et al. 2024. "New Connections of Medication Use and Polypharmacy with the Gut

Microbiota Composition and Functional Potential in a Large Population." *Scientific Reports* 14 (1): 23723. doi:10.1038/s41598-024-71571-4.

Li, Jinhe, Huilin Mou, and Ranran Yao. 2025. "Gut-Brain Axis and a Systematic Approach to Alzheimer's Disease Therapies." *Brain Science Advances* 11 (1). doi:10.26599/bsa.2024.9050031.

Lim, Byung-Gun, and Il-Ok Lee. 2019. "Anesthetic Management of Geriatric Patients." *Korean Journal of Anesthesiology* 73 (1): 8–29. doi:10.4097/kja.19391.

Liu, Lanxiang, Haiyang Wang, Xueyi Chen, Yangdong Zhang, Hanping Zhang, and Peng Xie. 2023. "Gut Microbiota and Its Metabolites in Depression: From Pathogenesis to Treatment." *EBioMedicine* 90 (March): 104527. doi:10.1016/j.ebiom.2023.104527.

Loh, Jian Sheng, Wen Qi Mak, Li Kar Stella Tan, Chu Xin Ng, Hong Hao Chan, Shiau Hueh Yeow, Jhi Biau Foo, Yong Sze Ong, Chee Wun How, and Kooi Yeong Khaw. 2024. "Microbiota–Gut–Brain Axis and Its Therapeutic Applications in Neurodegenerative Diseases." *Signal Transduction and Targeted Therapy* 9 (1): 37. doi:10.1038/s41392-024-01743-1.

Lu, Jiao, Yuan Wang, Lihong Hou, Zhenxing Zuo, Na Zhang, and Anle Wei. 2021. "Multimorbidity Patterns in Old Adults and Their Associated Multi-Layered Factors: A Cross-Sectional Study." *BMC Geriatrics* 21 (1). doi:10.1186/s12877-021-02292-w.

Maher, Robert L., Joseph Hanlon, and Emily R. Hajjar. 2013. "Clinical Consequences of Polypharmacy in Elderly." *Expert Opinion on Drug Safety* 13 (1): 57–65. doi:10.1517/14740338.2013.827660.

Mallick, Keya, Gopal Khodve, Ritika Ruwatia, and Sugato Banerjee. 2025. "Gut Microbes: Therapeutic Target for Neuropsychiatric Disorders." *Journal of Psychiatric Research* 184 (February): 27–38. doi:10.1016/j.jpsychires.2025.02.031.

Mangin, D., I. Heath, and M. Jamoulle. 2012. "Beyond Diagnosis: Rising to the Multimorbidity Challenge." *BMJ* 344 (June 2): e3526. doi:10.1136/bmj.e3526.

Marks, Jarrod, Iuliana Predescu, and Laura B. Dunn. 2021. "Ethical Issues in Caring for Older Adults." *Focus the Journal of Lifelong Learning in Psychiatry* 19 (3): 325–29. doi:10.1176/appi.focus.20210011.

Masutani, Rebecca, Ajinkya Pawar, Hemin Lee, Joel S. Weissman, and Dae Hyun Kim. 2020. "Outcomes of Common Major Surgical Procedures in Older Adults with and without Dementia." *JAMA Network Open* 3 (7): e2010395. doi:10.1001/jamanetworkopen.2020.10395.

Matsumoto, Ayaka, Yoshihiro Yoshimura, Hidetaka Wakabayashi, Fumihiko Nagano, Sayuri Shimazu, Yoshifumi Kido, Ai Shiraishi, Takenori Hamada, Kouki Yoneda, and Keisuke Maeda. 2025. "Polypharmacy Is Associated with Altered Gut Microbiota Diversity in Older Post-Stroke Inpatients." *Clinical Nutrition ESPEN* 70 (October): 427–33. doi:10.1016/j.clnesp.2025.10.015.

Moffa, Livia, and Claudio Tana. 2025. "Healthcare-Associated Infections (HAIS) in the Elderly: Molecular Mechanisms of Immunosenescence and Clinical, Nutritional and Therapeutic Implications." *International Journal of Molecular Sciences* 26 (19): 9649. doi:10.3390/ijms26199649.

Nas, Kemal, Levent Yazmalar, Volkan Şah, Abdulkadir Aydın, and Kadriye Öneş. 2015. "Rehabilitation of Spinal Cord Injuries." *World Journal of Orthopedics* 6 (1): 8. doi:10.5312/wjo.v6.i1.8.

National Institute for Health and Care Excellence (NICE). 2023. *Delirium: Prevention, Diagnosis and Management in Hospital and Long-Term Care.* NICE Clinical Guidelines, No. 103. https://www.ncbi.nlm.nih.gov/books/NBK553009/.

Obi-Azuike, Crystal, Ruona Ebiai, Taneil Gibson, Ariana Hernandez, Asma Khan, Gibson Anugwom, Alexsandra Urhi, Sakshi Prasad, Sara Ait Souabni, and Funso Oladunjoye. 2023. "A Systematic Review on Gut–brain Axis Aberrations in Bipolar Disorder and Methods of Balancing the Gut Microbiota." *Brain and Behavior* 13 (6): e3037. doi:10.1002/brb3.3037.

Olotu, Cynthia, Arved Weimann, Christian Bahrs, Wolfgang Schwenk, Martin Scherer, and Rainer Kiefmann. 2019. "The Perioperative Care of Older Patients." *Deutsches Ärzteblatt International* 116 (5): 63–69. doi:10.3238/arztebl.2019.0063.

Ortega, Miguel A., Miguel Angel Álvarez-Mon, Cielo García-Montero, ÓScar Fraile-MartíNez, Jorge Monserrat, Lucia Martinez-Rozas, Roberto Rodríguez-JiméNez, Melchor Álvarez-Mon, and Guillermo Lahera. 2023. "Microbiota–Gut–Brain Axis Mechanisms in the Complex Network of Bipolar Disorders: Potential Clinical Implications and Translational Opportunities." *Molecular Psychiatry* 28 (7): 2645–73. doi:10.1038/s41380-023-01964-w.

Pant, Usha, Krooti Vyas, and Elizabeth Papathanassoglou. 2025. "Psychosocial Support Interventions for Adult Critically Ill Patients during the Acute Phase of Their ICU Stay: A Scoping Review." *Healthcare* 13 (17): 2182. doi:10.3390/healthcare13172182.

Patil, Shradha, and SayedFarhan S Mehdi. 2025. "The Gut-Brain Axis and Mental Health: How Diet Shapes Our Cognitive and Emotional Well-Being." *Cureus* 17 (7): e88420. doi:10.7759/cureus.88420.

Potestio, Christopher, and Mary Katherine Carney. 2024. *Common Surgical Procedures in Geriatric Patients*. 366–95. Bentham Science Publishers. doi:10.2174/9789815238877124010023.

Prince, Martin J, Fan Wu, Yanfei Guo, Luis M Gutierrez Robledo, Martin O'Donnell, Richard Sullivan, and Salim Yusuf. 2014. "The Burden of Disease in Older People and Implications for Health Policy and Practice." *The Lancet* 385 (9967): 549–62. doi:10.1016/s0140-6736(14)61347-7.

Prusaczyk, Beth, Steven M. Cherney, Christopher R. Carpenter, and James M. DuBois. 2016. "Informed Consent to Research with Cognitively Impaired Adults: Transdisciplinary Challenges and Opportunities." *Clinical Gerontologist* 40 (1): 63–73. doi:10.1080/07317115.2016.1201714.

Qaffas, Mohammed F., Bandar S. Alenezi, Ahmed K. Zoghby, Ahmed A. Saad, Hanan S. Alkhelaiwi, Afnan M. Maashi, Abdulaziz A. Alshehri, et al. 2023. "Psychiatric Evaluation and Support for the Elderly Undergoing Elective Surgery." *International Journal of Community Medicine and Public Health* 11 (1): 506–10. doi:10.18203/2394-6040.ijcmph20233949.

Reynolds, Charles F., Dilip V. Jeste, Perminder S. Sachdev, and Dan G. Blazer. 2022. "Mental Health Care for Older Adults: Recent Advances and New Directions in Clinical Practice and Research." *World Psychiatry* 21 (3): 336–63. doi:10.1002/wps.20996.

Robinson, Thomas. 2008. "Postoperative Delirium in the Elderly: Diagnosis and Management." *Clinical Interventions in Aging* 3 (2): 351–5. doi:10.2147/cia.s2759.

Rojas-Solé, Catalina, Víctor Pinilla-González, José Lillo-Moya, Tommy González-Fernández, Luciano Saso, and Ramón Rodrigo. 2023. "Integrated Approach to Reducing Polypharmacy in Older People: Exploring the Role of Oxidative Stress and Antioxidant Potential Therapy." *Redox Report* 29 (1): 2289740. doi:10.1080/13510002.2023.2289740.

Ruiz, Nicolás Idárraga, Carlota G Herrera Giron, Carlos Alejandro Arragan Lezama, Sergio José Frias Redroban, Manuel Omar Ventura Herrera, and Gustavo Adolfo Sanic Coj. 2025. "Timing and Protocols for Microbiome Intervention in Surgical Patients: A Literature Review of Current Evidence." *Cureus* 17 (6): e86104. doi:10.7759/cureus.86104.

Sampson, Elizabeth L., Vivienne Gould, Dan Lee, and Martin R. Blanchard. 2006. "Differences in Care Received by Patients with and without Dementia Who Died during Acute Hospital Admission: A Retrospective Case Note Study." *Age and Ageing* 35 (2): 187–9. doi:10.1093/ageing/afj025.

Schapira, Lidia, James A. Tulsky, Robert Buckman, and Kathryn I. Pollak. 2008. "Communication: What Do Patients Want and Need?" 2008. *Journal of Oncology Practice* 4 (5): 249–53. doi:10.1200/jop.0856501.

Schenning, Katie J., and Stacie G. Deiner. 2015. "Postoperative Delirium in the Geriatric Patient." *Anesthesiology Clinics* 33 (3): 505–16. doi:10.1016/j.anclin.2015.05.007.

Schipa, Chiara, Ersilia Luca, Matteo Ripa, Liliana Sollazzi, and Paola Aceto. 2023. "Preoperative Evaluation of the Elderly Patient." *Saudi Journal of Anaesthesia* 17 (4): 482–90. doi:10.4103/sja.sja_613_23.

Schulz, Richard, and Paula R. Sherwood. 2008. "Physical and Mental Health Effects of Family Caregiving." *AJN American Journal of Nursing* 108 (9): 23–7. doi:10.1097/01.naj.0000336406.45248.4c.

Sessums, Laura L, Hanna Zembrzuska, and Jeffrey L Jackson. 2011. "Does This Patient Have Medical Decision-Making Capacity?" *JAMA* 306 (4): 420. doi:10.1001/jama.2011.1023.

Shaikh, Raja Gulfam, Animesh Dey, Vindhya Prakash Singh, Anita Khandagle, Baskaran M, Sunil Naik, and Asif Hasan. 2025. "Understanding the Impact of the Gut Microbiome on Mental Health: A Systematic Review." *Cureus* 17 (1): e78100. doi:10.7759/cureus.78100.

Sherman, Deborah Witt. 2019. "A Review of the Complex Role of Family Caregivers as Health Team Members and Second-Order Patients." *Healthcare* 7 (2): 63. doi:10.3390/healthcare7020063.

Stavrou, George. 2016. "Gut Microbiome, Surgical Complications and Probiotics." *Annals of Gastroenterology* 30 (1): 45–53. doi:10.20524/aog.2016.0086.

Steffens, David C., Gwenith G. Fisher, Kenneth M. Langa, Guy G. Potter, and Brenda L. Plassman. 2009. "Prevalence of Depression among Older Americans: The Aging, Demographics and Memory Study." *International Psychogeriatrics* 21 (5): 879. doi:10.1017/s1041610209990044.

Sulaimanova, Gulnura, and Shirin Talapbek Kyzy. 2025. "The Prevalence of Anxiety Disorders among Older Patients with COPD: Systematic Review and Meta-Analysis." *Geriatric Nursing* 64 (June): 103405. doi:10.1016/j.gerinurse.2025.103405.

Tabari, Masoomeh, Soheila Milani, and Malihe Aghasizadeh. 2025. "The Association of Gut Microbiome with Anesthesia Outcomes, Pain Management, and Patient Recovery." *Archives of Anesthesia and Critical Care*, December. doi:10.18502/aacc.v12i1.20546.

Tamene, Fasil Bayafers, Tirsit Ketsela Zeleke, Akalu Fetene Desalew, Getachew Yitayew Tarekegn, Ashenafi Kibret Sendekie, Selamawit Mengstu Tafere, Mekdes Kiflu, Tilaye Arega Moges, Fisseha Nigussie Dagnew, and Samuel Agegnew Wondm. 2025. "Polypharmacy and Associated Factors among Patients with Type Two Diabetes Mellitus with Comorbidity: A Multicenter Cross-Sectional Study in Northwest Ethiopia." *BMC Endocrine Disorders* 25 (1): 188. doi:10.1186/s12902-025-02011-1.

Tana, Claudio, Samanta Moffa, Marco Tana, Claudio Ucciferri, and Livia Moffa. 2025. "Gut Microbiota, Mild Cognitive Impairment and Dementia: A Systematic Review." *Neurology International* 17 (10): 155. doi: 10.3390/neurolint17100155.

Taylor, Mark E., and Beryl A. Oppenheim. 1998. "Hospital-Acquired Infection in Elderly Patients." *Journal of Hospital Infection* 38 (4): 245–60. doi:10.1016/s0195-6701(98)90074-3.

Teutem, Simon Van. 2024. "The Global Number of People Aged 65 Years and Older Is Set to Double within the Next Thirty Years." https://ourworldindata.org/data-insights/by-2060-the-number-of-people-aged-65-and-older-will-be-more-than-four-times-what-it-was-in-2000.

Theleritis, Christos, Maria-Ioanna Stefanou, Marina Demetriou, Evangelos Alevyzakis, Konstantinos Triantafyllou, Nikolaos Smyrnis, Demetrios Spandidos, and Emmanouil Rizos. 2024. "Association of Gut Dysbiosis with First-Episode Psychosis (Review)." *Molecular Medicine Reports* 30 (1). doi:10.3892/mmr.2024.13254.

Wang, Dexiu, Huiying Yao, Luoqi Wang, Bowen Lu, Wenkai Liu, Jinghan Li, Yujin Gong, et al. 2025. "Gut-Brain Axis and Vascular Dementia: A Review on Mechanisms and Chinese Herbal Medicine Therapeutics." *Frontiers in Microbiology* 16 (May): 1564928. doi:10.3389/fmicb.2025.1564928.

Wang, Xiao-Qing, He Li, Xiang-Nan Li, Cong-Hu Yuan, and Hang Zhao. 2021. "Gut-Brain Axis: Possible Role of Gut Microbiota in Perioperative Neurocognitive Disorders." *Frontiers in Aging Neuroscience* 13 (December): 745774. doi:10.3389/fnagi.2021.745774.

Wang, Yangyanqiu, Zhanbo Qu, Jian Chu, and Shuwen Han. 2024. "Aging Gut Microbiome in Healthy and Unhealthy Aging." *Aging and Disease* 16 (2): 980–1002. doi:10.14336/ad.2024.0331.

Wick, Jeannette. 2011. "Bedrest: Implications for the Aging Population." *Pharmacy Times* 77 (1): 54–6.

Yang, Yan, Zhipeng Xu, Jianrong Guo, Zhiqiang Xiong, and Baoji Hu. 2024. "Exploring the Gut Microbiome-Postoperative Cognitive Dysfunction Connection: Mechanisms, Clinical Implications, and Future Directions." *Brain Behavior & Immunity - Health* 38 (March): 100763. doi:10.1016/j.bbih.2024.100763.

Yu, Lingling, Xiaoling Li, Pengjun Zhang, and Guojun Zhang. 2020. "Ethical Guidance for Geriatric Clinical Research in China." *Aging Medicine* 3 (4): 218–23. doi:10.1002/agm2.12138.

Yu, Wanqiu, Zhaoqiong Zhu, and Fushan Tang. 2024. "Emerging Insights into Postoperative Neurocognitive Disorders: The Role of Signaling Across the Gut-Brain Axis." *Molecular Neurobiology* 61 (12): 10861–82. doi:10.1007/s12035-024-04228-y.

Zhang, Shi-Hua, Xiao-Yu Jia, Qing Wu, Jia Jin, Long-Sheng Xu, Lei Yang, Jun-Gang Han, and Qing-He Zhou. 2023. "The Involvement of the Gut Microbiota in Postoperative Cognitive Dysfunction Based on Integrated Metagenomic and Metabolomics Analysis." *Microbiology Spectrum* 11 (6): e0310423. doi:10.1128/spectrum.03104-23.

Zheng, Zhipeng, Yingnan Hu, Jingyi Tang, Wenjun Xu, Weihan Zhu, and Wei Zhang. 2023. "The Implication of Gut Microbiota in Recovery from Gastrointestinal Surgery." *Frontiers in Cellular and Infection Microbiology* 13 (February): 1110787. doi:10.3389/fcimb.2023.1110787.

Zhu, Sibo, Yanfeng Jiang, Kelin Xu, Mei Cui, Weimin Ye, Genming Zhao, Li Jin, and Xingdong Chen. 2020. "The Progress of Gut Microbiome Research Related to Brain Disorders." *Journal of Neuroinflammation* 17 (1). doi:10.1186/s12974-020-1705-z.

10 Epigenetics, Gene, and Microbiome Engineering

Shilpi Kumari, Abhijit Swain, Varshika Singh, and Vinay Kumar

10.1 INTRODUCTION

The gut–brain axis (GBA) is like the body's own sophisticated communication network, connecting our digestive system with our brain. This intricate system relies on a complex dance of neural, hormonal, and immune signals (Carabotti et al. 2015). At its heart, the GBA manages digestion, immune responses, and brain activity, which, in turn, affects our mood, thinking, and behavior. Key players in this communication include the vagus nerve (Bonaz et al. 2018), the hypothalamic–pituitary–adrenal (HPA) axis (Bertollo et al. 2025), and substances produced by gut bacteria. Recent breakthroughs in neuroscience and microbiology have shown that the GBA plays a crucial role in keeping our brains healthy, and when it's out of balance, it can be linked to various neurological and mental health disorders (Bakshi et al. 2024).

The vagus nerve acts as the main highway, sending signals back and forth between the gut and the brain (Figure 10.1). It picks up sensory information from the gut and affects the brain areas related to emotions and thinking (Bonaz et al. 2018). The HPA axis, which is essential for stress responses, interacts with gut bacteria to control cortisol levels, thus influencing our mood and behavior (Bertollo et al. 2025). Another important part of this axis is the immune system, which communicates with the brain through cytokines and other inflammatory substances (Kennedy and Silver 2016). This interaction between the immune system and the nervous system plays a significant

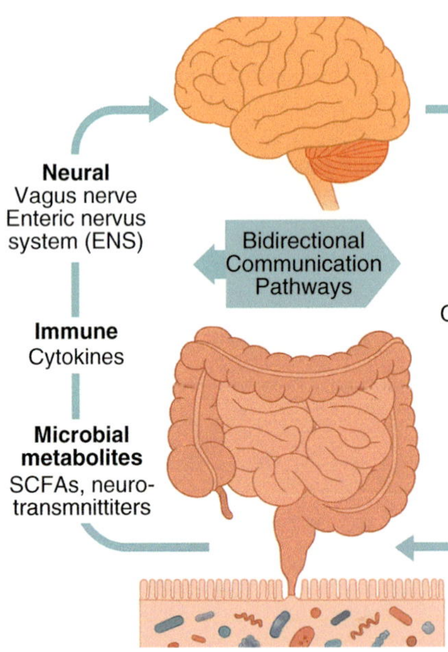

FIGURE 10.1 Diagram illustrating the gut–brain axis; highlighting the key components involved in the bidirectional communication system.

DOI: 10.1201/9781003610274-12

role in how the body responds to stress, inflammation, and infections, impacting brain development and mental health.

Emerging evidence indicates that when the balance of bacteria in our gut, known as gut dysbiosis, gets disrupted, it can mess with the communication between our gut and brain. This disruption may play a role in the development and progression of neurological disorders like anxiety, depression, autism spectrum disorder (ASD), Parkinson's disease, and Alzheimer's disease (Martin et al. 2018; Zhou et al. 2025). For example, when there are fewer good bacteria that fight inflammation and more bad bacteria that cause inflammation, it can lead to ongoing low-level inflammation. This inflammation affects the balance of neurotransmitters and the brain's ability to adapt and change (Ullah et al. 2023). This strong link between gut health and brain function suggests that treating issues in the GBA could offer new ways to help with neurological and psychiatric conditions (Ignácio et al. 2025).

The GBA has become a focal point of research, recognized as a vital pathway influencing neurological health. An increasing body of evidence underscores the complex interplay between epigenetics, gene expression, and the gut microbiome (Ma et al. 2023, Jung et al. 2025). These interrelated systems are crucial in neurodevelopment, affect cognitive abilities, and play a role in the mechanisms underlying neurological disorders. New studies indicate that epigenetic processes such as DNA methylation, histone modifications, and non-coding RNAs are essential for regulating gene expression in response to signals from gut microbes (Ma et al. 2023). When this regulatory framework is disrupted, it may lead to neurodegenerative diseases, mood disorders, and neurodevelopmental issues (Jung et al. 2025).

Recent innovations in microbiome manipulation, including the use of probiotics, prebiotics, and fecal microbiota transplantation (FMT), present new therapeutic avenues for restoring balance in the gut–brain connection. FMT specifically has demonstrated potential in altering the composition of gut microbiota, which can subsequently affect neurological outcomes (Ullah et al. 2023, Xu & Lu 2025).

At the same time, advancements in gene-editing techniques like CRISPR-Cas9 allow for precise corrections of genetic and epigenetic irregularities associated with neurological disorders. These technologies have shown promise in experimental models of neurodegenerative diseases, revealing possible therapeutic advantages (Khan et al. 2025).

This chapter consolidates the most recent research at the convergence of epigenomics, neurogenetics, and microbiome studies, emphasizing their combined effects on brain health. By reviewing both preclinical and clinical data, we investigate how strategic modifications to microbial populations and the host's epigenetic environment can be utilized for therapeutic purposes. Gaining a deeper insight into these mechanisms could lead to personalized treatment options, offering renewed hope for individuals suffering from neurological conditions.

10.2 THE RELEVANCE OF EPIGENETICS, GENE REGULATION, AND MICROBIOME INTERACTIONS IN THE GUT–BRAIN AXIS

Epigenetic modifications are heritable variations in gene expression that do not affect the underlying DNA sequence. They influence how genes are switched on or off in distinct cells and at different stages of development, acting as an additional layer of control on top of the genetic code. By altering the shape and accessibility of chromatin, these changes decide whether particular genes are tightly packed and silenced or open to being read by the cell's machinery. Epigenetic modifications, which are impacted by both internal and external variables, including developmental signals and environmental stimuli, stress, and food, aid organisms in adapting to new situations (Turner 2002). Importantly, epigenetic modifications are reversible, allowing cells to respond to both internal and external inputs while avoiding irreversible genetic changes. To guarantee that specialized cells maintain their identity over the course of an organism's life, they can also be handed on during cell division. Since anomalies in these regulatory systems can result in diseases including cancer, neurological problems, and metabolic syndromes, their significance in both health and disease is clear (Farooqui et al. 2012)

These modifications have a profound impact on the relationship between the gut and the brain. Consider them as pliable switches that assist the body in reacting to its surroundings. In this dynamic system, processes like DNA methylation, histone modification, and non-coding RNAs are essential because they impact how neurons work, adapt, and even impact mood, cognition, and general brain health. Essentially, epigenetics helps to refine the continuous dialogue between your stomach and brain (Stilling et al. 2014). Microbial metabolites may cause epigenetic alterations that lead to the development of complex neurological illnesses such as ASD, schizophrenia, and neurodegenerative diseases (Lozupone et al. 2020). For example, people with dysbiosis have been shown to have different DNA methylation patterns in genes linked to neurodevelopment. Similarly, microbial imbalances in individuals with Parkinson's and Alzheimer's diseases have been connected to alterations in histone modifications that impact neuroinflammation. These results highlight the function of gut bacteria as epigenetic regulators, able to affect gene expression by interacting with the host and microbial metabolites (Stilling et al. 2014)

They are sensitive to environmental factors such as nutrition, stress, and chemicals produced by gut bacteria, making epigenetics an important component of gut–brain communication. According to recent studies, gut bacteria are capable of producing butyrate and other short-chain fatty acids (SCFAs), which block the enzyme histone deacetylase (HDAC). This affects the brain's chromatin structure and gene transcription, suggesting that microbial metabolites have the ability to epigenetically control genes related to inflammation, degeneration, and brain development (Stilling et al. 2014)

These epigenetic modifications are closely linked to gene regulation in the GBA. Genes involved in neurotransmitter synthesis, nerve cell support, and synaptic plasticity are expressed in response to microbial metabolite-responsive transcription factors and signaling pathways (Stilling et al. 2014). For example, metabolites produced from intestinal tryptophan can influence serotonin synthesis, influencing mood and behavior. Furthermore, the host's endocannabinoid system can be modulated by the gut bacteria, which can affect emotional regulation and stress resilience. These results demonstrate how gut bacteria are potent epigenetic modulators that can affect patterns of gene expression crucial to neurological health (Table 10.1).

The microbiome's influence on the GBA extends beyond metabolism to include immune regulation and neuroinflammation. Numerous metabolites produced by gut bacteria, such as bile acids, SCFAs, and neurotransmitters, have an impact on the host's immune system and the blood–brain barrier's integrity (O'Riordan et al. 2022). For instance, SCFAs help preserve immunological homeostasis by promoting the development of regulatory T cells and having anti-inflammatory qualities. On the other hand, dysbiosis, or an imbalance of gut flora, can result in increased gut permeability and systemic inflammation, which can exacerbate neuroinflammatory diseases that are connected to mood disorders and cognitive decline. This reciprocal interaction between the host's genetic expression and the microbiome emphasizes how crucial it is to maintain a balanced gut microbiota for the best possible brain health (Nichols et al. 2021).

10.2.1 Epigenetic Modifications and Their Role in Gut–Brain Axis

In the GBA, epigenetic changes play a major role in determining whether genes are activated or inactive. They aid in controlling the intricate interactions that take place between the central nervous system, enteric nervous system (ENS), and gut bacteria. These minute modifications affect how genes function, which eventually impacts everything from mood to digestion, but they do not modify the DNA sequence itself (Figure 10.2). Below are some of the major types of epigenetic modifications involved in the gut-brain connection:

- **DNA methylation:** DNA methylation, a major epigenetic mechanism involving the addition of methyl groups to cytosine residues, is critical in regulating gene expression along the GBA (Stilling et al. 2014). This mechanism primarily happens at CpG sites and can silence genes associated with neuroinflammation, stress response, and neurotransmitter

TABLE 10.1
Epigenetic, Gene Regulatory, and Microbiome Interactions in the Gut–Brain Axis: Mechanisms and Implications for Brain Health

Category	Key Mechanisms	Specific Influence on Gut–Brain Axis	Impact on Brain Function & Health	References
Epigenetics	DNA methylation	Alters gene expression in response to environmental factors, including microbial metabolites	Regulates neurodevelopment and cognitive function; dysregulated methylation linked to autism and schizophrenia	Alam et al. (2017)
	Histone modification (acetylation/ deacetylation)	SCFAs (e.g., butyrate) inhibit histone deacetylase (HDAC), modifying chromatin structure and gene transcription	Affects neuroinflammation and neurodegeneration (e.g., Alzheimer's, Parkinson's)	Dong and Ciu (2022)
	Noncoding RNAs (e.g., microRNAs, lncRNAs)	Regulate gene expression post-transcriptionally in response to gut microbial signals	Modulate synaptic plasticity, neuroimmune responses, and mood disorders	Singh et al. (2021)
Gene regulation	Transcription factors (e.g., NF-κB, CREB)	Activated by microbial metabolites to regulate genes involved in neurotransmitter production and neuroprotection	Influences mood, cognition, and neuronal survival	Chen et al. (2021)
	Signaling pathways (e.g., TLRs, MAPK, mTOR)	Mediate immune responses and inflammation in reaction to gut microbiota	Regulate neuroinflammatory conditions and brain plasticity	Tang et al. (2021)
Microbiome interactions	SCFAs (butyrate, propionate, acetate)	Modulate histone modifications and promote regulatory T-cell differentiation, reducing inflammation	Support blood–brain barrier integrity and prevent neuroinflammatory diseases	Huang et al. (2021)
	Bile acids	Act as signaling molecules influencing gut and brain receptors	Affect metabolic pathways linked to neurodegenerative diseases	Needham et al. (2020)
	Tryptophan metabolites (e.g., serotonin, kynurenine pathway)	Influence neurotransmitter synthesis and immune signaling	Affect mood regulation, anxiety, and neurodegeneration	Teleanu et al. (2022)
	Gut hormones (e.g., GLP-1, ghrelin)	Communicate metabolic and inflammatory status between gut and brain	Modulate appetite, stress response, and cognitive functions	Van de Wouw et al. (2017)

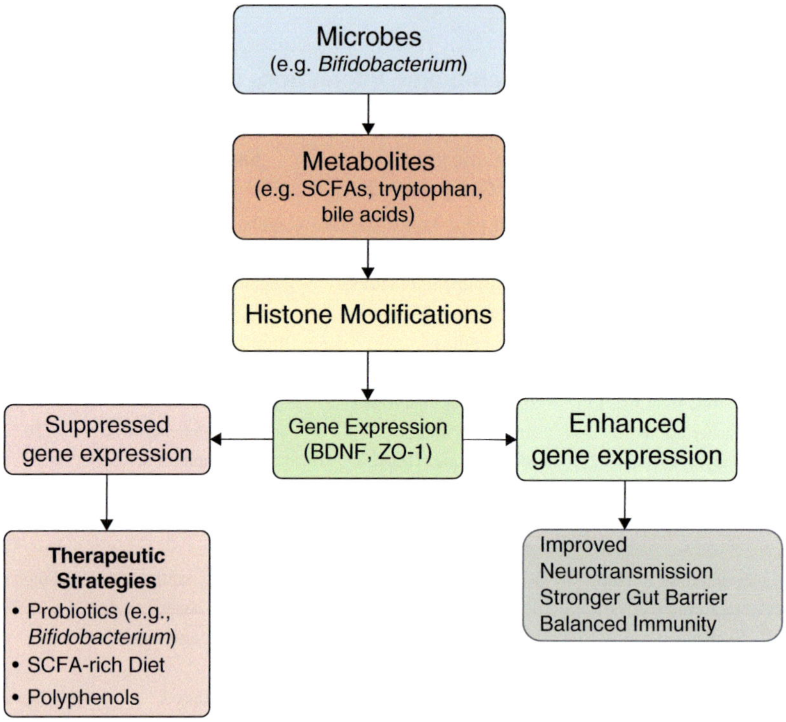

FIGURE 10.2 Gut microbiota influence on brain epigenetic signaling—microbes such as *Bifidobacterium* play a crucial role in metabolite production and modulate histone modifications; which leads to either enhanced gene expression of *BDNF* (critical for neuronal plasticity) and *ZO-1* (essential for gut barrier integrity) or if histone marks become dysregulated, they suppress gene expression, leading to reduced BDNF levels and increased vulnerability to conditions like depression and anxiety. Hence, therapeutic strategies such as interventions with probiotics, SCFA-rich diets, and polyphenols are advised to restore beneficial histone marks and support mental health.

function without changing the DNA sequence (Jang et al. 2017). Through metabolites such as SCFAs, which alter DNA methyltransferase (DNMT) activity, the gut microbiota affects these methylation patterns (Figure 10.3). Long-term epigenetic programming is shaped by the combined action of enzymes like DNMT1 that preserve preexisting methylation marks and DNMT3A/B that create new ones (Hervouet et al. 2018). Disruptions in this system, whether caused by gut dysbiosis, dietary factors, or environmental pollutants, have been linked to many disorders including irritable bowel syndrome (IBS), depression, and neurodegenerative diseases, emphasizing the delicate balance between microbial activity and host gene regulation (Maguire & Maguire 2019).

- **Histone modifications and chromatin remodeling:** Histone modifications and chromatin remodeling are critical epigenetic mechanisms that dynamically regulate gene expression in the GBA, facilitating communication between gut bacteria, the ENS, and the brain (Stilling et al. 2014). Post-translational modifications (PTMs) such as acetylation (H3K27ac), methylation (H3K4me3/H3K27me3), and phosphorylation (H3S10ph) change chromatin structure (Ramazi et al. 2020), affecting genes implicated in neuroinflammation, neurotransmitter production (e.g., serotonin), and gut barrier integrity. For example, gut microbial metabolites such as SCFAs block HDACs, enabling an open chromatin state that increases neuroprotective gene expression (O'Riordan et al. 2022). Conversely, repressive marks like H3K27me3 silence inflammatory pathways, while stress-induced

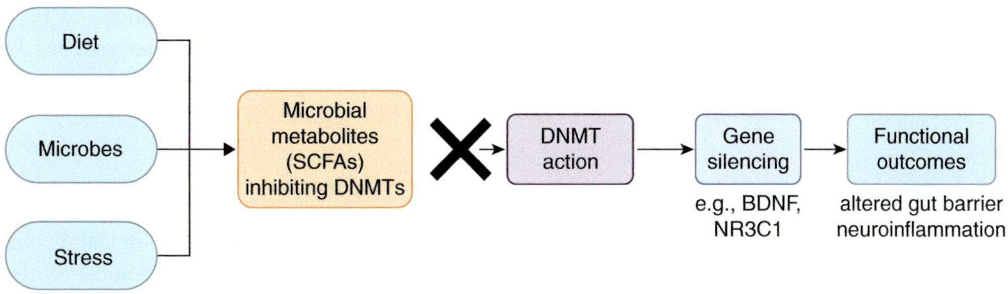

FIGURE 10.3 The diagram depicts the effect of environmental factors (diet, microbes, stress) on microbial metabolites production (SCFAs) that inhibits DNMTs action through epigenetics modifications (specifically DNA methylation via DNMTs) and thus affecting the gut–brain axis.

phosphorylation can activate immediate early genes in neurons. These modifications are enzymatically regulated (e.g., HATs, HDACs, and methyltransferases) and respond to dietary factors, dysbiosis, and environmental toxins, linking microbial activity to brain function and behavior (Farrelly et al. 2019).

- **Ubiquitination:** Ubiquitination (H2BK120ub) further fine-tune transcriptional programs, with SCFAs and tryptophan metabolites influencing these PTMs to regulate synaptic plasticity and immune responses. ADP-ribosylation (via PARPs) and citrullination (by PADs) modulate chromatin compaction in response to oxidative stress or infection, impacting neurodevelopmental and neurodegenerative processes (Hu et al. 2023). In GBA illnesses such as IBS, depression, and Parkinson's disease, dysregulation of these pathways is linked to abnormal histone marks that change gut permeability, glial activation, and neuronal signaling. The potential of therapeutic approaches that target these modifications like probiotics, polyphenol-rich diets, or HDAC inhibitors to restore GBA homeostasis is highlighted (Zheng et al. 2023).

10.3 GUT MICROBIOTA AND BRAIN EPIGENETIC SIGNALING

GBA is heavily influenced by various RNAs that regulate epigenetics and neurodevelopment. MicroRNAs (miRNAs) regulate post-transcriptional modifications through messenger RNA (mRNA) silencing and help in modulating neuroinflammation and stress responses (Begum et al. 2022). Long noncoding RNAs (lncRNAs) are involved in regulating chromatin remodeling and transcriptional regulation. lncRNAs are found to influence neurodegenerative diseases via gut microbiota-driven epigenetic changes (Ruffo et al. 2023). Circular RNAs (circRNAs) influence neuroprotection and gut microbiota interactions by preventing miRNAs from silencing target genes (Chen et al. 2021). circRNAs function as miRNA sponges. Further, mRNA also impacts brain signaling via neurotransmitter production. The stability and translation efficiency of mRNA are modulated by gut microbiota (Xu & Lu 2025). Transfer RNA (tRNA) helps in immune system modulation and neurodegenerative conditions (Winek & Soreq 2025). Understanding these RNA-mediated mechanisms could lead to novel therapeutic strategies, such as RNA-based interventions targeting gut microbiota to improve neurological health.

10.4 BIOINFORMATICS TOOLS FOR EPIGENETIC ANALYSIS

To study the epigenetic modifications, researchers use specialized bioinformatics tools that help decode the intricate regulatory networks by integrating data from various sources, such as DNA methylation and histone modification studies (Chenarani et al. 2021). They allow scientists to identify significant changes, visualize data, and understand the functional implications of these epigenetic marks. As the field of epigenetics grows, these advanced computational approaches are becoming increasingly important for understanding how the epigenome influences health and disease. A comprehensive account of these tools and techniques is summarized in Table 10.2.

10.5 ENGINEERING THE MICROBIOME FOR THERAPEUTIC INTERVENTIONS

The gut microbiome is a dynamic ecosystem that profoundly influences neurological health. Scientists are now uncovering ways to harness its potential for therapeutic interventions, aiming to restore balance and improve brain function. By carefully modulating the microbiome, we can enhance mental well-being, slow neurodegenerative diseases, and even alleviate mood disorders (Kip et al. 2023).

10.5.1 Microbiome Modulation Strategies

In the quest to unlock the gut–brain connection, researchers have identified several key strategies to optimize the microbiome. These include probiotics, FMT, and dietary interventions, all of which offer promising ways to support neurological health (Vendrik et al. 2020)

- **Probiotics and prebiotics: The microbial architects of brain health:** Not all bacteria are bad; in fact, some are essential for keeping our brains healthy. Probiotics, which are beneficial live bacteria, play a vital role in producing neurotransmitters like serotonin, dopamine, and gamma-aminobutyric acid (GABA) (Diez-Gutiérrez et al. 2020). Studies show that *Lactobacillus* and *Bifidobacterium* strains can reduce anxiety and depression by improving neurotransmitter balance and lowering stress hormone levels. Prebiotics, on the other hand, are like fertilizer for the good bacteria in our gut. Found in fiber-rich foods like garlic, onions, and bananas, prebiotics fuel the growth of beneficial microbes, leading to increased production of SCFAs potent molecules with neuroprotective and anti-inflammatory properties (Diez-Gutiérrez et al. 2020). By nurturing the right gut bacteria, prebiotics can create a more stable mental state and even enhance cognitive function.
- **FMT: Resetting the gut for a healthier brain**: FMT may sound unconventional, but this groundbreaking therapy is revolutionizing gut and brain health. It involves transplanting gut microbes from a healthy donor into a recipient suffering from microbiome imbalances (Kim & Gluck 2019). While FMT is best known for treating severe gastrointestinal infections, research now suggests that it may also help with neurological conditions such as Parkinson's disease, multiple sclerosis (MS), and ASD. By replenishing beneficial bacteria, FMT can reduce systemic inflammation, improve gut integrity, and restore gut–brain communication. Some Parkinson's patients have reported improvements in motor function following FMT, while studies on autism suggest that the therapy may alleviate behavioral symptoms by modifying gut microbial composition. Though still in its early stages, FMT holds immense potential for treating neurological disorders linked to dysbiosis (Table 10.3).

10.5.2 Bioinformatics in Microbiome Engineering

With advances in bioinformatics, researchers can now explore ways to engineer the microbiome to improve health outcomes, particularly in neurological disorders. By analyzing microbial genomes, designing engineered probiotics, and predicting microbial interactions through computational modeling, scientists are pioneering new frontiers in microbiome-based therapies (D'Urso & Broccolo 2024).

Understanding the genetic blueprint of microbes is the first step in engineering them for therapeutic use. Each microbe in the gut has a unique genetic code that determines its role in metabolism,

TABLE 10.2
Bioinformatics Tools and Techniques for DNA Methylation Analysis in the Gut–Brain Axis

Technique	Principle	Application in Gut–Brain Axis Studies	Advantages	Limitations	Reference
Bisulfite sequencing (BS-Seq)	Converts unmethylated cytosines to uracil while methylated cytosines remain unchanged	Identifies genome-wide DNA methylation patterns in gut microbiota and host tissues influencing neurological disorders	High resolution, single-base precision	High cost, requires deep sequencing	Park et al. (2016)
Reduced representation bisulfite sequencing (RRBS)	Enriches CpG-rich regions using restriction enzymes before bisulfite conversion	Targets differentially methylated regions (DMRs) associated with gut microbiome changes in neurodegenerative diseases	Cost-effective compared to whole-genome bisulfite sequencing	Bias toward CpG islands, limited coverage	Nakabayashi et al. (2022)
Methylated DNA immunoprecipitation sequencing (MeDIP-Seq)	Uses antibodies to capture methylated DNA for sequencing	Detects methylation changes in key regulatory genes linking gut microbiota to neuroinflammation	Efficient for large-scale studies, works with low DNA input	Lower resolution compared to BS-Seq, antibody specificity issues	Neary et al. (2020)
ELISA for global 5mC levels	Quantifies global 5-methylcytosine (5mC) levels in DNA samples	Measures overall epigenetic changes in host tissues in response to gut microbial metabolites	Simple, cost-effective	Lacks locus-specific resolution	Zhang et al. (2020)
Pyrosequencing	Measures DNA methylation at specific CpG sites through sequencing-by-synthesis	Used for targeted validation of gut–microbiome-induced methylation changes in neurological pathways	High accuracy, quantitative	Limited genomic coverage, expensive per sample	Ahmadian et al. (2006)
Nanopore-based methylation sequencing	Directly sequences native DNA, detecting methylation without bisulfite conversion	Enables real-time monitoring of host and microbial epigenetic modifications in gut–brain interactions	Long-read sequencing, single-molecule resolution	Higher error rates compared to BS-Seq	Liu et al. (2021)
Methylation-specific PCR (MSP)	Uses methylation-sensitive primers for amplification of methylated vs. unmethylated DNA	Useful for studying candidate genes affected by gut microbiota-derived epigenetic modifications	Sensitive, low cost	Limited to known methylation sites, not genome-wide	Derks et al. (2004)

TABLE 10.3
FMT Methods Applied for Gut and Brain Health Treatment

Fecal Microbial Therapy Focus	Description	References
Clostridioides difficile clearance	FMT restores gut microbiota by introducing beneficial bacteria to combat antibiotic-resistant *C. difficile* infections.	Smillie et al. (2018)
Gut-brain axis modulation	FMT may influence neurological functions by altering microbiota compositions linked to brain health.	Eslami et al. (2025)
Immune system regulation	Restores microbial diversity, reducing systemic inflammation and autoimmune responses.	Singh et al. (2024)
Autism spectrum disorder (ASD) microbial modulation	FMT alters gut microbiome composition, potentially alleviating behavioral symptoms in individuals with ASD.	Li et al. (2021)
Parkinson's disease symptom reduction	FMT enhances gut integrity and may improve motor function in Parkinson's patients.	Bruggeman et al. (2024)
Multiple sclerosis microbiota adjustment	Restoration of gut microbiome diversity helps manage immune dysregulation in multiple sclerosis.	Engen et al. (2020)

neurotransmitter production, and immune modulation. Bioinformatics technologies enable researchers to decode these genomes and uncover genes that produce beneficial substances including SCFAs, serotonin, dopamine, and other neuroactive molecules (Yan et al. 2022).

Genome annotation technologies like Prokka and RAST (Rapid Annotation using Subsystem Technology) assist scientists in analyzing microbial genomes and assigning functions to specific genes (Table 10.4). These tools enable researchers to identify probiotic microorganisms that have favorable metabolic pathways; detect antibiotic resistance genes to assess the safety of modified probiotics; understand how microbes interact inside the gut ecosystem; and optimize genetic changes to increase probiotic efficacy (Galanis et al. 2025). By mapping microbial genetic landscapes, bioinformatics enables researchers to create bacterial strains that blend easily into the gut microbiome, giving specific therapeutic benefits.

Probiotics have long been used to improve gut health, but bioinformatics is taking them to the next level by allowing for precise microbiome engineering. Instead, then depending on naturally occurring probiotics, scientists are developing bacterial strains with improved capabilities (Bober et al. 2018). Engineered probiotics can be tailored to produce specific neurotransmitters like GABA to reduce anxiety and depression; secrete anti-inflammatory compounds to combat neuroinflammation linked to Alzheimer's disease and Parkinson's disease; and improve gut barrier integrity by enhancing mucus production and tightening intestinal junctions, reducing leaky gut syndrome.

10.5.3 Modulating Epigenetic Markers to Restore Brain–Gut Homeostasis

The gut and the brain are intimately connected through the GBA, a complex communication network involving neurons, hormones, and immune signals. Recent research suggests that imbalances in the gut microbiome often caused by genetic factors, diet, or environmental stress can contribute to neurological disorders, including ASD, depression, and MS (Sahin & Sur. 2015).

Depression and anxiety disorders are increasingly linked to gut dysbiosis, where an imbalance in gut microbes disrupts the production of neurotransmitters like serotonin and GABA (Huang et al. 2021). While current treatments focus on antidepressants and therapy, gene editing offers a potentially longer-lasting solution by directly modifying the epigenetic regulators that influence stress responses and gut–brain communication. By using epigenome editing to adjust the activity of

TABLE 10.4
Bioinformatics Tools Applications in Microbiome Engineering

Tool	Method/Algorithm Used by Tool	Application in Microbiome Engineering	References
QIIME 2	Uses DADA2 for denoising, taxonomy classification via naive Bayes, and diversity analysis	Microbiome composition analysis, taxonomic profiling, functional inference	Bolyen et al. (2019)
MetaPhlAn	Marker-based metagenomic profiling to identify species presence and abundance	Identification of microbial communities, strain-level resolution in microbiomes	Truong et al. (2015)
HUMAnN	Hierarchical functional profiling using protein-coding gene annotation	Functional characterization of gut microbiota, predicting metabolic pathways	(Franzosa et al. (2018)
PICRUSt2	Infers the functional potential of microbiomes using marker gene-based prediction	Predicting metabolic pathways and microbiome interactions	Douglas et al. (2020)
Kraken2	Uses k-mer-based classification for rapid taxonomic assignment	Classification of microbial communities in metagenomic datasets	Wood et al. (2019)
MG-RAST	Automated metagenome analysis and annotation via similarity-based methods	Functional and taxonomic annotation of environmental microbiomes	Keegan et al. (2016)
anvi'o	Integrative visualization platform employing metagenomic binning techniques	Genome-resolved microbiome analysis, strain-level characterization	Eren et al. (2015)
MicrobiomeAnalyst	Web-based platform using statistical methods for microbiome data exploration	Comparative microbiome analysis, pathway prediction	Dhariwal (2017)

genes involved in neurotransmitter production and inflammation, researchers hope to develop new, more effective treatments for mood disorders that work by restoring gut–brain homeostasis at the molecular level (Alam et al. 2017). Unlike direct gene mutations, many of these conditions arise from epigenetic changes, which modify how genes are turned on or off without altering the DNA sequence itself. Gene-editing technologies like epigenome editing are now being explored to reverse these changes and restore healthy brain–gut interactions.

- **CRISPR guide RNA design: Precision in editing neurological disorders:** At the heart of gene editing is the guide RNA (gRNA), a short RNA molecule that directs CRISPR enzymes such as Cas9 to the precise region in the genome. The effectiveness and accuracy of gene editing are dependent on how well the gRNA is constructed. A badly constructed guide can cause off-target alterations, which can be disastrous in neurological applications. To optimize gRNA selection, several bioinformatics tools have been developed:
    - **CRISPOR**: Predicts the best gRNA sequences based on their specificity and probable off-target regions (Concordet et al. 2018).
    - **CHOPCHOP**: A widely used tool that analyzes CRISPR target sites in terms of efficiency and minimization of off-target effects (Labun et al. 2019)
    - **CCTop**: Improves editing success rates by rating based on mismatches and chromatin accessibility (Dobson et al. 2015)

  Precision is especially crucial in neurological illnesses since brain cells do not renew easily, making unintentional genetic changes potentially irreversible. Bioinformatics tools help reduce risks by identifying where CRISPR could accidentally edit nontarget genes, resulting in safer and more reliable interventions.

One of the biggest risks in gene and epigenome editing is off-target effects, unintended modifications that could lead to serious side effects. The challenge is even greater in neurological applications because neurons are highly interconnected, meaning even minor disruptions in gene regulation can have widespread consequences. To address this, bioinformatics plays a key role in predicting and minimizing risks before conducting lab experiments:

- **CRISPRoff and CRISPRon**: These predictive models estimate how CRISPR-based epigenetic modifications will affect gene expression levels (Shaffer et al. 2021)
- **DeepCRISPR**: Uses machine learning to analyze thousands of CRISPR experiments and predict the likelihood of off-target mutations (Chuai et al. 2018)
- **GUIDE-seq**: A computational approach that maps CRISPR-induced breaks in the genome, ensuring that edits occur in the intended locations (Tsai et al. 2015)

By running simulations before conducting actual experiments, bioinformatics allows researchers to test different gene-editing strategies in silico, ensuring the safest and most effective approach is chosen.

10.6 CONCLUSION

The intersection of epigenetics, gene regulation, and microbiome research has created a groundbreaking paradigm for comprehending the GBA in both health and illness. Epigenetic alterations—such as DNA methylation, modifications of histones, and the roles of non-coding RNAs—not only react to environmental and microbial signals but also play essential roles in regulating gene expression, which in turn influences neurological outcomes. Concurrently, the gut microbiota functions as both a modulator and a recipient of these epigenetic changes, forming a dynamic feedback mechanism that affects neurodevelopment, behavior, and neurodegenerative processes.

Disruptions within this complex system have been linked to various neurological conditions, including ASD, Parkinson's disease, and Alzheimer's disease. Recent advancements in microbiome manipulation and gene-editing technologies, such as CRISPR-Cas9 and epigenetic reprogramming methods, present promising new strategies for therapeutic intervention, potentially allowing for the reversal of disease mechanisms at their molecular origins.

Nonetheless, the application of these discoveries in clinical settings poses considerable challenges. The variability in individual microbiome compositions and epigenetic profiles highlights the necessity for tailored approaches. Additionally, ethical and safety issues related to gene and microbiome modifications must be thoroughly examined through continuous research and policy formulation.

The integration of multi-omics data—including epigenomics, metagenomics, transcriptomics, and metabolomics—coupled with sophisticated computational modeling, offers potential for clarifying the complexities of the GBA. Such interdisciplinary initiatives will be crucial for the development of next-generation, precision-based therapies for neurological disorders. As this domain evolves, it is set to enhance our understanding of brain–gut interactions significantly and to transform the landscape of clinical neuroscience and personalized medicine.

REFERENCES

Ahmadian, A., Ehn, M., & Hober, H. "Pyrosequencing: history, biochemistry and future". *Clinica Chimica Acta* (2006) 363(1–2), pp. 83–94.

Alam, R., Abdolmaleky, H. M., & Zhou, J. R. "Microbiome, inflammation, epigenetic alterations, and mental diseases". *American Journal of Medical Genetics Part B: Neuropsychiatric Genetics* (2017) 174(6), pp. 651–660.

Bakshi, I., Dey, S., Raut, A., Katta, S., & Sharma, P. "Exploring the gut-brain axis: a comprehensive review of interactions between the gut microbiota and the central nervous system". *International Journal for Multidisciplinary Research* (2024) 6, pp. 1–15

Begum, N., Mandhare, A., Tryphena, K. P., Srivastava, S., Shaikh, M. F., Singh, S. B., & Khatri, D. K. "Epigenetics in depression and gut-brain axis: a molecular crosstalk". *Frontiers in Aging Neuroscience* (2022) 14, p. 1048333.

Bertollo, A. G., Santos, C. F., Bagatini, M. D., & Ignácio, Z. M. "Hypothalamus-pituitary-adrenal and gut-brain axes in biological interaction pathway of the depression". *Frontiers in Neuroscience* (2025) 19, p. 1541075.

Bober, J. R., Beisel, C. L, & Nair, N. U. "Synthetic biology approaches to engineer probiotics and members of the human microbiota for biomedical applications". *Annual Review of Biomedical Engineering* (2018) 20, pp. 277–300.

Bolyen, E. et al. "Reproducible, interactive, scalable and extensible microbiome data science using QIIME 2". *Nature News* (2019) 37, pp. 852–857.

Bonaz, B., Bazin, T., & Pellissier, S. "The vagus nerve at the interface of the microbiota-gut-brain axis". *Frontiers in Neuroscience* (2018) 12, p. 336468.

Bruggeman, A., Vandendriessche, C., Hamerlinck, H., De Looze, D., Tate, D. J., Vuylsteke, M., De Commer, L., Devolder, L., Raes, J., Verhasselt, B., Laukens, D., Vandenbroucke, R. E., & Santens, P. "Safety and efficacy of faecal microbiota transplantation in patients with mild to moderate Parkinson's disease (GUT-PARFECT): a double-blind, placebo-controlled, randomised, phase 2 trial". *EClinical Medicine* (2024) 7, 102563.

Carabotti, M., Scirocco, A., Maselli, M. A., & Severi, C. "The gut-brain axis: interactions between enteric microbiota, central and enteric nervous systems". *Annals of Gastroenterology: Quarterly Publication of the Hellenic Society of Gastroenterology* (2015) 28(2), p. 203

Chen, M., Lai, X., Wang, X., Ying, J., Zhang, L., Zhou, B., Liu, X., Zhang, J., Wei, G., & Hua, F. "Long non-coding RNAs and circular RNAs: insights into microglia and astrocyte mediated neurological diseases". *Frontiers in Molecular Neuroscience* (2021) 14, p. 745066

Chen, Y., Xu, J., & Chen, Y. "Regulation of neurotransmitters by the gut microbiota and effects on cognition in neurological disorders". *Nutrients* (2021) 13(6), p. 2099.

Chenarani, N., Emamjomeh, A., Allahverdi, A., Mirmostafa, S., Afsharinia, M. H., & Zahiri, J. "Bioinformatic tools for DNA methylation and histone modification: a survey". *Genomics* (2021) 113(3), pp.1098–1113.

Chuai, G., Ma, H., Yan, J., Chen, M., Hong, N., Xue, D., Zhou, C., Zhu, C., Chen, K., Duan, B., & Gu, F. "DeepCRISPR: optimized CRISPR guide RNA design by deep learning". *Genome Biology* (2018) 19, pp. 1–18.

Concordet, J. P., & Haeussler, M. "CRISPOR: intuitive guide selection for CRISPR/Cas9 genome editing experiments and screens". *Nucleic Acids Research* (2018) 46(W1), pp. W242–W245.

D'Urso, F., & Broccolo, F. "Applications of artificial intelligence in microbiome analysis and probiotic interventions—an overview and perspective based on the current state of the art". *Applied Sciences* (2024) 14(19), p. 8627.

Derks, S., Lentjes, M. H. F. M., Hellebrekers, D. M. E. I., De Bruïne, A. P., Herman, J. G., & Van Engeland, M. "Methylation-specific PCR unraveled". *Analytical Cellular Pathology* (2004) 26(5–6), pp. 291–299.

DeSalle, R., & Perkins, S. L. *Welcome to the Microbiome: Getting to Know the Trillions of Bacteria and Other Microbes in, on, and Around You*, 2015, Yale University Press.

Dhariwal, A. *Statistical, Visual and Functional Analysis of Microbiome Data*. Master's thesis, McGill University, 2017.

Diez-Gutiérrez, L., San Vicente, L., Barrón, L. J. R., del Carmen Villarán, M., & Chávarri, M. "Gamma-aminobutyric acid and probiotics: multiple health benefits and their future in the global functional food and nutraceuticals market". *Journal of Functional Foods* (2020) 64, p. 103669.

Dobson, L., Reményi, I., & Tusnády, G. E. "CCTOP: a consensus constrained topology prediction web server". *Nucleic Acids Research* (2015) 43(W1), pp. W408–W412.

Dong, Y., & Cui, C. "The role of short-chain fatty acids in central nervous system diseases". *Molecular and Cellular Biochemistry* (2022) 477, pp. 2595–2607.

Douglas, G. M., Maffei, V. J., Zaneveld, J. R., Yurgel, S. N., Brown, J. R., Taylor, C. M., Huttenhower, C., & Langille, M. G. "PICRUSt2 for prediction of metagenome functions". *Nature Biotechnology* (2020) 38(6), 685–688.

Engen, P. A., Zaferiou, A., Rasmussen, H., Naqib, A., Green, S. J., Fogg, L. F., Forsyth, C. B., Raeisi, S., Hamaker, B., & Keshavarzian, A. "Single-arm, non-randomized, time series, single-subject study of fecal microbiota transplantation in multiple sclerosis". *Frontiers in Neurology* (2020) 11, p. 564830.

Eren, A. M., Esen, Ö. C., Quince, C., Vineis, J. H., Morrison, H. G., Sogin, M. L., & Delmont, T. O. "Anvi'o: an advanced analysis and visualization platform for 'omics data". *PeerJ* (2015) 3, p. e1319
Eslami, M., Zarifeh, A., Bahram, F. D., Shayan, Y., Faezeh, M. T., Valentyn, O., & Ramtin, N. "A novel frontier in gut–brain axis research: the transplantation of fecal microbiota in neurodegenerative disorders". *Biomedicines* (2025) 13(4), p. 915.
Farooqui, A. A., Farooqui, T., Panza, F., & Frisardi, V. "Metabolic syndrome as a risk factor for neurological disorders". *Cellular and Molecular Life Sciences* (2012) 69, pp. 741–762.
Farrelly, L. A., & Maze, I. "An emerging perspective on 'histone code'mediated regulation of neural plasticity and disease". *Current Opinion in Neurobiology* (2019) 59, pp. 157–163.
Franzosa, E. A., McIver, L. J., Rahnavard, G., Thompson, L. R., Schirmer, M., Weingart, G., Lipson, K. S., Knight, R., Caporaso, J. G., Segata, N., & Huttenhower, C. "Species-level functional profiling of metagenomes and metatranscriptomes". *Nature Methods* (2018) 15(11), pp. 962–968.
Galanis, A., Papadimitriou, K., & Moloney, G.M. "Editorial: Omics technologies and bioinformatic tools in probiotic research." *Frontiers in Microbiology* (2025) 16, 1577852.
Hervouet, E., Peixoto, P., Delage-Mourroux, R., Boyer-Guittaut, M., & Cartron, P.-F. "Specific or not specific recruitment of DNMTs for DNA methylation, an epigenetic dilemma". *Clin Epigenet* (2018) 10, p. 17.
Hu, M. L., Pan, Y. R., Yong, Y. Y., Liu, Y., Yu, L., Qin, D. L., Qiao, G., Law, B. Y. K., Wu, J. M., Zhou, X. G., & Wu, A. G. "Poly (ADP-ribose) polymerase 1 and neurodegenerative diseases: past, present, and future". *Ageing Research Reviews* (2023) 91, p. 102078.
Huang, X., Hussain, B., & Chang, J. "Peripheral inflammation and blood–brain barrier disruption: effects and mechanisms". *CNS Neuroscience & Therapeutics* (2021) 27(1), pp. 36–47.
Ignácio, Z. M., Clayton, J. B., Bagatini, M. D., & Veerappa, A. "Reviews in gut-brain axis: stress, dysregulation in gut-brain axis function and stress related disorders". *Frontiers in Neuroscience* (2025) 19, p. 1665282.
Jang, H. S., Shin, W. J., Lee, J. E., & Do, J. T. "CpG and non-CpG methylation in epigenetic gene regulation and brain function". *Genes* (2017) 8(6), p. 148.
Jung, S., Bae, H., Song, W. S., Chun, Y., Le, J., Alam, Y., Verlande, A., Chun, S. K., Kim, J., Kelly, M. E., Lopez, M. L., Park, S. H., Onofre, D., Baek, J., Jang, K.-H., Rubtsova, V. I., Anica, A., Masri, S., Lee, G., & Jang, C et al. "Dietary fibre-adapted gut microbiome clears dietary fructose and reverses hepatic steatosis". *Nature Metabolism* (2025) 7, pp. 1801–1818.
Keegan, K. P., Glass, E. M., & Meyer, F. "MG-RAST, a metagenomics service for analysis of microbial community structure and function". In: Martin, F., Uroz, S. (eds) *Microbial Environmental Genomics (MEG): Methods in Molecular Biology,* vol. 1399, 2016, Humana Press, New York, NY.
Kennedy, R. H., & Silver, R. "Neuroimmune signaling: Cytokines and the CNS." In: Pfaff, D. W., Volkow, N. D. (eds) *Neuroscience in the 21st century*, pp. 1–41, 2016. Springer, New York.
Khan, M. S., Qureshi, N., Khan, R., Son, Y., & Maqbool, T. "CRISPR/Cas9-Based therapeutics as a promising strategy for management of Alzheimer's disease: progress and prospects". *Frontiers in Cellular Neuroscience* (2025) 19, p. 1578138.
Kim, K. O., & Gluck, M. "Fecal microbiota transplantation: an update on clinical practice". *Clinical Endoscopy* (2019) 52(2), pp. 137–143.
Kip, E., & Parr-Brownlie, L. C. "Healthy lifestyles and wellbeing reduce neuroinflammation and prevent neurodegenerative and psychiatric disorders". *Frontiers in Neuroscience* (2023) 17, p. 1092537.
Labun, K., Montague, T. G., Krause, M., Torres Cleuren, Y. N., Tjeldnes, H., & Valen, E. "CHOPCHOP v3: expanding the CRISPR web toolbox beyond genome editing". *Nucleic Acids Research* (2019) 47(W1), pp. W171–W174.
Li, N., Chen, H., Cheng, Y., Xu, F., Ruan, G., Ying, S., Tang, W., Chen, L., Chen, M., Lv, L., Ping, Y., Chen, D., & Wei, Y. "Fecal microbiota transplantation relieves gastrointestinal and autism symptoms by improving the gut microbiota in an open-label." Study. *Frontiers in Cellular and Infection Microbiology* (2021) 11, 759435.
Liu, Y., Rosikiewicz, W., Pan, Z., Jillette, N., Wang, P., Taghbalout, A., Foox, J., Mason, C. E., Carroll, M., Cheng, A., & Li, S. "DNA methylation-calling tools for Oxford Nanopore sequencing: a survey and human epigenome-wide evaluation." *Genome Biology* (2021) 22, pp. 1–33.
Lozupone, M., D'Urso, F., Piccininni, C., Montagna, M. L., Sardone, R., Resta, E., Dibello, V., Daniele, A., Giannelli, G., Bellomo, A., & Panza, F. "The relationship between epigenetics and microbiota in neuropsychiatric diseases". *Epigenomics* (2020) 12(17), pp. 1559–1568.
Ma, J., Zhang, J., & Kuang, Z. "A microbiota-epigenetic circuit controls systematic circadian programs in the gut epithelium". *Frontiers in Systems Biology* (2023) 3, p. 1175306.

Maguire, M., & Maguire, G. "Gut dysbiosis, leaky gut, and intestinal epithelial proliferation in neurological disorders: towards the development of a new therapeutic using amino acids, prebiotics, probiotics, and postbiotics". *Reviews in the Neurosciences* (2019) 30(2), pp. 179–201.

Martin, C. R., Osadchiy, V., Kalani, A., & Mayer, E. A. "The brain-gut-microbiome axis." *Cellular and Molec ular Gastroenterology and Hepatology* (2018) 6, pp. 133–148.

Nakabayashi, K., Yamamura, M., Haseagawa, K., & Hata, K. "Reduced representation bisulfite sequencing (RRBS)." In: Smith, J. D., Chen, M. L. (eds) *Epigenomics: Methods and Protocols*, 2022, pp. 39–51, Springer.

Nakao, M. "Epigenetics: interaction of DNA methylation and chromatin". *Gene* (2001) 278(1–2), pp. 25–31.

Neary, J. L., & Melanie, A. C. "Methylated DNA immunoprecipitation sequencing (MeDIP-seq): principles and applications." In: Tollefsbol, T. O. (ed) *Epigenetics Methods*, 2020, pp. 157–179, Academic Press.

Needham, B. D., Kaddurah-Daouk, R., & Mazmanian, S. K. "Gut microbial molecules in behavioural and neurodegenerative conditions". *Nature Reviews Neuroscience* (2020) 21(12), pp. 717–731.

Nichols, R. G., & Davenport, E. R. "The relationship between the gut microbiome and host gene expression: a review". *Human genetics* (2021) 140(5), pp. 747–760.

O'Riordan, K. J., Collins M. K., Moloney, G. M., Knox, E. G., Aburto, M. R., Fulling, C., Morley, S. J., Clarke, G., Schellekens, H., & Cryan, J. F. "Short chain fatty acids: microbial metabolites for gut-brain axis signalling". *Molecular and Cellular Endocrinology* (2022) 546, p. 111572.

Park, Y., & Hao, W. "Differential methylation analysis for BS-seq data under general experimental design". *Bioinformatics* (2016) 32(10), pp. 1446–1453.

Ramazi, S., Allahverdi, A., & Zahiri, J. "Evaluation of post-translational modifications in histone proteins: a review on histone modification defects in developmental and neurological disorders". *Journal of Biosciences* (2020) 45(1), p. 135.

Ruffo, P., De Amicis, F., Giardina, E., & Conforti, F.L. "Long-noncoding RNAs as epigenetic regulators in neurodegenerative diseases". *Neural Regeneration Research* (2023) 18(6), pp. 1243–1248.

Sahin, M., & Sur, M. "Genes, circuits, and precision therapies for autism and related neurodevelopmental disorders". *Science* (2015) 350(6263), p. aab3897.

Shaffer, C. "CRISPR: new techniques and complex models: tools for altering gene expression—CRISPRi, CRISPRa, CRISPRoff, and CRISPRon—and CRISPR-compatible cellular models are opening new research possibilities". *Genetic Engineering & Biotechnology News* (2021) 41(S4), pp. S6–S9.

Singh, R., Zogg, H., & Ro, S. "Role of microRnas in disorders of gut–brain interactions: clinical insights and therapeutic alternatives". *Journal of Personalized Medicine* (2021) 11(10), p. 1021.

Singh, V., Choi, S. D., Mahra, K., Son, H., Lee, H., Lee, Y.-J., Kim, E. S., & Shin, J.-H. "Cultured fecal microbial community and its impact as fecal microbiota transplantation treatment in mice gut inflammation". *Applied Microbiology and Biotechnology* (2024), 108, p. 463.

Smillie, C. S., Sauk, J., Gevers, D., Friedman, J., Sung, J., Youngster, I., Hohmann, E. L., Staley, C., Khoruts, A., Sadowsky, M. J., Allegretti, J. R., Smith, M. B., Xavier, R. J., Alm, E. J. "Strain tracking reveals the determinants of bacterial engraftment in the human gut following fecal microbiota transplantation". *Cell Host Microbe* (2018) 23(2), pp. 229–240.e5.

Stilling, R. M., Dinan, T. G., & Cryan, J. F. "Microbial genes, brain & behaviour–epigenetic regulation of the gut–brain axis". *Genes, Brain and Behaviour* (2014) 13(1), pp. 69–86.

Taka, N., Yoshida, W. "Quantification of global DNA methylation level using 5-methylcytosine dioxygenase". *Analytical and Bioanalytical Chemistry* (2020) 412, pp. 5299–5305.

Tang, W., Meng, Z., Li, N., Liu, Y., Li, L., Chen, D., & Yang, Y. "Roles of gut microbiota in the regulation of hippocampal plasticity, inflammation, and hippocampus-dependent behaviors". *Frontiers in Cellular and Infection Microbiology* (2021) 10, p. 611014.

Teleanu, R. I., Niculescu, A. G., Roza, E., Vladâcenco, O., Grumezescu, A. M., & Teleanu, D. M. "Neurotransmitters—key factors in neurological and neurodegenerative disorders of the central nervous system". *International Journal of Molecular Sciences* (2022) 23(11), p. 5954.

Truong, D. T., Eric, A. F., Timothy, L. T., Matthias, S., George, W., Edoardo, P., Adrian, T., Curtis, H., & Nicola, S. "MetaPhlAn2 for enhanced metagenomic taxonomic profiling". *Nature Methods* (2015) 12(10), pp. 902–903.

Tsai, S. Q., Zheng, Z., Nguyen, N. T., Liebers, M., Topkar, V. V., Thapar, V., Wyvekens, N., Khayter, C., Iafrate, A. J., Le, L. P., & Aryee, M. J. "GUIDE-seq enables genome-wide profiling of off-target cleavage by CRISPR-Cas nucleases". *Nature Biotechnology* (2015) 33(2), pp. 187–197.

Turner, B. M. *Chromatin and Gene Regulation: Molecular Mechanisms in Epigenetics*, John Wiley & Sons, 2002.

Ullah, H., Arbab, S., Tian, Y., Liu, C.-Q., Chen, Y., Qijie, L., Khan, M. I. U., Hassan, I. U., & Li, K. "The gut microbiota–brain axis in neurological disorder". *Frontiers in Neuroscience* (2023) 17, p. 1225875

Ullah, H., Arbab, S., Tian, Y., Liu, C., Chen, Y., Qijie, L., Khan, M. I., Hassan, I. U., & Li, K. "The gut microbiota–brain axis in neurological disorder". *Frontiers in Neuroscience* (2023) 17, p. 1225875.

Van de Wouw, M., Schellekens, H., Dinan, T. G., & Cryan, J. F. "Microbiota-gut-brain axis: modulator of host metabolism and appetite". *The Journal of Nutrition* (2017) 147(5), pp. 727–745.

Vendrik, K. E., Ooijevaar, R. E., De Jong, P. R., Laman, J. D., Van Oosten, B. W., Van Hilten, J. J., Ducarmon, Q. R., Keller, J. J., Kuijper, E. J., & Contarino, M. F. "Fecal microbiota transplantation in neurological disorders". *Frontiers in Cellular and Infection Microbiology* (2020) 10, p. 98.

Winek, K., & Soreq, H. "Emerging roles of transfer RNA fragments in the CNS". *Brain* (2025) 148(8), pp. 2631–2645.

Wood, D. E., & Salzberg, S. L. "Kraken: ultrafast metagenomic sequence classification using exact alignments". *Genome Biology* (2014) 15, pp. 1–12.

Xu, J., & Lu, Y. "The microbiota-gut-brain axis and central nervous system diseases: from mechanisms of pathogenesis to therapeutic strategies". *Frontiers in Microbiology* (2025) 16, p. 1583562

Yakoo, S. *Comparative Analysis of Gene Annotation Tools*. Master's thesis, San Diego State University, 2018.

Yan, J., Feng, X., Zhou, X., Zhao, M., Xiao, H., Li, R., & Shen, H. "Identification of gut metabolites associated with Parkinson's disease using bioinformatic analyses". *Frontiers in Aging Neuroscience*, (2022) 14, 927625.

Zhang, H., Luan, J., He, L., Pan, X., Zhang, H., Li, Y., & Li, H. "Role of the gut-brain axis in neurological diseases: molecular connections and therapeutic implications (Review)". *International Journal of Molecular Medicine* (2025) 56, 192.

Zheng, Y., Bonfili, L., Wei, T., & Eleuteri, A. M. "Understanding the gut–brain axis and its therapeutic implications for neurodegenerative disorders". *Nutrients* (2023) 15(21), 4631.

Zhou, M., Niu, B., Ma, J., Ge, Y., Han, Y., Wu, W., & Yue, C. "Intervention and research progress of gut microbiota-immune-nervous system in autism spectrum disorders among students". *Frontiers in Microbiology* (2025) 16, 1535455.

Zhou, S., Yu, X., Wang, M., Meng, Y., Song, D., Yang, H., Wang, D., Bi, J., & Xu, S. "Long non-coding RNAs in pathogenesis of neurodegenerative diseases". *Frontiers in Cell and Developmental Biology* (2021) 9, p. 719247.

Section C

Microbiome, Aging, and Environment

11 Built-Up Environment Microbes and Mental Health

Sucheta Das, Shamba Chatterjee, and Subrata Trivedi

11.1 INTRODUCTION

Population expansion has been one of the major factors in recent decades, along with urbanization that affect the lives globally. According to estimates, the proportion of people living in cities worldwide is expected to rise from 55% to 67% between 2020 and 2050 (Pesantez 2018). Both the size and population of cities are increasing; by 2030, 706 cities are predicted to have a population of at least 1 million. At the same time, we are seeing an increase in autoimmune and chronic noncommunicable diseases (NCDs) (Miller 2023). Along with the "Old Friends Hypothesis" (Rook 2023), some other theories, like Reconciling Competing Theories and Hygiene Hypothesis, allowed for a new understanding of NCDs, which are caused by our contemporary lifestyle and significantly impact modern residents. The human gut microbiome is altered by current lifestyle choices. A recent study examined the differences in the bacterial microbiota between children in a rural hamlet in Burkina Faso, West Africa, and children in urban Europe, and the result showed significant differences among the microbiomes tested (Rook 2012). Given that NCDs account for 74% of deaths worldwide, according to WHO reports (WHO 2025), significant improvements in living conditions and their effects on health are necessary.

In recent years, the study of indoor microbes, known as the microbiome of the built environment (MoBE), has flourished due to recent developments in DNA sequencing methods, mainly metagenomic and general omics. It is believed that 90% of people in industrialized countries today live indoors, spending the majority of their time in human-designed, human-built, and human-managed places and structures, such as homes, workplaces, and educational institutions (Hoisington et al. 2015). People are constantly surrounded by germs when they travel in cars, on public transportation, and from their homes to their places of employment. Even within this subset of structures, there is a great deal of variation not just in the systems, materials, and designs of the buildings, but also in the social and economic traits, densities, and behaviors of the occupants, as well as in the relationships between the buildings and the external site and infrastructure. A diverse collection of organisms, including viruses, prokaryotes (bacteria and archaea), and microbial eukaryotes (fungi, microscopic algae, and protozoa), is present in a specific area (National Academies of Sciences et al. 2017). In addition to living microorganisms and their genetic material, these microbes also comprise proteins, metabolic products, dormant or inactive bacteria, and other cell components or fragments that may have a positive, negative, or neutral effect on human occupiers. Furthermore, humans engage in a cycle of exposure, absorption, and shedding with the indoor microbiome; these interactions can have various unknowable effects on human health.

Again, the human body's micro-ecosystem is made up of viruses, bacteria, fungi, protists, archaea, and other trillions of microbes. The entire microbial gene content, sometimes known as the human microbiome, is essential to human immunity and function. A complex link, mediated by the immune system, exists between the endogenous human microbiome and the surrounding microbial world, or environment. The health and illness of humans can be significantly impacted by breaking this link (Sharma and Gilbert 2018). The development of new microbial therapies may be significantly impacted by our ability to comprehend how the immune response is influenced by microbial exposure and how this feedback affects endogenous microbial metabolic activity. Many facets of

DOI: 10.1201/9781003610274-14

host physiology, including the development and operation of the immune system, are regulated by gut microorganisms (Fung et al. 2017). Furthermore, there is mounting evidence that the gut flora controls behavior, brain development, and function (Cryan and Dinan 2012; Collins et al. 2012; Sampson and Mazmanian 2015). Immunomodulation by the microbiota is being recognized as a significant mechanism that coordinates microbiota–gut–brain communication.

According to the World Health Organization (WHO), there is "No Health without Mental Health," and on the other hand, mental health is frequently harmed when physical health is lacking (https://www.who.int/westernpacific/activities/promoting-mental-health). Numerous studies have examined the reciprocal connection between NCDs and mental illnesses (WHO 2025). One possible link between the MoBE and mental health is that it may have an impact on the human microbiome. The transfer of microorganisms from the built environment to occupants must be demonstrated for that influence to occur. In this chapter, the integration of the MoBE and mental health domains is covered in greater detail. Specifically, the impact of the MoBE on the human microbiome and potential modifications to the MoBE to improve mental health will be examined.

11.2 MICROBIAL ECOLOGY OF THE BUILT-UP ENVIRONMENT

Numerous studies have been conducted in recent decades on the microbial communities found in natural settings, including soil, freshwater systems, marine systems, and the microbiomes of different animals, including humans. According to one definition, human build environment (BEs) are "man-made structures, features, and facilities viewed collectively as the places in which people live and work" (CDC 2024). The microbial populations present in the built environment, including indoor spaces, public areas, and air and water handling systems, are frequently overlooked. Studying the communities present in this environment, rather than just individual species of microbes, helps us better understand the ecological and evolutionary factors that influence them as well as how they may interact with other animals and ourselves (Ciric 2022). The majority of microorganisms cannot grow in our houses and other buildings because of the low relative humidity and temperature that humans typically maintain for comfort and to inhibit microbial growth. Moreover, in modern apartments, most of the rooms are sealed and isolated from the outside, which restricts the colonization of indoor places by environmental species. Additionally, this sealing aids in the buildup of our own bodily associates in the areas where we spend the most time (Lax et al. 2017). Different construct materials and household furniture retain unique characteristics, including surface structure, moisture content and potential, and different antimicrobial components. The great bulk of the variation in microbial communities within a constructed environment is therefore frequently explained by surface type (Hsu et al. 2016; González et al. 2020; Hu et al. 2019). One of the main factors influencing the variety and kinds of microorganisms present in homes and other buildings is the people who live there. The quantity and compositional variety of indoor-associated microbial profiles are influenced by the types of inhabitants (e.g., people vs. dogs), their illness status, length of stay, and occupancy density. Body-associated bacteria are more likely to be present in a place with higher occupant densities and longer stays (Qian et al. 2012), and sick people spread disease-causing organisms throughout their surroundings. The presence of pathogenic microorganisms can change the makeup of other microbial taxa present on surfaces, in addition to increasing the number and potential for transmission of diseases (Marotz et al. 2018). Additionally, having pets in or near a home boosts microbial diversity and encourages the colonization of taxa that may shield kids from developing autoimmune and allergy illnesses (Stein et al. 2016). The environment's influence on the organization of microbial communities can be seen in the microbiome found in dust that has been collected from different locations within a building (Zhou et al. 2023). Nygaard et al. (2020) used high-throughput sequencing to demonstrate the differences in bacterial microbiomes observed on floor dust and on HVAC (heating, ventilation, and air conditioning) input and exhaust filter dust. They demonstrated the abundance of outside organisms linked to the natural environment, like *Sphingomonas* and *Pseudomonas*, in the HVAC input filter populations. *Streptococcus* and *Staphylococcus*, two taxa linked to human skin and the respiratory system, predominated in the floor dust and dust from the HVAC exhaust filters. Their research demonstrates how occupancy affects the microbiota in indoor environments. Tanaka et al. (2020) has shown that location and aerosol size had an impact on the microbiomes of bioaerosols collected from outdoor urban and suburban areas in Japan. It was discovered that whilst soil and plant-associated bacteria dominated suburbia samples, human skin-associated bacteria dominated urban bioaerosols. Furthermore, bacteria detected in smaller particles differed from those found in bigger, coarser particles. The study suggests that local environmental factors, including indoor spaces, may have an impact on outdoor bacterial communities. The ventilation method used in a place might affect the microbial communities in the nasal passages (Chen et al. 2019). Additionally, bacteria detected on the surfaces' around hospitalized patients were more likely to be present, underscoring the significance of the patient's surroundings. Several factors may influence the microbiome of an indoor place, like the openness of a room, preference for indoor plants, drug usage habits of inhabitants, and even the behavioral patterns of those individuals (Kembel et al. 2012; Dockx et al. 2022). Antibiotic-resistant bacteria are becoming more prevalent in BEs as a result of the overuse and misuse of antibiotic drugs; this is a serious health issue identified by the WHO (https://www.who.int/news-room/spotlight/ten-threats-to-global-health-in-2019). Through direct deposition from feces or indirect selection following consumption and drug release into the environment, horizontal transfer of resistance gene to other group of indoor or even human microbiota, antibiotic-resistant bacteria can grow (Wang et al. 2023). Bacteria detected on the surfaces around hospitalized patients were more likely to be present, underscoring the significance of the patient's surroundings (Russotto et al. 2015). Like hospitals, homes may serve as a reservoir for the spread of antibiotic-resistant bacteria to people. Cleanliness of the built environment is effective for the removal of pathogenic strains from the surroundings, but it has also resulted in the loss of numerous positive relationships that are critical to human health. The rise and maintenance of antibiotic resistance, as observed during the COVID-19 pandemic, has also been linked to increased cleaning techniques. Reduced rates of SARS-CoV-2 and seasonal flu infection have been linked to the sharp rise in air and surface cleaning (Hayashi and Konishi 2021); however, these changed cleaning practices have also been linked to a rise in ARG abundance and hospital healthcare-acquired infections (O'Toole 2021). It has also been observed that due to the variations that exist in the architectural style, building materials, cultural variations, and even seasonal variations in weather and light availability, vegetation pattern, geographical characteristics, etc. can influence the microbial population of any built environment (Hill and Gilbert 2023). A stronger link to microbiomes outside of BEs leads to greater overall microbial diversity "inside" BEs, which in turn increases the diversity of human microbiomes in ways associated with increased human health, though it is still unclear how such structures affect the human microbiome.

11.3 IMPACT OF THE BUILT-UP ENVIRONMENT MICROBIOME ON THE HUMAN MICROBIOME

Human health and disease may be significantly impacted by the MoBE through both direct (i.e., exposure to pathogenic and helpful bacteria) and indirect (i.e., changing the makeup and structure of the human microbiota) methods. Built environments have various uses and unique operating circumstances that affect occupant health and the indoor microbiome. An increase in the number of microorganisms in the BE causes pathogenesis as well as increased immunity. The dynamics and diversity of indoor microbial communities can influence human occupants' skin- and stool-associated bacterial communities and are shaped by shared occupancy of the indoor environment. Each individual has a unique microflora and can acquire some components of it from their surroundings, which can again quickly spread across the immediate surroundings and may even reach other persons who are in the same location (Adams et al. 2015; Franzosa et al. 2015; Meadow et al. 2015). Leung et al. (2018) found that the percentage of skin microorganisms shared by cohabiting households ranged from 7% to 94%. In comparison to non-physically engaging residents who share the same space, Lax et al. (2014)

showed that with time, physically interacting occupants share more skin bacterial taxa. So as the occupants are continuously exposed to the indoor microbiome, it follows that the microbiomes found in the indoor environment could also have a profound effect on human health.

Microbes contribute to both human health and the health of the planet by being a member of a vast and complex ecosystem. On and within humans, commensal microorganisms have occupied ecological niches for almost 3 million years. First, it's critical to quickly examine the ways in which microbiota normally contributes to health in order to gain a better understanding of how the changed microbiota may jeopardize human health. They include, but are not limited to, immune system "education," defense against infections, preservation of environmental barriers (such as intestinal mucosa and cutaneous structures), nutrient extraction, vitamin synthesis, conversion of dietary phytochemicals, lipid metabolism, supply of short-chain fatty acids (SCFAs), and various other potentially bioactive metabolites, as well as the metabolism of xenobiotics (Hou et al. 2022). Some research using germ-free (GF) and specific-pathogen-free (SPF) animals has long shown that diverse ecosystems and microbiological colonization can affect behavior, the immune system, and even the availability of neurotransmitters in the brain (Sharon et al. 2016; Delgado-Ocaña and Cuesta 2024). A Japanese study from 2004 found that the hippocampal and cerebral cortex of GF animals had lower levels of gene expression of brain-derived neurotrophic factor than those of SPF animals (Sharvin et al. 2023). The study has also shown that after acute stress, GF mice had increased activity of the hypothalamus–pituitary–adrenal (HPA) axis. Restitution with the advantageous microbe *Bifidobacterium infantis* significantly reduced the GF mice's heightened HPA stress response. As a result, research showed how common commensal microorganisms affect the anatomy and function of the brain. Table 11.1 summarises the effect of environment on gut microbiome.

TABLE 11.1
Effect of Environment and Gut Microbiome

Place	Experiment	Observation	References
Farm	Microbial diversity checked in farm workers and nonworkers	High number of Proteobacteria and Clostridiaceae and less species diversity in workers.	Sun et al. (2020)
Office with shifting duty	Stool samples of 10 male security officers doing rotational day/night shifting duty	Increased abundance in Firmicutes and Bacteroidetes in the day shift. In the night shift: Bacteroidetes lowered; increased actinobacteria and Firmicutes.	Mortaş et al. (2020)
Hospitals	Microbial diversity compared between hospital workers and nonworkers	Hypervirulent strains of *Clostridium difficile* are prevalent in healthcare workers.	Friedman et al. (2013)
	Microbial diversity checked within COVID 19 frontline workers and nonworkers	High number of *Bacteroides eggerthii* and less number of *Eubacterium hallii*.	Gao et al. (2022)
	Medical workers with long time staying at hospitals	Increase in *Bifidobacterium, Butyrivibrio, Clostridiaceae, Clostridium, Dialister, Desulfovibrio, Lactobacillus, Odoribacter, Ruminococcus,* and reduction in *Bacteroides* and *Blautia* number.	Zheng et al. (2021)
Tunnel	Microbial diversity checked before 3 weeks and after 8 hours/day exposure (up to 3 weeks)	Increase in *Bifidobacterium, Romboutsia, Clostridium,* and *Leucobacter* and decrease in the count of Reduced *Faecalibacterium* and *Roseburia* in the lower gut.	Lu et al. (2021)

11.4 RELATION BETWEEN MENTAL HEALTH AND HUMAN MICROBIOME

Mental health is intricately linked with the nervous, immune, and endocrine systems, forming a complex web of interactions. Psycho-neuro-immuno-endocrinology, initially defined in 1936, is the study of the relationship between immunological responses, neurological and endocrine processes, and the psyche (mental) (González-Díaz et al. 2017). These systems communicate through various pathways, including hormonal signals, neurotransmitters, and immune cell communication, impacting mood, behavior, and overall health. The neurological, endocrine, and immune systems can all be greatly impacted by microbes found in the built environment. Early exposure to specific bacteria can affect how these systems develop and work, which can have an impact on mental health and general well-being.

11.4.1 Intestinal Immune System and Microbes

Gut microbial interactions, especially those involving the gut microbiome, are important for neurodevelopment and can affect behavior, brain function, and possibly the etiology of neurodegenerative illnesses. Neurogenerative processes such as myelination, neurogenesis, and the establishment of the blood–brain barrier (BBB) depends on the gut microbiota (Ogbonnaya et al. 2015; Figure 11.1). Research on the gut–brain axis shows that the gut microbiota plays a crucial part in coordinating behavior and brain development, and the immune system is becoming recognized as a key modulator of these relationships. In the central nervous system (CNS), intestinal bacteria influence the development and activity of tissue-resident immune cells. Additionally, microbes affect

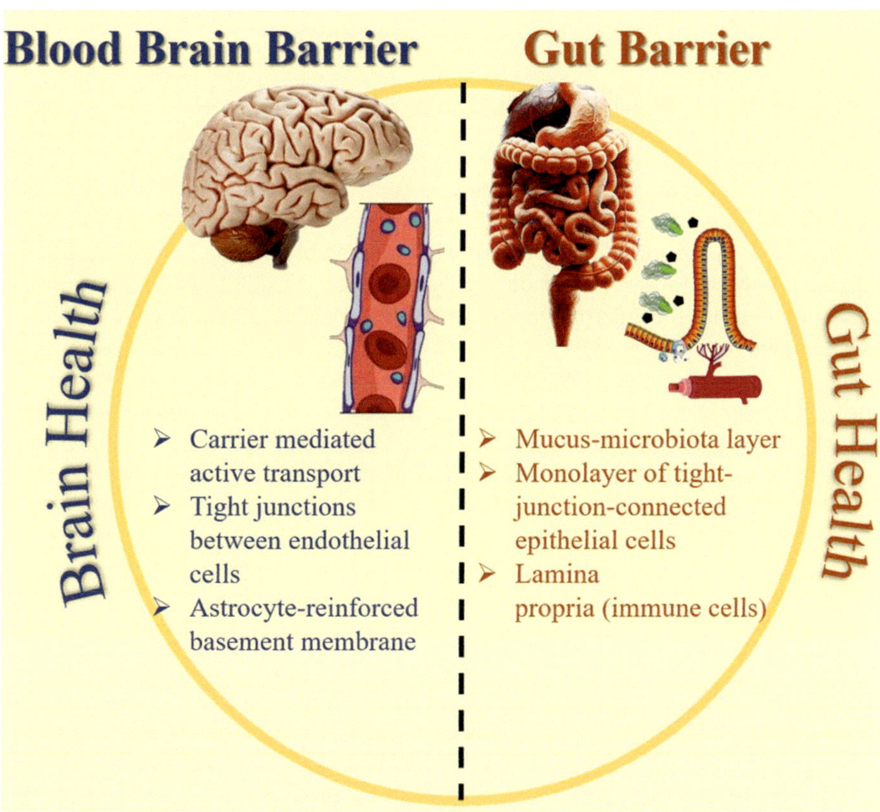

FIGURE 11.1 Two types of biological barrier, blood–brain barrier (BBB) and the gut barrier.

the activation of peripheral immune cells, which control reactions to brain damage, autoimmunity, neurogenesis, and neuroinflammation. Via the vagal and pelvic connections between the CNS, enteric nervous system, and sympathetic prevertebral ganglia, the gastrointestinal tract can regulate microbial density via neurally regulating gut motility, gut permeability, release of gut hormones, and immune function (Tobias and Sadiq 2025). Enterocytes, secretory cells, and gut-associated lymphoid tissue (GALT) make up the majority of the gut's epithelial lining, which is covered with mucosal surface layers (Pott and Hornef 2012). Enterocytes have the ability to release cytokines and chemokines and exhibit innate immune receptors, but the GALT uses lymphocytes to generate an adaptive immune response when necessary. The secretory cells aid in the release of mucus from goblet cells, the regulation of bacterial diversity through the secretion of antimicrobials from Paneth cells, and the regulation of the nervous system through the release of chemicals from enteroendocrine cells, including ghrelin, somatostatin, cholecystokinin, peptide YY, and serotonin. In order to survive in a relatively hostile environment where the host intestinal barrier is constantly under immune monitoring, the gastrointestinal microbiota had to evolve (Rea et al. 2016).

Gut barrier (GB) mediates the communication between the environment and the body in both human and rodent models. The inner left part is the BBB. It is present between the endothelial cells lining the brain's blood arteries. Occludins linked to the cytoskeleton by scaffold proteins, such as zonula occludens (ZO), and specific tight junction proteins, such as claudin-5, offer particular barrier capabilities. The astrocytes, mural cells, pericytes, and microglia form the neurovascular unit, which is a complex network surrounding the BBB. The inner right part is the GB. It is made of a layer of intestinal lumen lining epithelial cells called enterocytes. With the help of claudins, desmosomes, and gap junctions, these cells control the flow of chemicals from the gut lumen to the body. Most importantly, because of its intimate relationship with the gut microbiome, the GB is susceptible to alterations in bacterial activity. The GB also functions as a major signaling hub, interacting with immune cells, mast cells, blood vessels, and enteric neurons

11.4.2 Communication between the Microbiota and the Epithelium

When activated, epithelial pattern recognition receptors can activate and recognize pathogen-associated molecular patterns, which are specific to bacteria and other microorganisms (Duerkop et al. 2009). According to De Souza et al. (2005), continued cytokine production in response to invasive pathogens or as a result of a recurrently high inflammatory reaction can weaken the intestinal barrier and mucosal layer; as a result, increased bacterial infiltration occurs across the gut. This "leaky gut" may let the microbiota and inflammatory tone communicate, which could be the cause of the persistent low-grade inflammation frequently seen in mood disorders like depression. Among the strains of bacteria present in the gut (an average of more than 700) Bacteriodetes and Firmicute phylotypes are predominant (Ley et al. 2008). Many neurotransmitters and neuromodulators can be synthesized and released by these commensal bacteria, or they can stimulate enteroendocrine cells to synthesize and release neuropeptides. Gamma-aminobutyric acid can be produced by *Bifidobacterium* and *Lactobacillus* species: *Candida, Streptococcus, Escherichia*, and *Enterococcus* species can create 5-HT, while *Bacillus, Saccharomyces*, and *Escherichia* species can make norepinephrine. *Lactobacillus* can create acetylcholine, while *Bacillus* can produce dopamine (Lyte 2013). These microbiologically produced neurotransmitters have the ability to penetrate the intestinal mucosa and potentially mediate effects locally within the gut, or they can enter the bloodstream and influence brain physiological processes.

Bacteriocins, bile acids, choline, and SCFAs, including butyric acid, acetic acid, propionic acid, and lactic acid, are among the bioactive substances that gut bacteria can release and are essential to both host health and illness. It has been revealed that these fatty acids have neuroactive qualities, can reach the bloodstream and activate free fatty acid receptors (FFARs), or mediate epigenetic events (Kasubuchi et al. 2015).

11.5 CONSEQUENCES OF MICROBIAL INTERACTION

A healthy, resting inflammatory state is present under homeostatic settings, when the microbiota stimulates the release of cytokines and chemokines, which in turn control the local concentrations of bacteria in the gut. Microbiota can produce and release neurotransmitters and SCFAs, affect the release of neuropeptides and hormones from intestinal enteroendocrine cells, and affect local immune responses at the epithelium. Gut peptides are believed to affect peripheral neuronal connectivity and have the ability to operate centrally to affect behavior. According to current theories, these circulating cytokines, chemokines, endocrine messengers, and microbial byproducts can enter the blood and lymphatic systems or affect the messages sent by the vagal and spinal afferent neurons, which can then affect centrally mediated processes like neuroinflammation and the control of HPA axis activity. These substances can reach the CNS through the choroid plexus, meninges, perivascular space, and parenchymal tissue (Kasubuchi et al. 2015).

11.6 CROSSING THE BLOOD–BRAIN BARRIER

Though the CNS lacks the lymphatic drainage system, the brain possesses it. Yet the BBB, a highly capillarized structure made up of astrocytes and endothelial cells, tightly regulates the migration of immune cells, cytokines, chemokines, endocrine messengers, and microbial byproducts within the brain. Some biological conditions, like stress, infection, or injury, can reduce this barrier's permeability, letting messenger molecules enter and leave and influencing the brain's homeostatic environment. According to Ransohoff and Engelhardt (2012), these chemicals can enter the CNS through the choroid plexus, meninges, perivascular space, and parenchymal tissue. Resident microglia and macrophages always surveil the process (Nayak et al. 2012). These cells are the first to notice changes in their surroundings, and once they become active, they support the brain both locally and by attracting monocytes to the area, activating the neuroinflammation.

11.6.1 Peripheral Influence of Microbiota Regulating Neuroinflammation and HPA Axis

According to new data, the peripheral immune system recruits inflammatory monocytes to the brain through a unique neuroimmune circuit that involves microglia activation and altered sympathetic neuronal tone (Wohleb et al. 2014). Monocyte trafficking from the spleen has a role in mediating several behavioral abnormalities that result from chronic HPA activation. Spleen, an area where FFAR2 receptors are highly concentrated, these FFAR2 receptors are Gi/Go G protein-coupled receptors that are found in lymphocytes. In the absence of a complex microbiota, it has been demonstrated that SCFAs, the natural ligands for FFARs, can counteract the negative effects on microglia density, morphology, and maturity (Erny et al. 2015). When combined, these results imply that the gastrointestinal microbiota may indirectly control centrally mediated events by controlling the movement of monocytes to the brain, which in turn activates microglia and activates the HPA axis, potentially through mechanisms mediated by SCFA. HPA is a system of glands and hormones that controls the body's response to stress. It is the main neuroendocrine system that links stress to physiological changes.

According to the findings, the bacteria that live in our stomach throughout the early stages of life affect our cerebral development via activating microglia. Because of this, disturbances (like stress or antibiotics) to our microbiota during these crucial times of dynamic microbiota–host interactions can significantly change gut–brain signaling, impact health throughout life, and raise the risk of neurodevelopmental disorders. The BE is made up of interactional zones that are marked by mobility and exchange with the surroundings rather than just physical boundaries. These interactions alter, and sociospatial disparities persist on different time and space scales (Figure 11.2).

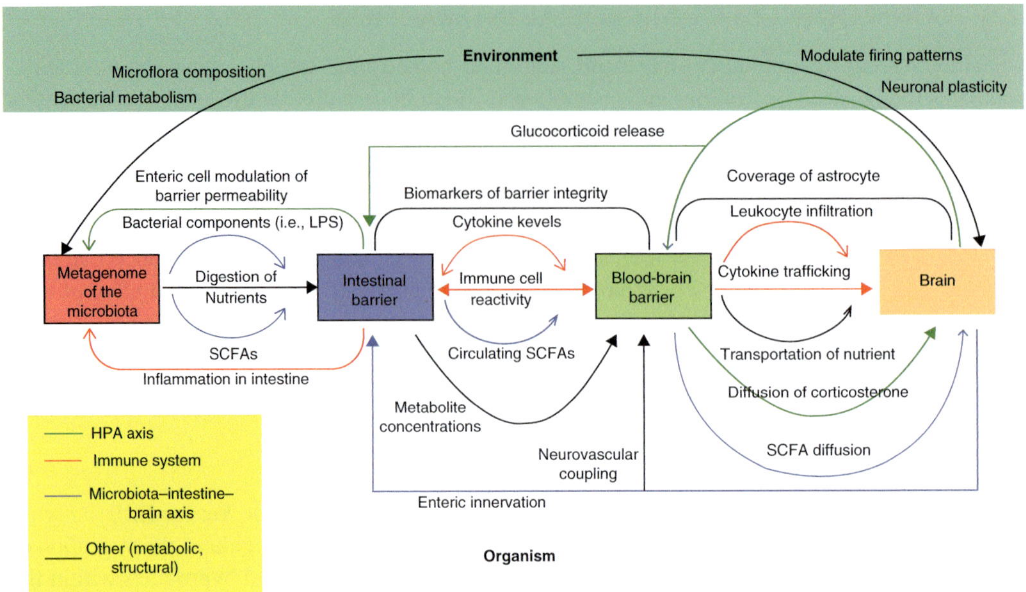

FIGURE 11.2 The channel and communication style that connects barrier communication to the environment (BBB, blood–brain barrier; GB, gut barrier; HPA, hypothalamic–pituitary–adrenal; SCFA, short-chain fatty acid; LPS, lipopolysaccharide; Paton et al. 2023). (Published by Canadian Science Publishing under Open access.)

11.6.2 Modifications of the MoBE to Improve Mental Health

The significance of the BE for microbiomes is not well understood till date. There is a gap in knowledge in this context. With a better understanding of the interactions between the MoBE and the host-associated microbiota of human occupants, research could proceed to the identification of environmental microorganisms that are either beneficial or harmful to the mental health of individuals, thereby mitigating potential diatheses or stressors. From the beginning, human beings have manipulated, modified, and changed their environment for their benefit, survival, protection, convenience, aesthetic purposes, or some other reasons. For getting better health, modification of the built environment is a newer thought. Better mental health conditions might arise from purposeful changes to the built environment that boost microbial biodiversity either by exposure to more immunoregulatory antigens or by introducing probiotics. Furthermore, research on mental health and MoBE logically extends to neurodevelopmental illnesses addressed from a human microbiome perspective, including schizophrenia and autism spectrum disorders. As BEs continue to migrate into less permeable borders between the inside and external, shifting microbial richness and composition may represent a new hazard to human health. Since hospitals' BEs are committed to restoring health, they offer valuable case studies for the built environment–microbiome focus. Aside from decreased diversity, we still don't fully understand what a sick (dysbiotic) microbiome is. Furthermore, totally sterile settings are not ideal for organ function and developmental programs that maintain health and longevity. The concept of "sick building syndrome" is very old (Murphy 2006). Since their creation, buildings have had undesired health effects, and despite human efforts to make them safer and healthier, they frequently simply increase the risks of disease and other health issues. Although the issue is old, there are novel toxicities in materials and systems today. In the built environment, there are a number of potential treatments to lower exposures to harmful bacteria. These interventions include the use of different surface disinfection techniques, raising outdoor ventilation rates and improving air filtration efficiency. Another more recent intervention strategy is to introduce healthy bacteria indoors. Manipulation of physical environments such as local or systemic variations in temperature, ventilation, moisture, and light (particularly ultraviolet [UV]) are commonly included in these.

It has been found that the rate and method of ventilation may influence the composition of bacterial communities in indoor air. A certain amount of indoor–outdoor air exchange occurs in every building, some of it intentional (ventilation) and some of it accidental (infiltration). Natural or hybrid mechanical ventilation methods are better than infiltration because they allow for better ventilation control, the removal of outdoor pollutants and dehumidification of outdoor air, and the potential to lower energy costs by recovering or releasing heat from outgoing air.

A useful measure to get rid of airborne bacteria is the use of particle filtration to purge the air in building HVAC systems. More recently, the health benefits of lowering indoor concentrations of fine and ultrafine particles have been taken into account, which has led to the need for the implementation of higher filter efficiency levels in buildings. Filters are categorized using the so-called Minimum Efficiency Reporting Value (MERV) scale after particle filtration efficiencies are determined by testing using American Society of Heating, Refrigerating and Air-Conditioning Engineers (ASHRAE) Standard 52.2 (ASHRAE, 2017). MERV ratings range from 1 to 20, with higher values corresponding to more effective particle removal (Huttenhower et al. 2012).

UV-A, -B, and -C light exposure, which ranges from 100 to 400 nanometers (nm), can harm living things' DNA and prevent them from replicating, making a cell noninfectious. For almost 50 years, UVGI has been used to disinfect indoor air and surfaces. It is usually produced by low-pressure mercury vapor lamps with spectral power distribution concentrated at 254 nm (Reed 2010). UV light can be used for the sterilization of surfaces as well as indoor air. Upper-level UVGI is typically applied in healthcare settings to prevent the transmission of *Mycobacterium tuberculosis*. The use of mobile UV units in hospitals to disinfect surfaces is a widely accepted technique. Introduction of aerosols containing oxidants and surfactants inside built environments for the express purpose of indoor disinfection, chemicals like ozone, vaporized hydrogen peroxide, chlorine derivatives (vaporized hypochlorous acid and chlorine dioxide), and peracetic acid microdroplets are used. Otherwise, cleaning of surfaces with chemicals also reduces the microbial load of any BE.

Most recently, different active agents like metal nanoparticles, quaternary ammonium salts, polyhexamethylene biguanide, triclosan, chitosan, dyes, and regenerable halamine compounds have been used to treat surfaces and textiles against microbes (Hasan et al. 2013). Applying liquid substances, such as biocidal paints, that give surfaces long-lasting antimicrobial action and activating antimicrobial surfaces with metals like copper or silver are two further tactics that need to be verified and studied further for use in the built environment.

More recently, a new concept, "environmental probiotics", has been popularized, which promotes exposure to potentially advantageous microorganisms. The addition of beneficial organisms to the building environment has several advantages. Environmental probiotics may lessen human exposure to infectious or otherwise hazardous agents by preventing opportunistic pathogens from colonizing and growing in the environment. Beneficial bacteria in buildings may serve as a source and modulator of the human microbiome, and the resulting alterations in the human microbiome may prevent or lessen disease; however, these relationships have not yet been well investigated. Cross-sectional epidemiological studies have shown that having pets, particularly dogs, lowers the likelihood of developing allergic sensitization (Ownby et al. 2002). Humans may come into contact with commensal bacteria that are spread by their family members, other people, animals, and the environment less frequently due to modern lifestyle and decreased exposure to outdoor areas, plants, and animals. A tendency toward immunoregulation failure and poorly controlled inflammatory responses in humans may be linked to the rise in certain diseases in affluent, urbanized civilisations. It has been suggested that these immune system malfunctions result from a lack of exposure to species from the environment during growth time (Figure 11.3).

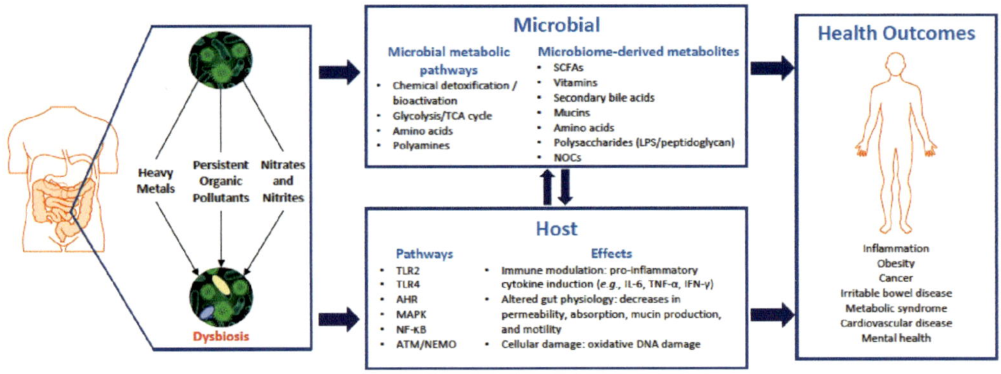

FIGURE 11.3 Chemical pollution's impact on the gut microbiota and inflammatory response (Teffera et al. 2024). (Published by Elsevier under Creative Commons license).

11.7 CONCLUSION

MoBE is one of the critical factors influencing microbiome diversity among human beings. It modulates the immune system function and mental well-being by influencing the host-associated microbiome. In brief, it can be suggested that MoBE can alter the gut microbiota, occupants' immune system, wellness, and mental health. Critical observations are needed to determine the regulatory framework and factors associated with MoBE that not only help in controlling the human microbiota but also make it easier to manage moods, emotions, and mental health for both the general public (prevention) and people with psychiatric diseases (intervention).

REFERENCES

Adams, Rachel I., Ashley C. Bateman, Holly M. Bik, and James F. Meadow. 2015. "Microbiota of the Indoor Environment: A Meta-Analysis." *Microbiome* 3 (1): 49. https://doi.org/10.1186/s40168-015-0108-3.

CDC. 2024. "Built Environment Assessment Tool and Manual." *CDC Physical Activity*, November 7. https://www.cdc.gov/physical-activity/php/built-environment-assessment/index.html.

Chen, Chang-Hua, Ming-Li Liou, Cheng-Yang Lee, Ming-Chuan Chang, Han-Yueh Kuo, and Tzu-Hao Chang. 2019. "Diversity of Nasal Microbiota and Its Interaction with Surface Microbiota among Residents in Healthcare Institutes." *Scientific Reports* 9 (1): 6175. https://doi.org/10.1038/s41598-019-42548-5.

Ciric, Lena. 2022. "Microbes in the Built Environment." *Scientific Reports* 12 (1): 8732. https://doi.org/10.1038/s41598-022-12254-w.

Collins, Stephen M., Michael Surette, and Premysl Bercik. 2012. "The Interplay between the Intestinal Microbiota and the Brain." *Nature Reviews Microbiology* 10 (11): 735–42. https://doi.org/10.1038/nrmicro2876.

Cryan, John F., and Timothy G. Dinan. 2012. "Mind-Altering Microorganisms: The Impact of the Gut Microbiota on Brain and Behaviour." *Nature Reviews Neuroscience* 13 (10): 701–12. https://doi.org/10.1038/nrn3346.

De Souza, Cláudio T., Eliana P. Araujo, Silvana Bordin et al. 2005. "Consumption of a Fat-Rich Diet Activates a Proinflammatory Response and Induces Insulin Resistance in the Hypothalamus." *Endocrinology* 146 (10): 4192–99. https://doi.org/10.1210/en.2004-1520.

Delgado-Ocaña, Susana, and Santiago Cuesta. 2024. "From Microbes to Mind: Germ-Free Models in Neuropsychiatric Research." *mBio* 15 (10): e0207524. https://doi.org/10.1128/mbio.02075-24.

Dockx, Yinthe, Martin Täubel, Esmée M. Bijnens et al. 2022. "Indoor Green Can Modify the Indoor Dust Microbial Communities." *Indoor Air* 32 (3): e13011. https://doi.org/10.1111/ina.13011.

Duerkop, Breck A., Shipra Vaishnava, and Lora V. Hooper. 2009. "Immune Responses to the Microbiota at the Intestinal Mucosal Surface." *Immunity* 31 (3): 368–76. https://doi.org/10.1016/j.immuni.2009.08.009.

Erny, Daniel, Anna Lena Hrabě de Angelis, Diego Jaitin et al. 2015. "Host Microbiota Constantly Control Maturation and Function of Microglia in the CNS." *Nature Neuroscience* 18 (7): 965–77. https://doi.org/10.1038/nn.4030.

Franzosa, Eric A., Katherine Huang, James F. Meadow et al. 2015. "Identifying Personal Microbiomes Using Metagenomic Codes." *Proceedings of the National Academy of Sciences* 112 (22). https://doi.org/10.1073/pnas.1423854112.

Friedman, N. Deborah, James Pollard, Douglas Stupart et al. 2013. "Prevalence of Clostridium Difficile Colonization among Healthcare Workers." *BMC Infectious Diseases* 13 (1): 459. https://doi.org/10.1186/1471-2334-13-459.

Fung, Thomas C., Christine A. Olson, and Elaine Y. Hsiao. 2017. "Interactions between the Microbiota, Immune and Nervous Systems in Health and Disease." *Nature Neuroscience* 20 (2): 145–55. https://doi.org/10.1038/nn.4476.

Gao, Fengjie, Ruijin Guo, Qingyan Ma et al. 2022. "Stressful Events Induce Long-Term Gut Microbiota Dysbiosis and Associated Post-Traumatic Stress Symptoms in Healthcare Workers Fighting against COVID-19." *Journal of Affective Disorders* 303 (April): 187–95. https://doi.org/10.1016/j.jad.2022.02.024.

González, Lina M., Nikita Mukhitov, and Christopher A. Voigt. 2020. "Resilient Living Materials Built by Printing Bacterial Spores." *Nature Chemical Biology* 16 (2): 126–33. https://doi.org/10.1038/s41589-019-0412-5.

González-Díaz, Sandra Nora, Alfredo Arias-Cruz, Bárbara Elizondo-Villarreal, and Olga Patricia Monge-Ortega. 2017. "Psychoneuroimmunoendocrinology: Clinical Implications." *World Allergy Organization Journal* 10 (January). https://doi.org/10.1186/s40413-017-0151-6.

Hasan, Jafar, Russell J. Crawford, and Elena P. Ivanova. 2013. "Antibacterial Surfaces: The Quest for a New Generation of Biomaterials." *Trends in Biotechnology* 31 (5): 295–304. https://doi.org/10.1016/j.tibtech.2013.01.017.

Hayashi, Takuma, and Ikuo Konishi. 2021. "Significant Decrease in Seasonal Influenza in the COVID-19 Era: Impact of Global Movement Restrictions?" *Journal of Clinical Medicine Research* 13 (3): 191–94. https://doi.org/10.14740/jocmr4450 (https://www.jocmr.org/index.php/JOCMR/article/view/4450/25893383).

Hill, Megan S., and Jack A. Gilbert. 2023. "Microbiology of the Built Environment: Harnessing Human-Associated Built Environment Research to Inform the Study and Design of Animal Nests and Enclosures." *Microbiology and Molecular Biology Reviews* 87 (4): e00121–21. https://doi.org/10.1128/mmbr.00121-21.

Hoisington, Andrew J., Lisa A. Brenner, Kerry A. Kinney, Teodor T. Postolache, and Christopher A. Lowry. 2015. "The Microbiome of the Built Environment and Mental Health." *Microbiome* 3 (December): 60. https://doi.org/10.1186/s40168-015-0127-0.

Hou, Kaijian, Zhuo-Xun Wu, Xuan-Yu Chen et al. 2022. "Microbiota in Health and Diseases." *Signal Transduction and Targeted Therapy* 7 (1): 135. https://doi.org/10.1038/s41392-022-00974-4.

Hsu, Tiffany, Regina Joice, Jose Vallarino et al. 2016. "Urban Transit System Microbial Communities Differ by Surface Type and Interaction with Humans and the Environment." *mSystems* 1 (3): e00018–16. https://doi.org/10.1128/mSystems.00018-16.

Hu, Jinglin, Sarah Ben Maamar, Adam J. Glawe, Neil Gottel, Jack A. Gilbert, and Erica M. Hartmann. 2019. "Impacts of Indoor Surface Finishes on Bacterial Viability." *Indoor Air* 29 (4): 551–62. https://doi.org/10.1111/ina.12558.

Huttenhower, Curtis, Dirk Gevers, Rob Knight et al. 2012. "Structure, Function and Diversity of the Healthy Human Microbiome." *Nature* 486 (7402): 207–14. https://doi.org/10.1038/nature11234.

Kasubuchi, Mayu, Sae Hasegawa, Takero Hiramatsu, Atsuhiko Ichimura, and Ikuo Kimura. 2015. "Dietary Gut Microbial Metabolites, Short-Chain Fatty Acids, and Host Metabolic Regulation." *Nutrients* 7 (4): 2839–49. https://doi.org/10.3390/nu7042839.

Kembel, Steven W., Evan Jones, Jeff Kline et al. 2012. "Architectural Design Influences the Diversity and Structure of the Built Environment Microbiome." *The ISME Journal* 6 (8): 1469–79. https://doi.org/10.1038/ismej.2011.211.

Lax, Simon, Naseer Sangwan, Daniel Smith et al. 2017. "Bacterial Colonization and Succession in a Newly Opened Hospital." *Science Translational Medicine* 9 (391): eaah6500. https://doi.org/10.1126/scitranslmed.aah6500.

Lax, Simon, Daniel P. Smith, Jarrad Hampton-Marcell et al. 2014. "Longitudinal Analysis of Microbial Interaction between Humans and the Indoor Environment." *Science* 345 (6200): 1048–52. https://doi.org/10.1126/science.1254529.

Leung, Marcus H. Y., Xinzhao Tong, David Wilkins, Hedwig H. L. Cheung, and Patrick K. H. Lee. 2018. "Individual and Household Attributes Influence the Dynamics of the Personal Skin Microbiota and Its Association Network." *Microbiome* 6 (1): 26. https://doi.org/10.1186/s40168-018-0412-9.

Ley, Ruth E., Micah Hamady, Catherine Lozupone et al. 2008. "Evolution of Mammals and Their Gut Microbes." *Science* 320 (5883): 1647–51. https://doi.org/10.1126/science.1155725.

Lu, Zhen-Hua, Yi-Wen Liu, Zhao-Hua Ji et al. 2021. "Alterations in the Intestinal Microbiome and Mental Health Status of Workers in an Underground Tunnel Environment." *BMC Microbiology* 21 (1): 7. https://doi.org/10.1186/s12866-020-02056-3.

Lyte, Mark. 2013. "Microbial Endocrinology in the Microbiome-Gut-Brain Axis: How Bacterial Production and Utilization of Neurochemicals Influence Behavior." *PLOS Pathogens* 9 (11): e1003726. https://doi.org/10.1371/journal.ppat.1003726.

Marotz, Clarisse A., Jon G. Sanders, Cristal Zuniga, Livia S. Zaramela, Rob Knight, and Karsten Zengler. 2018. "Improving Saliva Shotgun Metagenomics by Chemical Host DNA Depletion." *Microbiome* 6 (1): 42. https://doi.org/10.1186/s40168-018-0426-3.

Meadow, James F., Adam E. Altrichter, Ashley C. Bateman et al. 2015. "Humans Differ in Their Personal Microbial Cloud." *PeerJ* 3 (September): e1258. https://doi.org/10.7717/peerj.1258.

Miller, Frederick W. 2023. "The Increasing Prevalence of Autoimmunity and Autoimmune Diseases: An Urgent Call to Action for Improved Understanding, Diagnosis, Treatment, and Prevention." *Current Opinion in Immunology* 80 (February): 102266. https://doi.org/10.1016/j.coi.2022.102266.

Mortaş, Hande, Saniye Bilici, and Tarkan Karakan. 2020. "The Circadian Disruption of Night Work Alters Gut Microbiota Consistent with Elevated Risk for Future Metabolic and Gastrointestinal Pathology." *Chronobiology International* 37 (7): 1067–81. https://doi.org/10.1080/07420528.2020.1778717.

Murphy, M. 2006. *Sick Building Syndrome and the Problem of Uncertainty: Environmental Politics, Technoscience, and Women Workers*. Duke University Press. https://www.dukeupress.edu/sick-building-syndrome-and-the-problem-of-uncertainty.

National Academies of Sciences, Engineering, and Medicine. 2017. "Microorganisms in Built Environments: Impacts on Human Health." In: *Microbiomes of the Built Environment: A Research Agenda for Indoor Microbiology, Human Health, and Buildings*. National Academies Press. https://www.ncbi.nlm.nih.gov/books/NBK458822/.

Nayak, Debasis, Bernd H. Zinselmeyer, Kara N. Corps, and Dorian B. McGavern. 2012. "In Vivo Dynamics of Innate Immune Sentinels in the CNS." *Intravital* 1 (2): 95–106. https://doi.org/10.4161/intv.22823.

Nygaard, Anders B., Hege S. Tunsjø, Roger Meisal, and Colin Charnock. 2020. "A Preliminary Study on the Potential of Nanopore MinION and Illumina MiSeq 16S rRNA Gene Sequencing to Characterize Building-Dust Microbiomes." *Scientific Reports* 10 (1): 3209. https://doi.org/10.1038/s41598-020-59771-0.

Ogbonnaya, Ebere S., Gerard Clarke, Fergus Shanahan, Timothy G. Dinan, John F. Cryan, and Olivia F. O'Leary. 2015. "Adult Hippocampal Neurogenesis Is Regulated by the Microbiome." *Biological Psychiatry* 78 (4): e7–9. https://doi.org/10.1016/j.biopsych.2014.12.023.

O'Toole, Ronan F. 2021. "The Interface between COVID-19 and Bacterial Healthcare-Associated Infections." *Clinical Microbiology and Infection* 27 (12): 1772–76. https://doi.org/10.1016/j.cmi.2021.06.001.

Ownby, Dennis R., Christine Cole Johnson, and Edward L. Peterson. 2002. "Exposure to Dogs and Cats in the First Year of Life and Risk of Allergic Sensitization at 6 to 7 Years of Age." *JAMA* 288 (8): 963–72. https://doi.org/10.1001/jama.288.8.963.

Paton, Sam E. J., José L. Solano, François Coulombe-Rozon, Manon Lebel, and Caroline Menard. 2023. "Barrier–Environment Interactions along the Gut–Brain Axis and Their Influence on Cognition and Behaviour throughout the Lifespan." *Journal of Psychiatry and Neuroscience* 48 (3): E190–208. https://doi.org/10.1503/jpn.220218.

Pesantez, Gerardo. 2018. *68% of the World Population Projected to Live in Urban Areas by 2050*. United Nation. https://www.un.org/development/desa/en/news/population/2018-revision-of-world-urbanization-prospects.html.

Pott, Johanna, and Mathias Hornef. 2012. "Innate Immune Signalling at the Intestinal Epithelium in Homeostasis and Disease." *EMBO Reports* 13 (8): 684–98. https://doi.org/10.1038/embor.2012.96.

Qian, J., D. Hospodsky, N. Yamamoto, W. W. Nazaroff, and J. Peccia. 2012. "Size-Resolved Emission Rates of Airborne Bacteria and Fungi in an Occupied Classroom." *Indoor Air* 22 (4): 339–51. https://doi.org/10.1111/j.1600-0668.2012.00769.x.

Ransohoff, Richard M., and Britta Engelhardt. 2012. "The Anatomical and Cellular Basis of Immune Surveillance in the Central Nervous System." *Nature Reviews Immunology* 12 (9): 623–35. https://doi.org/10.1038/nri3265.

Rea, Kieran, Timothy G. Dinan, and John F. Cryan. 2016. "The Microbiome: A Key Regulator of Stress and Neuroinflammation." *Neurobiology of Stress and Inflammation*, vol. 4 (October): 23–33. https://doi.org/10.1016/j.ynstr.2016.03.001.

Reed, Nicholas G. 2010. "The History of Ultraviolet Germicidal Irradiation for Air Disinfection." *Public Health Reports* 125 (1): 15–27. https://doi.org/10.1177/003335491012500105.

Rook, Graham. 2012. "A Darwinian View of the Hygiene or 'Old Friends' Hypothesis." *Microbe Magazine* 7 (April): 173–80. https://www.grahamrook.net/OldFriends/oldfriends.html.

Rook, Graham A. W. 2023. "The Old Friends Hypothesis: Evolution, Immunoregulation and Essential Microbial Inputs." *Frontiers in Allergy* 4 (September). https://doi.org/10.3389/falgy.2023.1220481.

Russotto, Vincenzo, Andrea Cortegiani, Santi Maurizio Raineri, and Antonino Giarratano. 2015. "Bacterial Contamination of Inanimate Surfaces and Equipment in the Intensive Care Unit." *Journal of Intensive Care* 3 (1): 54. https://doi.org/10.1186/s40560-015-0120-5.

Sampson, Timothy R., and Sarkis K. Mazmanian. 2015. "Control of Brain Development, Function, and Behavior by the Microbiome." *Cell Host & Microbe* 17 (5): 565–76. https://doi.org/10.1016/j.chom.2015.04.011.

Sharma, Anukriti, and Jack A. Gilbert. 2018. "Microbial Exposure and Human Health." *Current Opinion in Microbiology* 44 (August): 79–87. https://doi.org/10.1016/j.mib.2018.08.003.

Sharon, Gil, Timothy R. Sampson, Daniel H. Geschwind, and Sarkis K. Mazmanian. 2016. "The Central Nervous System and the Gut Microbiome." *Cell* 167 (4): 915–32. https://doi.org/10.1016/j.cell.2016.10.027.

Sharvin, Brendan L., Maria Rodriguez Aburto, and John F. Cryan. 2023. "Decoding the Neurocircuitry of Gut Feelings: Region-Specific Microbiome-Mediated Brain Alterations." *Neurobiology of Disease* 179 (April): 106033. https://doi.org/10.1016/j.nbd.2023.106033.

Stein, Michelle M., Cara L. Hrusch, Justyna Gozdz et al. 2016. "Innate Immunity and Asthma Risk in Amish and Hutterite Farm Children." *New England Journal of Medicine* 375 (5): 411–21. https://doi.org/10.1056/NEJMoa1508749.

Sun, Jian, Xiao-Ping Liao, Alaric W. D'Souza et al. 2020. "Environmental Remodeling of Human Gut Microbiota and Antibiotic Resistome in Livestock Farms." *Nature Communications* 11 (1): 1427. https://doi.org/10.1038/s41467-020-15222-y.

Tanaka, Daisuke, So Fujiyoshi, Fumito Maruyama et al. 2020. "Size Resolved Characteristics of Urban and Suburban Bacterial Bioaerosols in Japan as Assessed by 16S rRNA Amplicon Sequencing." *Scientific Reports* 10 (1): 12406. https://doi.org/10.1038/s41598-020-68933-z (https://www.nature.com/articles/s41598-020-68933-z).

Teffera, Menna, Alex C. Veith, Sean Ronnekleiv-Kelly et al. 2024. "Diverse Mechanisms by Which Chemical Pollutant Exposure Alters Gut Microbiota Metabolism and Inflammation." *Environment International* 190 (August): 108805. https://doi.org/10.1016/j.envint.2024.108805.

Tobias, Abraham, and Nazia M. Sadiq. 2025. *Physiology, Gastrointestinal Nervous Control*. StatPearls Publishing. https://www.ncbi.nlm.nih.gov/books/NBK545268/.

Wang, Weitao, You Weng, Ting Luo, Qiang Wang, Guiling Yang, and Yuanxiang Jin. 2023. "Antimicrobial and the Resistances in the Environment: Ecological and Health Risks, Influencing Factors, and Mitigation Strategies." *Toxics* 11 (2): 185. https://doi.org/10.3390/toxics11020185.

WHO. 2025. *Noncommunicable Diseases*. World Health Organization. https://www.who.int/news-room/fact-sheets/detail/noncommunicable-diseases.

Wohleb, Eric S., Daniel B. McKim, Daniel T. Shea et al. 2014. "Re-Establishment of Anxiety in Stress-Sensitized Mice Is Caused by Monocyte Trafficking from the Spleen to the Brain." *Biological Psychiatry* 75 (12): 970–81. https://doi.org/10.1016/j.biopsych.2013.11.029.

Zheng, Ning, Sheng-Hui Li, Bo Dong et al. 2021. "Comparison of the Gut Microbiota of Short-Term and Long-Term Medical Workers and Non-Medical Controls: A Cross-Sectional Analysis." *Clinical Microbiology and Infection* 27 (9): 1285–92. https://doi.org/10.1016/j.cmi.2020.10.033.

Zhou, Jia-Cheng, Yi-Fei Wang, Dong Zhu, and Yong-Guan Zhu. 2023. "Deciphering the Distribution of Microbial Communities and Potential Pathogens in the Household Dust." *Science of The Total Environment* 872 (May): 162250. https://doi.org/10.1016/j.scitotenv.2023.162250.

12 Traditional Foods
Global Coverage and Processing Methods

Deepak Kumar, Sakshi, Ritik Kumar, and Anil Kumar

12.1 INTRODUCTION

Traditional food products (TFPs) express historical and geographically distinct culture, identity, and history (Galli 2018). Traditional cuisine must be tied to a place and a custom to last. Traditional foods are defined by the European Commission as those passed down through generations, having a minimum of one human generation of at least 25 years (Guerrero et al. 2009). Country cultures are reflected in land and sea-based cuisine and drinks. Locals have traditionally regarded these dishes as daily or occasional. Traditional foods affect diet, health, and status (Al-Habsi, Al-Khusaibi, and Rahman 2019). Japan's rich eating culture, such as Washoku cuisine, promotes peace and longevity through rice, seasonal vegetables, and fish (Kulawik and Dordević 2022). Miso, a cell wall-bound fermented soybean paste, enhances umami in Japanese dishes by adding heat or sweetness. Indian cuisine relies on rice, wheat, and fermented dairy products including butter, curd, and milk (Satish Kumar et al. 2013). Lightly roasting turmeric improves flavor and health. Contemporary fine dining uses polenta and "broa" to highlight cultural history and vegetarian or plant-based alternatives (Jiménez-Fernández et al. 2023). Squash blossoms and prickly pear cactii embellish Mexican dishes, merging indigenous and Spanish traditions with health (Baldivia et al. 2018). Chia seeds and cocoa, rich in omega-3s and polyphenols, highlight Mexico's variety and cuisine. Human diets have historically included fermentation, drying, roasting, malting, and salting (Hotz and Gibson 2007). These procedures improve food's taste, shelf life, and nutrition. Traditional foods' mental health benefits are getting attention (Aslam et al. 2020). Through the gut–brain axis, fermentable foods improve mental health (Selhub, Logan, and Bested 2014).

12.2 TRADITIONAL FOODS

The ancient civilization across the globe is rich in traditional foods with enormous diversity in preparation and component selection. East and Southeast Asia use filamentous fungi like *Aspergillus* to malt rice and beans in their warm, humid climate, while Western countries use malt and fruit juices. Table 12.1 summarizes different fermented foods in diverse geographical domains:

12.2.1 Traditional Foods of Japan

Koji, a rice, wheat, and soybean mold, ferments Japanese food. Koji makes miso, soy sauce, vinegar, mirin, pickles, sake, and spirits (Kusumoto et al. 2021). Japan appreciates sushi, similar to the French croissant, for its raw fish eaten with vinegar rice or sashimi (thinly sliced raw fish over ice) (Kulawik and Dordević 2022). Miso, created from rice, barley, and soy, makes cooking tasty. Rice miso is fermented and seasoned with *Zygosaccharomyces rouxii* and *Tetragenococcus halophilus*, where the cooked soybeans are processed with rice koji (Kusumoto et al. 2021). The Ministry of Foreign Affairs recognizes Washoku as part of Japan's cultural inheritance, while the Ministry of Education and Culture promotes Japanese values and spirit in cultural events abroad. Washoku

DOI: 10.1201/9781003610274-15

TABLE 12.1
Traditional Foods of Different Countries

S. No.	Countries	Traditional Foods or Beverages	References
1.	Japan	Miso, sushi, tofu, tempura	Saeed et al. (2022), Nawa, Hatz, and Blum (2005)
2.	India	Idli, dosa, ambali, dhokla, selroti	Sarkar et al. (2015)
3.	Mexico	Tepache, piznate, tejuino, taberna	Ojeda-Linares et al. (2021)
4.	Africa	Kilishi, banda or kundi, ndariko and jirge	Aworh (2023)
5.	China	Green bean dessert, barbecue pork bun, red-bean dessert, moon cakes	Chen (2010)
6.	Germany	Bruhwurst, kochwurst, sauerkraut	Lücke and Zangerl (2014)
7.	Greece	Leek sausages, chickenpea soup, Cherry tomato of Santorini	Skalkos et al. (2021)
8.	Belgium	Belgium waffles, meat loaf, meat balls	Trichopoulou, Soukara, and Vasilopoulou (2007)

keeps families close by having shared mealtimes, practicing cultural traditions like chopstick use and conscious eating, and avoiding modern eating patterns by using fresh, minimally processed raw materials and low-calorie intake, which contributes to Japan's low BMI and high life expectancy. Therefore, UNESCO recognizes Japan as an intangible cultural heritage.

12.2.2 Traditional Foods of India

Traditional Indian diets contain antioxidants, probiotics, and fiber. They supply critical nutrients, immunity boosters, weight control, and regional and traditional Indian cuisine ingredients. Fermented idli, dosa, and dhokla promote good bacterial growth and have antibacterial, cancer-fighting properties (Satish Kumar et al. 2011). Indians make puffed rice by soaking, boiling, and roasting rice in hot sandy conditions. Grain expansion occurs due to internal steam pressure, with amylose, moisture, and heat parameters affecting product lightness and crispiness (Kumar and Prasad 2018). Millets, such as sorghum, pearl millet, and finger millet, are dried at a high temperature that preserves proteins, vitamins, and minerals while also improving their digestibility and texture. During the process of milling, the grains develop a kind of puff and expand (Choudhury, Das, and Baroova 2011). Indian food has historically used turmeric, which has a unique flavor and nutrient and is used as an antioxidant, anti-atherosclerotic, tumor-blocking, and natural remedy. Traditional medicine has used it to treat excessive cholesterol, blood circulation, urinary tract infections, Alzheimer's, and cancer (Mata et al. 2021).

12.2.3 Traditional Foods of Mexico

Over 100 natural species provide tastes, nutrients, and biological components, making Mexican food flowers the most diverse. History confirms edible flowers' culinary and health benefits. Baking or boiling edible flowers in Mexican cuisine helps improve regional nutrition (Mulík and Ozuna 2020). Sweets, soups, and tamales use squash blossoms. *Erythrina americana* coral tree blossoms, boiled to eliminate toxins, are eaten during Lent for their medicinal properties (Oliva Hernández, López Herrera, and Castillo Linares 2020). The funeral tree blooms (*Quararibea funebris*) are used to prepare Mexican-inspired drinks and pastries (Jiménez-Fernández et al. 2023). Research reveals that corn, fruit, and vegetable diets boost metabolism. Nutrigenetics and nutrigenomics research focus on how food bioactives impact gene expression and metabolism (Mulík et al. 2022). Prickly pear (*Opuntia ficus-indica*), a cactus native to Mexico, is among the indispensable fruit, seed, and pad varieties that are enriched with different vitamins, antioxidants, fiber, and essential fatty acids. These studies found inconsistent glucose and insulin effects. Ancient Mesoamerican pseudo-cereal chia (*Salvia hispanica*) controls glucose and weight with omega-3s, proteins, flavonoids, and minerals (Baldivia et al. 2018). Cocoa polyphenols like epicatechin may benefit cardiovascular health, insulin sensitivity, and glucose management (Sojecka, Drozd-Rzoska, and Rzoska 2024).

12.2.4 Continental

Cultural and historical roots formed the basis for traditional foods to be of paramount importance when considering health advantages. The Mediterranean diet, for instance, emphasizes the use of plant-based foods, olive oil, and moderate wine consumption. As humanity becomes increasingly interconnected due to globalization, dietary habits are changing. This endeavor recognizes that traditional foods are indispensable for health and the preservation of cultural heritage, despite regional dietary shifts (Pasqualone et al. 2022). Maize has dominated European cuisine for years. Maize-based "broa" bread is popular in Portugal. Portugal and others may make wheat and rye bread with 2 g of amino acids from maize cultivar GP-1183 (Bento-Silva et al. 2021). Multiple countries used maize as a currency, altering eating habits, farming, cuisine, and culture. Basque Country maize displaced millet and other grains, making talo a mainstay in the harsh climate. Canary Islands high-energy-density foods include berber-influenced roasted maize flour gofio with milk and frangollo (Revilla et al. 2022). Portuguese "broa" contains approximately 50% maize flour, wheat or rye, and proteins. This combo's taste and texture appeal to healthy bakery clients. In the mid-20th century, wheat displaced maize as Southeastern Europe's bread grain. Heat-treated maize maintains baked goods' moisture without enhancing baking quality. Corn flour and salty water make Italian polenta creamy, eaten with butter, cheese, sausages, and mushrooms (Motti et al. 2020).

12.3 PROCESSING METHODS OF TRADITIONAL FOODS

Food preservation relies on food formulation and processing technology to give millions of people healthy, safe, affordable, tasty, and sustainable food (Forde and Decker 2022). Some classic food processing methods are as follows.

12.3.1 Fermentation

Pioneering biotechnology focuses on indigenous microbes in natural food sources for fermentation. First, microbiological stability was improved to enhance fruit and vegetable-based food shelf life at ambient temperatures (Paul Ross, Morgan, and Hill 2002). Fermented meals are foods or beverages manufactured through regulated microbial growth and the transformation of food components through enzymatic action. Historically, a wide variety of foods, such as meat, fish, dairy products, vegetables, soybeans, other legumes, grains, and fruits, have experienced fermentation. Numerous factors, such as the microbes, nutritional components, and ambient circumstances, influence the fermentation process, resulting in thousands of distinct types of fermented foods (Marco et al. 2017). Vitamins, prebiotics/probiotics, increased digestibility, food security (allelopathic activity against harmful microbial contamination), and new fragrances, taste, and texture were immediately discovered in food fermentation. Fermenting plant and animal sources with bacteria, yeasts, and fungi has produced great food for millennia. *Neurospora, Bacillus, Mucor, Aspergillus, Actinomucor, Rhizopus, Monascus, Saccharomyces, Acetobacter*, and *Lactobacillus* ferment traditionally. Boring vegetarian diets like rice need flavor. In the Orient, fermented fish is a popular cuisine and might have been the first fermentation-based product. In vegetarian diets that consist of bland foods like rice, flavor is particularly crucial. The most popular and well-known fermentation in Oriental cuisine is shoyu, which is also used extensively as a seasoning. The world is home to various traditional fermented foods and beverages, including yogurt, cheeses, fermented sausages, sourdough bread, soy sauce, fish sauce, and fermented vegetables, such as "miso" (fermented soybeans), "kimchi" (fermented spicy cabbage), "sauerkraut" (fermented cabbage), and "surströmming" (fermented herring), as well as "kefir" (fermented milk), "kombucha" (fermented tea), beer, wine, or "kvass".

12.3.2 Drying

Drying is used in direct preparation and processing. The conservation strategy is used to provide year-round food and medication availability. Drying solids involves heating liquid and solid materials to eliminate water. Hot-drying fresh food involves many steps. Fresh materials are energized by heat drying. The cellular structure's water particles migrate to the product's surface and evaporate when free water evaporates (Calín-Sánchez et al. 2020). Fresh food with less moisture decreases bacterial growth and extends shelf life. Drying affects microbial growth, sensory factors, and enzyme activity (Özbek and Dadali 2007). To reduce weight and volume, fruits, vegetables, and herbs are dried to save on packing, storing, and shipping. Drying may provide novel healthy snacking and commercial items (Sehrawat, Nema, and Kaur 2018).

12.3.3 Malting

Nowadays, it is acknowledged that whole grains constitute a significant source of fiber as well as other nutrients like vitamins and trace minerals. It has been demonstrated that grains can help prevent and lower the risk of obesity, cancer, coronary heart disease, and Type 2 diabetes (Misra et al. 2008; Baranwal 2017). Soaking, malting, fermenting, and roasting increase plant-based nutrient bioavailability at home. Malting improves plant-based diet macro- and micronutrient bioavailability (Cornwell 2004; Nkundabombi, Nakimbugwe, and Muyonga 2016). Anti-nutritional factor decrease is the fundamental reason malting promotes bioavailability (Alowo, Muggaga, and Ongeng 2018). Steeping, germination, and drying are the primary malting steps. The kernels steep to absorb water and eliminate contaminants. Germination breaks proteins and carbohydrates into amino acids and sugars, causing enzyme synthesis. Drying stops kernel growth and lowers moisture and because of its enzymes, malted flour is utilized in various foods, especially baby formulae and weaning meals (Baranwal 2017).

12.3.4 Salting

People have historically preserved food using various techniques, including smoking, chilling, and salting, which is among the oldest (Sidhi et al. 2018). Salting alters texture, color, flavor, and aroma. Salt prevents food spoilage by being bacteriostatic (Zobell, Anderson, and Smith 1937). Food preservation involves dry, wet, or mixed salting. Dry salting fermented foods preserves them (Agirman and Erten 2018). Saturating cheese and pickles with salt is wet salting. Salt softens vegetables and ferments lactic acid bacteria. Salt inhibits hazardous germs but boosts dairy lactic acid bacteria (Nurmilah et al. 2022). It affects taste, texture, viscosity, and water-binding. Salt affects meat flavor, hardness, color, water retention, and microbiological growth (Nurmilah et al. 2022).

12.4 FOOD PRESERVATION

Several methods have been devised for food preservation with the main aim to secure the nutrient quality and composition. The following methods have been exploited in the past as well as in recent times for household and commercial purposes (Figure 12.1).

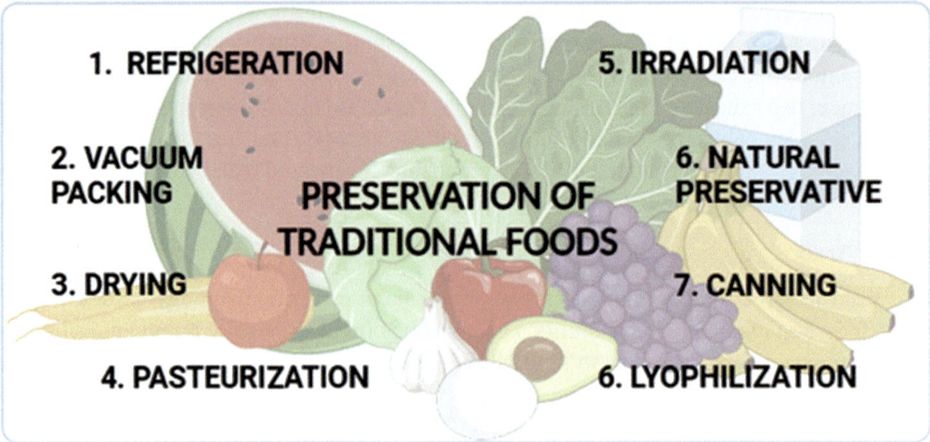

FIGURE 12.1 Preservation methods.

12.4.1 Refrigeration

To preserve nutrients and limit microbial activity, food must be refrigerated. Technologies aim to keep food cold and preserve quality (Aste, Del Pero, and Leonforte 2017). The Potler effect reduces freezer mobility. Strength, portability, and cheap maintenance make ancient and modern refrigerators ideal for off-grid or humanitarian use. Natural cooling systems must be more efficient than conventional freezers (He et al. 2017). Food freezes, hydrogen is produced, and closed-loop energy consumption is activated (Pina, Lozano, and Serra 2018). Adsorption and absorption in open- and closed-cycle refrigeration save energy. This green technology can enable off-grid air cooling, freezing, and refrigeration in remote places without electricity (Sarbu and Sebarchievici in 2015). Conservation, environmental friendliness, or solar radiation and low-grade heat substrates cool efficiently and cheaply. Solid-gas cycles' low thermal conductivity and advanced heat recovery design enable food preservation, pharmaceutical storage, and precise temperature control (Gao et al. 2018). To stay fresh and safe from storage to consumption, food needs refrigeration. For food preservation, refrigeration technology improves thermal stability, energy efficiency, and waste reduction.

12.4.2 Pasteurization

Pasteurization, created by Louis Pasteur, warms food to kill harmful bacteria and enzymes, extending its shelf life without affecting its chemical or physical properties. Food pH impacts pathogen removal in low-acid foods (pH>4.5) and spoilage prevention in acidic meals (pH<4.5). Certain carbonated and still juice methods lower infection risk and improve product safety. It also stops spore-forming microbes and heat-resistant enzymes like peroxidase from deteriorating acidic and canned foods (Amit et al. 2017). High-Temperature Short-Time (HTST) pasteurization destroys pathogens, while Ultra-High-Temperature (UHT) kills heat-resistant spores and improves shelf life through aseptic packaging (Arslanoglu et al. 2019). Pasteurization might lower vitamins and minerals, but food safety is worth it. Pasteurization vats heat and chill raw feedstock after a certain time. Some systems partially heat and cool the pasteurizer during filling and chilling. These are slow and need continual hand care to avoid overheating. They have poor regenerative heating, making heating and cooling more expensive, but they can be connected with many vats to maintain process continuity at staff expense (Arslanoglu et al. 2019). HTST pasteurizers use plate heat exchangers and continuous flow. These pasteurize uniformly and more efficiently than batch systems using regenerative heating and cooling (Indumathy et al. 2022). Metering pumps, pressure switches, and centrifugal booster pumps regulate flow and pressure, a 7 kPa pressure difference between both sides of the pasteurizer, and diverting underheated products to custom make them or ensure an accurate holding time to avoid undercooking.

12.4.3 Irradiation

Soft food preservation solutions address foodborne illnesses and high post-harvest losses. Cold plasma, high-pressure processing, and pulsed electric field replace energy-intensive, low-quality procedures (Delorme et al. 2020). Not all foods inactivate spores efficiently. Food irradiation by UV light, gamma rays, and electron beams deep into foods is the most effective way to kill viruses while preserving their appearance, flavor, and nutrition. Food taste, nutrition, and safety are preserved by UV/gamma radiation (Nair and Sharma 2016). Gamma irradiation increases fruit and vegetable quality, storability, and nutrition. According to Maraei and Elsawy (2017), white onions, green bell peppers, and tomatoes kept freshness, vitamins, and soluble fiber during cold storage when irradiated for up to 5 kGy, with 2 kGy being more effective. Gamma, X-ray, and UV-C radiation inactivate microbial Domain Name System (DNS) without altering food nutrition or flavor. To extend the shelf life of agricultural produce, UV-C wavelengths lower microbial loads and protect antioxidants. Maraei and Elsawy (2017) suggest that ionizing radiation sterilizes most bulky products. Safe, efficient, and scalable food processing may benefit from UV-LED microbial control. Gamma and X-ray irradiation are state-of-the-art methods of food preservation, facilitating microbial reduction, prolonged shelf life, and the retention of sensory and nutritional viability in fruits and vegetables (Ashtari et al. 2019). Gamma irradiation preserves sugar, inhibits degradation, and blocks fungal growth. High-mitigation bacteria die without radiation from X-rays (Park and Ha 2019). Many of their approaches increase food preservation system safety, post-harvest waste, and customer appeal (Zhang et al. 2020).

12.4.4 Natural Preservatives

Streptomyces natalensis' FDA-approved, nonsynthetic preservative natamycin prevents mycotoxins and fungal pollution (Meena et al. 2021). Natural antifungal preservative natamycin kills mold and yeast in cheese and yogurt to improve shelf life without affecting taste. Preservatives are sprayed, dipped, or coated with finely divided alginate or zein films to meet regulatory requirements and avoid spoilage, mycotoxin, and microbiological contamination (Lee and Paik 2016). Natamycin prolongs the shelf life and flavor of strawberries, apples, grapes, mushrooms, olives, and mulberries by preventing mold, spore germination, and fruit rot. Natamycin inhibits mold and yeast growth, extending the shelf life of baked goods without compromising taste or texture (Tsiraki et al. 2018). Plant, animal, and microorganism antimicrobial and antioxidant chemicals reduce microbial growth to prevent spoilage, infections, and synthetic preservatives (Hintz, Matthews, and Di 2015). Because they preserve food without altering taste or nutrients, natural preservatives are popular. Mint, cinnamon, and anise oils inhibit bacterial growth, improving food shelf life and quality (El-Saber Batiha et al. 2021). Chitosan, lysozyme in chicken and cow milk, and lactoferrin in animal-produced products restrict bacteria growth, reduce water, and break the bacterial cell wall in solid color foods like fruit and dairy during storage. Phenolic substances, vitamins C and E, and antioxidants delay food oxidative rancidity, degradation, and discoloration. They neutralize free radicals and protect fresh fruit, poultry, meat, and fish from oxidation. Honey and chitosan reduce lipid oxidation and food spoilage by removing free radicals. Honey contains phenolic acids, flavonoids, and vitamins, while crustacean chitosan reduces food oxidation (Himanshu and Khangwal 2024). Antioxidants from spices, citrus peels, and oil seeds protect food. Flavonoids, polyphenols, terpenoids, and essential oils prevent lipid oxidation and browning in meats, fish, fruits, and vegetables.

12.4.5 Lyophilization

Lyophilization preserves food by freezing it and then reducing pressure, which causes the water to go straight from the solid phase to the gaseous phase. Oven drying, while effective for small batches, is slower, consumes more energy, and is not suitable for large-scale preservation (Nowak and Jakubczyk 2020). Receiving and preparing raw materials, blanching, grinding, combining with preservatives and seasonings, packing in vacuum cans, sealing, heat processing, and cooling are all important steps to ensure food safety and quality. Canned meats are pickled, heat processed, and pressure cooked for safety, flavor, and lifespan, with checks on processing and clear labeling (Rahman et al. 2023).

12.5 ROLE OF TRADITIONAL FOOD IN MENTAL HEALTH

Higher rates of depression and other mental health issues have been linked to changing lifestyles (Colla et al. 2006; Hidaka 2012; Logan and Selhub 2012). Diet may help prevent mental health issues, including depression (Jacka et al. 2013). As our understanding of the human microbiome grows, so does our understanding of its relationships with physiology, including mental health issues like depression and anxiety (Selhub, Logan, and Bested 2014).

12.5.1 Gut–Brain Connection

Depression and anxiety often coincide with gut difficulties, suggesting a bidirectional relationship between gut health and mental well-being (Aslam et al. 2020). Fermented foods, which may be healthy, are found in many traditional diets including the Mediterranean diet (Jacka et al. 2010). Fermentation involves adding or naturally occurring bacteria to food. Fermented foods contain probiotics, prebiotics, and biogenics. Fermented foods like yogurt, soy sauce, cheese, miso, kimchi, kefir, tempeh, kombucha, and sauerkraut contain different levels of these three beneficial components. The gut flora, macronutrient digestion and absorption, gut permeability, and gut immune cells can be affected by any or all of the above factors (Rao et al. 2009). Fermented foods also have anti-inflammatory, immunomodulatory, and brain-modulatory properties (Choi et al. 2008). Thus, fermented foods may alleviate sadness and anxiety by altering the molecular pathways that cause them (Aslam et al. 2020).

12.5.2 Nutritional Value

Dysbiosis is a condition defined by an imbalance in gut bacteria, metabolic abnormalities, or gastrointestinal tract bacteria dispersion. The typical Western diet lacks fiber and prebiotics needed to develop and diversify gut flora. Instead, the diet is high in sugar, fat, and salt, which can promote dangerous germs and fungi. This can enhance intestinal permeability, inflammation, and bacterial endotoxins such as LPS entering blood vessels (Hrncir 2022). Endogenous hormones, neuropeptides, neurotransmitters, the microbiota–gut–brain axis, and brain composition, structure, and function are affected by diet. These are essential for stress reduction, inflammation reduction, and cognitive performance (Muscaritoli 2021). Traditional meals are more than just cultural; they provide critical nutrients. These foods are rich in vitamins, minerals, antioxidants, and fiber, which improve health.

12.6 CONCLUSION

Traditional food processing and preservation methods have preserved cultures and food security worldwide. Traditional procedures including fermenting, drying, malting, and salting improve food's flavor, shelf life, and nutrition. Modern advances like pasteurization, vacuum packaging, and refrigeration ensure food safety and quality. Growing understanding of the link between traditional foods and mental health highlights the importance of diet for overall health, especially through the gut–brain connection. Fermented foods boost gut flora, making them a good mental health strategy. Maintaining and respecting historic culinary practices while using modern preservation methods can nourish future generations physically and psychologically.

REFERENCES

Agirman, Bilal, and Huseyin Erten. 2018. "The Influence of Various Chloride Salts to Reduce Sodium Content on the Quality Parameters of Şalgam (Shalgam): A Traditional Turkish Beverage Based on Black Carrot." *Journal of Food Quality* 2018 (June): 1–11. doi:10.1155/2018/3292185.

Al-Habsi, Nasser, Mohammed Al-Khusaibi, and Mohammad Shafiur Rahman. 2019. "GCC Traditional Foods: Preparation and Processing." In: *Traditional Foods: History, Preparation, Processing and Safety*, edited by Mohammed Khalfan Al-Khusaibi, Nasser Abdullah Mohammed Al-Habsi, and Mohammad Shafiur Rahman, 37–49. Cham, Switzerland: Springer Nature. doi:10.1007/978-3-030-24620-4_3.

Alowo, Docus, Christopher Muggaga, and Duncan Ongeng. 2018. "The Effect of Traditional Malting Technology Practiced by an Ethnic Community in Northern Uganda on In-vitro Nutrient Bioavailability and Consumer Sensory Preference for Locally Formulated Complementary Food Formulae." *Food Science & Nutrition* 6 (8): 2491–98. doi:10.1002/fsn3.856.

Amit, Sadat Kamal, Md. Mezbah Uddin, Rizwanur Rahman, S. M. Rezwanul Islam, and Mohidus Samad Khan. 2017. "A Review on Mechanisms and Commercial Aspects of Food Preservation and Processing." *Agriculture & Food Security* 6 (1): 51. doi:10.1186/s40066-017-0130-8.

Arslanoglu, Sertac, Clair-Yves Boquien, Caroline King, Delphine Lamireau, Paola Tonetto, Debbie Barnett, Enrico Bertino et al. 2019. "Fortification of Human Milk for Preterm Infants: Update and Recommendations of the European Milk Bank Association (EMBA) Working Group on Human Milk Fortification." *Frontiers in Pediatrics* 7 (March). doi:10.3389/fped.2019.00076.

Ashtari, Meysam, Orang Khademi, Mahmoud Soufbaf, Hamideh Afsharmanesh, and Mohammad Ali Askari Sarcheshmeh. 2019. "Effect of Gamma Irradiation on Antioxidants, Microbiological Properties and Shelf Life of Pomegranate Arils Cv. 'Malas Saveh'." *Scientia Horticulturae* 244 (January): 365–71. doi:10.1016/j.scienta.2018.09.067.

Aslam, Hajara, Jessica Green, Felice N. Jacka, Fiona Collier, Michael Berk, Julie Pasco, and Samantha L. Dawson. 2020. "Fermented Foods, the Gut and Mental Health: A Mechanistic Overview with Implications for Depression and Anxiety." *Nutritional Neuroscience* 23 (9): 659–71. doi:10.1080/1028415X.2018.1544332.

Aste, Niccolò, Claudio Del Pero, and Fabrizio Leonforte. 2017. "Active Refrigeration Technologies for Food Preservation in Humanitarian Context: A Review." *Sustainable Energy Technologies and Assessments* 22 (August): 150–60. doi:10.1016/j.seta.2017.02.014.

Aworh, Ogugua Charles. 2023. "African Traditional Foods and Sustainable Food Security." *Food Control* 145 (March): 109393. doi:10.1016/j.foodcont.2022.109393.

Baldivia, Anacleto Sosa, Guadalupe Ruiz-Ibarra, Raúl René Robles de la Torre, Reyna Robles López, and Aurora Montufar López. 2018. "The Chia (Salvia Hispanica): Past, Present and Future of an Ancient Mexican Crop." *Australian Journal of Crop Science* 12 (10): 1626–32. doi:10.21475/ajcs.18.12.10.p1202.

Baranwal, Deepika. 2017. "Malting: An Indigenous Technology Used for Improving the Nutritional Quality of Grains- A Review." *Asian Journal of Dairy and Food Research* 36 (3). doi:10.18805/ajdfr.v36i03.8960.

Bento-Silva, Andreia, Noélia Duarte, Elsa Mecha, Maria Belo, Ana Teresa Serra, Maria Carlota Vaz Patto, and Maria Rosário Bronze. 2021. "Broa, an Ethnic Maize Bread, as a Source of Phenolic Compounds." *Antioxidants* 10 (5): 672. doi:10.3390/antiox10050672.

Bourrie, Benjamin C. T., Benjamin P. Willing, and Paul D. Cotter. 2016. "The Microbiota and Health Promoting Characteristics of the Fermented Beverage Kefir." *Frontiers in Microbiology* 7 (May). doi:10.3389/fmicb.2016.00647.

Calín-Sánchez, Ángel, Leontina Lipan, Marina Cano-Lamadrid, Abdolreza Kharaghani, Klaudia Masztalerz, Ángel A. Carbonell-Barrachina, and Adam Figiel. 2020. "Comparison of Traditional and Novel Drying Techniques and Its Effect on Quality of Fruits, Vegetables and Aromatic Herbs." *Foods* 9 (9): 1261. doi:10.3390/foods9091261.

Chen, Ya-Jun. 2010. "Glycemic Index and Glycemic Load of Selected Chinese Traditional Foods." *World Journal of Gastroenterology* 16 (12): 1512. doi:10.3748/wjg.v16.i12.1512.

Chilton, Stephanie, Jeremy Burton, and Gregor Reid. 2015. "Inclusion of Fermented Foods in Food Guides around the World." *Nutrients* 7 (1): 390–404. doi:10.3390/nu7010390.

Chizoba Ekezie, Flora-Glad, Jun-Hu Cheng, and Da-Wen Sun. 2018. "Effects of Nonthermal Food Processing Technologies on Food Allergens: A Review of Recent Research Advances." *Trends in Food Science & Technology* 74 (April): 12–25. doi:10.1016/j.tifs.2018.01.007.

Choi, Yong Hwan, Hyun Lim, Moon Young Heo, Dae Young Kwon, and Hyun Pyo Kim. 2008. "Anti-Inflammatory Activity of the Ethanol Extract of Chungkukjang, Korean Fermented Bean: 5-Lipoxygenase Inhibition." *Journal of Medicinal Food* 11 (3): 539–43. doi:10.1089/jmf.2007.0125.

Choudhury, Manisha, Pranati Das, and Basanti Baroova. 2011. "Nutritional Evaluation of Popped and Malted Indigenous Millet of Assam." *Journal of Food Science and Technology* 48 (6): 706–11. doi:10.1007/s13197-010-0157-3.

Colla, Judith, Stephan Buka, David Harrington, and Jane M. Murphy. 2006. "Depression and Modernization." *Social Psychiatry and Psychiatric Epidemiology* 41 (4): 271–79. doi:10.1007/s00127-006-0032-8.

Cornwell, Trudy. 2004. "Dietary Phytoestrogens and Health." *Phytochemistry* 65 (8): 995–1016. doi:10.1016/j.phytochem.2004.03.005.

Coton, Monika, Audrey Pawtowski, Bernard Taminiau, Gaëtan Burgaud, Franck Deniel, Laurent Coulloumme-Labarthe, Abdoulaye Fall, Georges Daube, and Emmanuel Coton. 2017. "Unraveling Microbial Ecology of Industrial-Scale Kombucha Fermentations by Metabarcoding and Culture-Based Methods." *FEMS Microbiology Ecology* 93 (5). doi:10.1093/femsec/fix048.

Delorme, Mariana M., Jonas T. Guimarães, Nathália M. Coutinho, Celso F. Balthazar, Ramon S. Rocha, Ramon Silva, Larissa P. Margalho et al. 2020. "Ultraviolet Radiation: An Interesting Technology to Preserve Quality and Safety of Milk and Dairy Foods." *Trends in Food Science & Technology* 102 (August): 146–54. doi:10.1016/j.tifs.2020.06.001.

El-Saber Batiha, Gaber, Diaa E. Hussein, Abdelazeem M. Algammal, Toyosi T. George, Philippe Jeandet, Ali Esmail Al-Snafi, Achyut Tiwari et al. 2021. "Application of Natural Antimicrobials in Food Preservation: Recent Views." *Food Control* 126 (August): 108066. doi:10.1016/j.foodcont.2021.108066.

Forde, Ciarán G., and Eric A. Decker. 2022. "The Importance of Food Processing and Eating Behavior in Promoting Healthy and Sustainable Diets." *Annual Review of Nutrition* 42 (1): 377–99. doi:10.1146/annurev-nutr-062220-030123.

Galli, Francesca. 2018. "Traditional Food." In *Case Studies in the Traditional Food Sector*, edited by Alessio Cavicchi and Cristina Santini, 3–24. Cambridge, MA: Woodhead Publishing. doi:10.1016/B978-0-08-101007-5.00002-6.

Gao, Jia, Li-Wei Wang, Peng Gao, Guo-Liang An, and He-Ti Lu. 2018. "Design and Analysis of a Gas Heating/Cooling Sorption Refrigeration System with Multi-Salt Solid Sorbent of CaCl2 and MnCl2." *International Journal of Heat and Mass Transfer* 126 (November): 39–47. doi:10.1016/j.ijheatmasstransfer.2018.05.154.

Guerrero, Luis, Maria Dolors Guàrdia, Joan Xicola, Wim Verbeke, Filiep Vanhonacker, Sylwia Zakowska-Biemans, Marta Sajdakowska et al. 2009. "Consumer-Driven Definition of Traditional Food Products and Innovation in Traditional Foods. A Qualitative Cross-Cultural Study." *Appetite* 52 (2): 345–54. doi:10.1016/j.appet.2008.11.008.

He, Chang, Zhanquan Zhang, Boqiang Li, Yong Xu, and Shiping Tian. 2019. "Effect of Natamycin on Botrytis Cinerea and Penicillium Expansum—Postharvest Pathogens of Grape Berries and Jujube Fruit." *Postharvest Biology and Technology* 151 (May): 134–41. doi:10.1016/j.postharvbio.2019.02.009.

He, Rong-Rong, Huai-Yu Zhong, Yang Cai, Di Liu, and Fu-Yun Zhao. 2017. "Theoretical and Experimental Investigations of Thermoelectric Refrigeration Box Used for Medical Service." *Procedia Engineering* 205: 1215–22. doi:10.1016/j.proeng.2017.10.356.

Hidaka, Brandon H. 2012. "Depression as a Disease of Modernity: Explanations for Increasing Prevalence." *Journal of Affective Disorders* 140 (3): 205–14. doi:10.1016/j.jad.2011.12.036.

Himanshu, and Ishu Khangwal. 2024. "Honey as a Potential Preservative in the Food Industry." In *Honey in Food Science and Physiology*, 181–95. Singapore: Springer Nature Singapore. doi:10.1007/978-981-97-3565-5_8.

Hintz, Tana, Karl K. Matthews, and Rong Di. 2015. "The Use of Plant Antimicrobial Compounds for Food Preservation." *BioMed Research International* 2015: 1–12. doi:10.1155/2015/246264.

Hotz, Christine, and Rosalind S. Gibson. 2007. "Traditional Food-Processing and Preparation Practices to Enhance the Bioavailability of Micronutrients in Plant-Based Diets." *The Journal of Nutrition* 137 (4): 1097–1100. doi:10.1093/jn/137.4.1097.

Hrncir, Tomas. 2022. "Gut Microbiota Dysbiosis: Triggers, Consequences, Diagnostic and Therapeutic Options." *Microorganisms* 10 (3): 578. doi:10.3390/microorganisms10030578.

Indumathy, Marimuthu, Sobana Sobana, Bidisha Panda, and Rames C. Panda. 2022. "Modelling and Control of Plate Heat Exchanger with Continuous High-Temperature Short Time Milk Pasteurization Process: A Review." *Chemical Engineering Journal Advances* 11 (August): 100305. doi:10.1016/j.ceja.2022.100305.

Jacka, Felice N., Julie A. Pasco, Arnstein Mykletun, Lana J. Williams, Allison M. Hodge, Sharleen Linette O'Reilly, Geoffrey C. Nicholson, Mark A. Kotowicz, and Michael Berk. 2010. "Association of Western and Traditional Diets With Depression and Anxiety in Women." *American Journal of Psychiatry* 167 (3): 305–11. doi:10.1176/appi.ajp.2009.09060881.

Jacka, Felice N., Eivind Ystrom, Anne Lise Brantsaeter, Evalill Karevold, Christine Roth, Margaretha Haugen, Helle Margrete Meltzer, Synnve Schjolberg, and Michael Berk. 2013. "Maternal and Early Postnatal Nutrition and Mental Health of Offspring by Age 5 Years: A Prospective Cohort Study." *Journal of the American Academy of Child & Adolescent Psychiatry* 52 (10): 1038–47. doi:10.1016/j.jaac.2013.07.002.

Jiménez-Fernández, Maribel, Naida Juárez-Trujillo, María Remedios Mendoza-López, Juan Luis Monribot-Villanueva, and José Antonio Guerrero-Analco. 2023. "Nutraceutical Potential, and Antioxidant and Antibacterial Properties of Quararibea Funebris Flowers." *Food Chemistry* 411 (June): 135529. doi:10.1016/j.foodchem.2023.135529.

Kulawik, Piotr, and Dani Dordević. 2022. "Sushi Processing: Microbiological Hazards and the Use of Emerging Technologies." *Critical Reviews in Food Science and Nutrition* 62 (5): 1270–83. doi:10.1080/10408398.2020.1840332.

Kumar, Shiv, and Kamlesh Prasad. 2018. "Effect of Parboiling and Puffing Processes on the Physicochemical, Functional, Optical, Pasting, Thermal, Textural and Structural Properties of Selected Indica Rice." *Journal of Food Measurement and Characterization* 12 (3): 1707–22. doi:10.1007/s11694-018-9786-4.

Kusumoto, Ken-Ichi, Youhei Yamagata, Rina Tazawa, Manabu Kitagawa, Taeko Kato, Kenji Isobe, and Yutaka Kashiwagi. 2021. "Japanese Traditional Miso and Koji Making." *Journal of Fungi* 7 (7): 579. doi:10.3390/jof7070579.

Lee, Na-Kyoung, and Hyun-Dong Paik. 2016. "Status, Antimicrobial Mechanism, and Regulation of Natural Preservatives in Livestock Food Systems." *Korean Journal for Food Science of Animal Resources* 36 (4): 547–57. doi:10.5851/kosfa.2016.36.4.547.

Logan, Alan C., and Eva M. Selhub. 2012. "Vis Medicatrix Naturae: Does Nature 'Minister to the Mind'?" *BioPsychoSocial Medicine* 6 (1): 11. doi:10.1186/1751-0759-6-11.

Lucera, Annalisa, Cristina Costa, Amalia Conte, and Matteo A. Del Nobile. 2012. "Food Applications of Natural Antimicrobial Compounds." *Frontiers in Microbiology* 3. doi:10.3389/fmicb.2012.00287.

Lücke, Friedrich-Karl, and Peter Zangerl. 2014. "Food Safety Challenges Associated with Traditional Foods in German-Speaking Regions." *Food Control* 43 (September): 217–30. doi:10.1016/j.foodcont.2014.03.014.

Maraei, Rabab W., and Khaled M. Elsawy. 2017. "Chemical Quality and Nutrient Composition of Strawberry Fruits Treated by γ-Irradiation." *Journal of Radiation Research and Applied Sciences* 10 (1): 80–87. doi:10.1016/j.jrras.2016.12.004.

Marco, Maria L., Dustin Heeney, Sylvie Binda, Christopher J. Cifelli, Paul D. Cotter, Benoit Foligné, Michael Gänzle et al. 2017. "Health Benefits of Fermented Foods: Microbiota and Beyond." *Current Opinion in Biotechnology* 44 (April): 94–102. doi:10.1016/j.copbio.2016.11.010.

Mata, Isabella Rosa da, Sheila Rosa da Mata, Rafaella Camara Rocha Menezes, Larissa Slongo Faccioli, Kathleen Krüger Bandeira, and Simone Morelo Dal Bosco. 2021. "Benefits of Turmeric Supplementation for Skin Health in Chronic Diseases: A Systematic Review." *Critical Reviews in Food Science and Nutrition* 61 (20): 3421–35. doi:10.1080/10408398.2020.1798353.

Meena, Mahima, Priyanka Prajapati, Chandrakala Ravichandran, and Rachna Sehrawat. 2021. "Natamycin: A Natural Preservative for Food Applications—A Review." *Food Science and Biotechnology* 30 (12): 1481–96. doi:10.1007/s10068-021-00981-1.

Misra, Anoop, Lokesh Khurana, Sumit Isharwal, and Swati Bhardwaj. 2008. "South Asian Diets and Insulin Resistance." *British Journal of Nutrition* 101 (4): 465–73. doi:10.1017/S0007114508073649.

Motti, Riccardo, Giuliano Bonanomi, Virginia Lanzotti, and Raffaele Sacchi. 2020. "The Contribution of Wild Edible Plants to the Mediterranean Diet: An Ethnobotanical Case Study Along the Coast of Campania (Southern Italy)." *Economic Botany* 74 (3): 249–72. doi:10.1007/s12231-020-09504-1.

Mulík, Stanislav, María Hernández-Carrión, Sara E. Pacheco-Pantoja, Natalia Aguilar-Ruiz, and César Ozuna. 2022. "Culinary Uses of Mexican Edible Flowers: Recipe Analysis." *International Journal of Gastronomy and Food Science* 28 (June): 100539. doi:10.1016/j.ijgfs.2022.100539.

Mulík, Stanislav, and César Ozuna. 2020. "Mexican Edible Flowers: Cultural Background, Traditional Culinary Uses, and Potential Health Benefits." *International Journal of Gastronomy and Food Science* 21 (October): 100235. doi:10.1016/j.ijgfs.2020.100235.

Muscaritoli, Maurizio. 2021. "The Impact of Nutrients on Mental Health and Well-Being: Insights from the Literature." *Frontiers in Nutrition* 8 (March). doi:10.3389/fnut.2021.656290.

Nagao, Fumiko, Masafumi Nakayama, Takashi Muto, and Kookumura. 2000. "Effects of a Fermented Milk Drink Containing *Lactobacillus Casei* Strain Shirota on the Immune System in Healthy Human Subjects." *Bioscience, Biotechnology, and Biochemistry* 64 (12): 2706–8. doi:10.1271/bbb.64.2706.

Nair, P.M., and Arun Sharma. 2016. "Food Irradiation." In Innovative Food Processing Technologies: A Comprehensive Review, edited by Kai Knoerzer and Kai-S. Muthukumarappan, 19–29. Oxford: Elsevier. doi:10.1016/B978-0-12-815781-7.02950-4.

Nawa, Yukifumi, Christoph Hatz, and Johannes Blum. 2005. "Sushi Delights and Parasites: The Risk of Fishborne and Foodborne Parasitic Zoonoses in Asia." *Clinical Infectious Diseases* 41 (9): 1297–1303. doi:10.1086/496920.

Nkundabombi, Marie Grace, Dorothy Nakimbugwe, and John H. Muyonga. 2016. "Effect of Processing Methods on Nutritional, Sensory, and Physicochemical Characteristics of Biofortified Bean Flour." *Food Science & Nutrition* 4 (3): 384–97. doi:10.1002/fsn3.301.

Nowak, Dorota, and Ewa Jakubczyk. 2020. "The Freeze-Drying of Foods—The Characteristic of the Process Course and the Effect of Its Parameters on the Physical Properties of Food Materials." *Foods* 9 (10): 1488. doi:10.3390/foods9101488.

Nurmilah, Siti, Yana Cahyana, Gemilang Lara Utama, and Abderrahmane Aït-Kaddour. 2022. "Strategies to Reduce Salt Content and Its Effect on Food Characteristics and Acceptance: A Review." *Foods* 11 (19): 3120. doi:10.3390/foods11193120.

Ojeda-Linares, César, Gonzalo D. Álvarez-Ríos, Carmen Julia Figueredo-Urbina, Luis Alfredo Islas, Patricia Lappe-Oliveras, Gary Paul Nabhan, Ignacio Torres-García, Mariana Vallejo, and Alejandro Casas. 2021. "Traditional Fermented Beverages of Mexico: A Biocultural Unseen Foodscape." *Foods* 10 (10): 2390. doi:10.3390/foods10102390.

Oliva Hernández, Jorge, María Aurelia López Herrera, and Erika Belem Castillo Linares. 2020. "Composición Química y Producción de Follaje de Erythrina Americana (Fabaceae) En Cercos Vivos Durante Dos Épocas Climáticas." *Revista de Biología Tropical* 69 (1). doi:10.15517/rbt.v69i1.41822.

Özbek, Belma, and Gökçe Dadali. 2007. "Thin-Layer Drying Characteristics and Modelling of Mint Leaves Undergoing Microwave Treatment." *Journal of Food Engineering* 83 (4): 541–49. doi:10.1016/j.jfoodeng.2007.04.004.

Park, Ji-Sook, and Jae-Won Ha. 2019. "X-Ray Irradiation Inactivation of Escherichia Coli O157:H7, Salmonella Enterica Serovar Typhimurium, and Listeria Monocytogenes on Sliced Cheese and Its Bactericidal Mechanisms." *International Journal of Food Microbiology* 289 (January): 127–33. doi:10.1016/j.ijfoodmicro.2018.09.011.

Pasqualone, Antonella, Francesca Vurro, Carmine Summo, Mokhtar H. Abd-El-Khalek, Haneen H. Al-Dmoor, Tomislava Grgic, Maria Ruiz et al. 2022. "The Large and Diverse Family of Mediterranean Flat Breads: A Database." *Foods* 11 (15): 2326. doi:10.3390/foods11152326.

Ross, R. Paul, Sheila Morgan, and Colin Hill. 2002. "Preservation and Fermentation: Past, Present and Future." *International Journal of Food Microbiology* 79 (1–2): 3–16. doi:10.1016/S0168-1605(02)00174-5.

Pina, Eduardo A., Miguel A. Lozano, and Luis M. Serra. 2018. "Thermoeconomic Cost Allocation in Simple Trigeneration Systems Including Thermal Energy Storage." *Energy* 153 (June): 170–84. doi:10.1016/j.energy.2018.04.012.

Rahman, Mizanur M., Mohammad Abul Hashem, Mohammad Abul Kashem Azad, Mohammad Shah Haidar Choudhury, and Mohammad Khaja Junaid Bhuiyan. 2023. "Techniques of Meat Preservation- A Review." *Meat Research* 3 (3). doi:10.55002/mr.3.3.55.

Rao, A. Venket, Alison C. Bested, Tracey M. Beaulne, Martin A. Katzman, Christina Iorio, John M. Berardi, and Alan C. Logan. 2009. "A Randomized, Double-Blind, Placebo-Controlled Pilot Study of a Probiotic in Emotional Symptoms of Chronic Fatigue Syndrome." *Gut Pathogens* 1 (1): 6. doi:10.1186/1757-4749-1-6.

Revilla, Pedro, Mara Lisa Alves, Violeta Andelković, Carlotta Balconi, Isabel Dinis, Pedro Mendes-Moreira, Rita Redaelli et al. 2022. "Traditional Foods from Maize (Zea Mays L.) in Europe." *Frontiers in Nutrition* 8 (January). doi:10.3389/fnut.2021.683399.

Rezac, Shannon, Car Reen Kok, Melanie Heermann, and Robert Hutkins. 2018. "Fermented Foods as a Dietary Source of Live Organisms." *Frontiers in Microbiology* 9 (August). doi:10.3389/fmicb.2018.01785.

Saeed, Farhan, Muhammad Afzaal, Yasir Abbas Shah, Mujahid Hassan Khan, Muzzamal Hussain, Ali Ikram, Huda Ateeq, Muhammad Noman, Shamaail A. Saewan, and Ashraf O. Khashroum. 2022. "Miso: A Traditional Nutritious & Health-endorsing Fermented Product." *Food Science & Nutrition* 10 (12): 4103–11. doi:10.1002/fsn3.3029.

Salminen, Seppo, Constant Bouley, Marie-Christine Boutron, John H. Cummings, Albert Franck, Glenn R. Gibson, Erika Isolauri, Marie-Claude Moreau, Marcel Roberfroid, and Ian Rowland. 1998. "Functional Food Science and Gastrointestinal Physiology and Function." *British Journal of Nutrition* 80 (S1): S147–71. doi:10.1079/BJN19980108.

Sarbu, Ioan, and Calin Sebarchievici. 2015. "General Review of Solar-Powered Closed Sorption Refrigeration Systems." *Energy Conversion and Management* 105 (November): 403–22. doi:10.1016/j.enconman.2015.07.084.

Sarkar, Preetam, Lohith Kumar Dasappa Hanumappa, Chanda Dhumal, Shubham Subrot Panigrahi, and Ruplal Choudhary. 2015. "Traditional and Ayurvedic Foods of Indian Origin." *Journal of Ethnic Foods* 2 (3): 97–109. doi:10.1016/j.jef.2015.08.003.

Satish Kumar, Rajendran, Palanisamy Kanmani, Narayanan Yuvaraj, Kothari Alagu Paari, Vairamani Pattukumar, and Venkatesan Arul. 2013. "Traditional Indian Fermented Foods: A Rich Source of Lactic Acid Bacteria." *International Journal of Food Sciences and Nutrition* 64 (4): 415–28. doi:10.3109/09637486.2012.746288.

Satish Kumar, Rajendran, Palanisamy Kanmani, Narayanan Yuvaraj, Kothari Alagu Paari, Vairamani Pattukumar, and Venkatesan Arul. 2011. "Lactobacillus Plantarum AS1 Binds to Cultured Human Intestinal Cell Line HT-29 and Inhibits Cell Attachment by Enterovirulent Bacterium Vibrio Parahaemolyticus." *Letters in Applied Microbiology* 53 (4): 481–87. doi:10.1111/j.1472-765X.2011.03136.x.

Sehrawat, Rachna, Prabhat K. Nema, and Barjinder Pal Kaur. 2018. "Quality Evaluation and Drying Characteristics of Mango Cubes Dried Using Low-Pressure Superheated Steam, Vacuum and Hot Air Drying Methods." *LWT* 92 (June): 548–55. doi:10.1016/j.lwt.2018.03.012.

Selhub, Eva M., Alan C. Logan, and Alison C. Bested. 2014. "Fermented Foods, Microbiota, and Mental Health: Ancient Practice Meets Nutritional Psychiatry." *Journal of Physiological Anthropology* 33 (1): 2. doi:10.1186/1880-6805-33-2.

Sidhi, Sigit Deddy Purnomo, Andreas Pujianto, Djoko Prasetyo, Akhmad Nurfauzi, and Muhfizar. 2018. "Experimental Study of Salted Fish Drying Under Greenhouse Dryer." *Russian Journal of Agricultural and Socio-Economic Sciences* 77 (5): 281–87. doi:10.18551/rjoas.2018-05.33.

Skalkos, Dimitris, Ioanna S. Kosma, Areti Vasiliou, and Raquel P. F. Guine. 2021. "Consumers' Trust in Greek Traditional Foods in the Post COVID-19 Era." *Sustainability* 13 (17): 9975. doi:10.3390/su13179975.

Sojecka, Agata Angelika, Aleksandra Drozd-Rzoska, and Sylwester J. Rzoska. 2024. "Food Preservation in the Industrial Revolution Epoch: Innovative High Pressure Processing (HPP, HPT) for the 21st-Century Sustainable Society." *Foods* 13 (19): 3028. doi:10.3390/foods13193028.

Stanton, Catherine, R Paul Ross, Gerald F Fitzgerald, and Douwe Van Sinderen. 2005. "Fermented Functional Foods Based on Probiotics and Their Biogenic Metabolites." *Current Opinion in Biotechnology* 16 (2): 198–203. doi:10.1016/j.copbio.2005.02.008.

Sun, Hao, Xueqin Ni, Xu Song, Bin Wen, Yi Zhou, Fuqin Zou, Mingyue Yang et al. 2016. "Fermented Yupingfeng Polysaccharides Enhance Immunity by Improving the Foregut Microflora and Intestinal Barrier in Weaning Rex Rabbits." *Applied Microbiology and Biotechnology* 100 (18): 8105–20. doi:10.1007/s00253-016-7619-0.

Tillisch, Kirsten, Jennifer Labus, Lisa Kilpatrick, Zhiguo Jiang, Jean Stains, Bahar Ebrat, Denis Guyonnet et al. 2013. "Consumption of Fermented Milk Product With Probiotic Modulates Brain Activity." *Gastroenterology* 144 (7): 1394–1401.e4. doi:10.1053/j.gastro.2013.02.043.

Trichopoulou, Antonia, Stavroula Soukara, and Effie Vasilopoulou. 2007. "Traditional Foods: A Science and Society Perspective." *Trends in Food Science & Technology* 18 (8): 420–27. doi:10.1016/j.tifs.2007.03.007.

Tsiraki, Maria I., Tahra El-Obeid, Hany M. Yehia, Layal Karam, and Ioannis N. Savvaidis. 2018. "Effects of Chitosan and Natamycin on Vacuum-Packaged Phyllo: A Pastry Product." *Journal of Food Protection* 81 (12): 1982–87. doi:10.4315/0362-028X.JFP-18-236.

Xiang, Huan, Dongxiao Sun-Waterhouse, Geoffrey I.N. Waterhouse, Chun Cui, and Zheng Ruan. 2019. "Fermentation-Enabled Wellness Foods: A Fresh Perspective." *Food Science and Human Wellness* 8 (3): 203–43. doi:10.1016/j.fshw.2019.08.003.

Zhang, Hongfei, Thi Mai Hoa Ha, Hon Luen Seck, and Weibiao Zhou. 2020. "Inactivation of Escherichia Coli O157:H7 and Salmonella Typhimurium in Edible Bird's Nest by Low-Energy X-Ray Irradiation." *Food Control* 110 (April): 107031. doi:10.1016/j.foodcont.2019.107031.

Zobell, Claude E., D. Quentin Anderson, and W. Whitney Smith. 1937. "The Bacteriostatic and Bactericidal Action of Great Salt Lake Water." *Journal of Bacteriology* 33 (3): 253–62. doi:10.1128/jb.33.3.253-262.1937.

13 Traditional Foods
Components and Biological Effects

Dimple Kothari, Ruchi Jariwala, and Niralee Patel

13.1 INTRODUCTION

Over the past few decades, there has been a consistent upward trend in the demand for premium food products, accompanied by a heightened interest in food quality. This shift is attributed to market forces, including the expectations of increasingly sophisticated and knowledgeable consumers and growing health and environmental awareness (Mascarello et al. 2015). Contemporary consumers are particularly discerning, emphasising the quality and health benefits of their food choices (Sajdakowska et al. 2018). In the current food economy, quality has emerged as a central concern (Grunert 2005), with recent trends indicating that consumers' aspirations for healthier lifestyles and environmental responsibility are significantly influencing their food purchasing decisions and quality perceptions. This contribution is predicated on the understanding that food quality is shaped by consumer perceptions (Baiardi, Puglisi, and Scabrosetti 2016) and proposes that a critical aspect of comprehending food quality evaluation lies in identifying the cues that consumers utilise in this process. Understanding these perceptions is crucial, as they form the basis for consumers' purchasing decisions (van Rijswijk and Frewer 2008).

Food preservation, alternatively referred to as food processing, encompasses the techniques and treatments applied to food to prevent deterioration. This involves inhibiting the proliferation of pathogenic microorganisms, averting fat oxidation (rancidity), and preserving the nutritional content, texture, and flavour of food items (Lianou, Panagou, and Nychas 2016; Necidová et al. 2019). It is generally accepted that food products are susceptible to degradation when exposed to chemicals, environmental microorganisms, and intrinsic food enzymes. During transit, food may experience deterioration, loss of visual appeal, and diminished nutritional value. According to the World Health Organisation, approximately 600 million individuals—equivalent to 1 in 10 people globally—contract illnesses from consuming contaminated food annually. This results in 420,000 fatalities each year and a loss of 33 million healthy life years. Furthermore, diarrhoeal illnesses claim the lives of 2.2 million children in developing countries annually (WHO 2018).

As time progresses, various modern techniques have emerged to preserve diverse food types for differing periods. Whilst certain processes pose disadvantages, some have benefits that are being supplanted by alternative methods, allowing for extended storage of food items without compromising their flavour, nutritional content, quality, aroma, taste, or freshness. Numerous approaches that have been utilised for years have undergone assessment (Benlloch-Tinoco et al. 2015; Ashie et al. 1996). Basic food preservation techniques such as freezing, drying, chemical preservation, chilling, and pasteurisation are extensively employed globally and continue to gain traction. Scientific advancements and innovations enhance the development of current technologies and the development of novel technologies, including obstacle technology, irradiation, and high-pressure technology (Amit et al. 2017; Deborah Blum 2012). Some bacterial genera can be viewed as advantageous symbionts engaged in a beneficial relationship with human beings. Certain bacterial species have been identified as possible pathogens, and an imbalance in their ratios may make the host more susceptible to disease. Though this may differ based on the particular host situation, *Lactobacilli* and *Bifidobacteria* species are generally considered "beneficial" bacteria and are commonly

 DOI: 10.1201/9781003610274-16

incorporated into probiotic supplements. Conversely, organisms such as *Escherichia coli*, certain strains of the lipopolysaccharide-producing taxa and *Clostridium* genus, like *Enterobacteriaceae*, have been linked to pathological conditions and associated symptoms (Tojo 2014; Wang et al. 2017; Ogunrinola et al. 2020).

The correlation between the two predominant phyla, expressed as the *Firmicutes:Bacteroidetes* ratio, has been associated with numerous clinical disorders (Magne et al. 2020; Sweeney and Morton 2013), although its connection to obesity remains contentious (Sze and Schloss 2016). A significant element that underscores the difficulties in identifying a healthy microbiome is the considerable heterogeneity observed among people. The existence of essential microbial functions, known as the "functional core," may, therefore, be more important in distinguishing a healthy microbial state than the presence of particular microorganisms, which is the basis for defining a healthy microbiome (Redondo-Useros et al. 2020; Bäckhed et al. 2012). This suggests that various microorganisms can carry out distinct metabolic tasks, allowing people with varied microbiota compositions to display the same microbial behaviours. Moreover, the current uncertainties regarding the human microbiome hinder the characterisation of a healthy ecosystem. Current microbiota research has yet to fully explore the variety and functionality of certain taxa and strains, despite the tremendous advancements in sequencing technologies over the last 10 years (Thomas and Segata 2019). A topic of discussion within the *Prevotella* genus, strain-level diversity may play a crucial role in establishing the relationships between a particular bacterial genus and health or illness (De Filippis et al. 2019).

While the fundamental gut microbiota in adults exhibits relative stability, environmental factors have been shown to influence the composition of the gut microbial community (Rinninella et al. 2019; Martínez, Muller, and Walter 2013; Yatsunenko et al. 2012). Research has demonstrated that dietary habits, whether short-term (David et al. 2014) or long-term (Wu et al. 2011), play a crucial role in shaping microbial composition and diversity. Scholars have thoroughly investigated the effects of both specific nutrients and overall dietary patterns on the microbiome. A comprehensive overview of the dietary elements affecting the gastrointestinal microbial population is presented in Table 13.1. Though certain broad conclusions can be drawn regarding the influence of diet on microbiota composition, recent investigations indicate that the interplay between diet and microbes is highly individualised and contingent upon one's existing microbial profile (Wu et al. 2016). This suggests that dietary supplements may require modification to accommodate each individual's baseline microbiome (Johnson et al. 2019).

13.2 BIOACTIVE COMPOUNDS IN TRADITIONAL FOODS AND THEIR NEUROPROTECTIVE EFFECTS

The human gut microbiota is a complex community of archaea, bacteria, fungi, protozoans, and viruses that live in symbiosis within the gastrointestinal system (Faruqui et al. 2021). In the human intestine, there are 1,000 of species and 7,000 strains of bacteria, with Bacteroidetes (including Prevotella and Bacteroides) and Firmicutes (such as *Clostridium, Eubacterium,* and *Lactobacillus*) being the most common phyla (Askarova et al. 2020; Faruqui et al. 2021; Szablewski 2018). Recent research highlights the critical role of keeping the gut microbiome healthy, as the amount, biological traits, composition, and distribution of the natural gut flora are constantly changing. There is a strong correlation between intestinal permeability and impairments in gut barrier function and disruption of gut flora. Microbial metabolites may enter the bloodstream as a result of a compromised intestinal barrier. Numerous proinflammatory cytokines may enter the central nervous system through a breach in the blood–brain barrier, activating astrocytes and microglia and resulting in neuroinflammation (Rutsch, Kantsjö, and Ronchi 2020; Tang et al. 2020).

TABLE 13.1
Impact of Dietary Components on Gut Microbiota and Health

S. No.	Dietary Components	Effects	Bacteria	References
1.	Fermented Foods	• Beneficial outcomes resulting from the consumption of microorganisms and their metabolites • Increase beneficial microbiome	*Bifidobacterium*	Kaczmarek et al. (2019)
2.	Nuts	• "Prebiotic effect" at the genus level	*Firmicutes* genera, including some butyrate producers (e.g., *Faecalibacterium* and *Roseburia*), and *Dialister*	Taylor et al. (2020)
3.	Fibres and prebiotics	• Bacterial diversity and the quantity of helpful microorganisms often rise with the kind of dietary fibre. • Increase beneficial microbes • Decreases potentially pathogenic bacteria	*Bifidobacterium*, *Lactobacillus*, *Akkermansia*, *Faecalibacterium*, and *Enterobacteriaceae*	Dhillon, Li, and Ortiz (2019)
4.	Plant-based proteins	• Increase beneficial microbes • Decreases Pathogens	*Bifidobacterium*, *Roseburia*, *Lactobacillus*, and *C. perfringens*	Kim et al. (2020)
5.	MUFA/PUFAs	• Increase beneficial microbes	*Lactobacillus*, *Lachnospira*, *Roseburia*, and *Bifidobacterium*	Watson et al. (2018)
6.	Polyphenols	• "Prebiotic"-like effect has been documented • Decrease potential pathogens and increase symbionts		González et al. (2020)

The enteric and central nervous systems communicate in two directions, linking the emotional and cognitive functions of the brain to the intestinal system (Szablewski 2018). Current studies underscore the crucial role of gut bacteria in gastrointestinal–brain axis function, affecting endocrine, neurological, and immunological functions (Rutsch, Kantsjö, and Ronchi 2020). The gut microbiota synthesises diverse neuroactive compounds, including bacterial metabolites, neurotransmitters, and short-chain fatty acids, which influence cognitive functions and neuronal activity (Rutsch, Kantsjö, and Ronchi 2020; Dalile et al. 2019; Sun et al. 2020; Saresella et al. 2010; Clarke et al. 2019). Gut bacteria make these compounds by fermenting dietary fibre and other chemicals (including ammonia, nitric oxide, and ethanol) (Dalile et al. 2019; Rowland et al. 2018). However, Alzheimer's disease, anxiety, autism spectrum disorders, and sadness have all been connected to gut dysbiosis (Wang et al. 2017; Santos et al. 2019). As a result, the gut–brain axis has become a viable target for therapies meant to improve brain function and treat mental and neurological conditions. Altering the gut microbiota by various interventions, including dietary modifications, antibiotics, prebiotics, faecal microbiota transplantation, and probiotics, can improve brain health and treat illnesses connected to the brain (Sun et al. 2020; Hill et al. 2014; Khanna et al. 2017).

Diet can have either beneficial or harmful effects on health, and it may influence microbial groups (Aaldijk and Vermeiren 2022; Govindarajan et al. 2011). The microbiota composition can be significantly altered by various factors, such as eating patterns, microbiome-specific interventions, and natural supplements. As a consequence, the gut–brain axis may be affected, alleviating

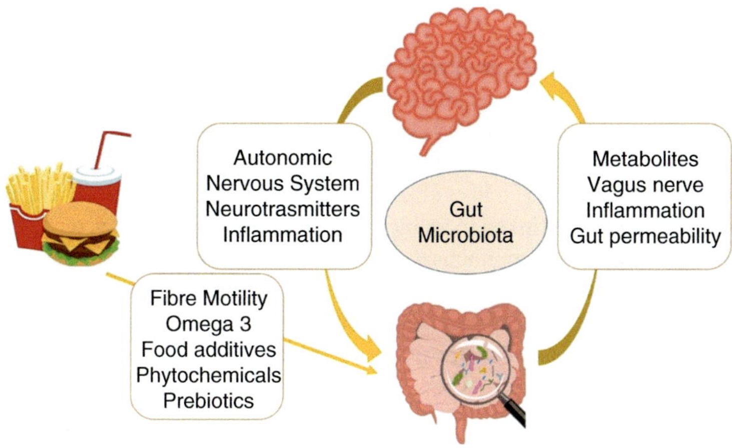

FIGURE 13.1 Gut–brain axis and dietary influence on gut microbiome.

Alzheimer's disease-related pathology (Solfrizzi et al. 2017; Moore et al. 2018). Figure 13.1 depicts the gut–brain axis and dietary influence on gut microbiota. This image illustrates the intricate relationship between diet, brain function, and gut microbiota, highlighting the gut–brain axis. Various physiological processes are regulated by the gut microbiota, depicted at the centre, which is influenced by both dietary and neural factors.

13.2.1 Dietary Proteins

Protein is a crucial macronutrient that is essential for maintaining human health. The source and quantity of protein, whether derived from animals or plants, can influence overall health and cognitive abilities (Sun et al. 2021). Consistent protein consumption over time may affect the likelihood of cognitive deterioration, with a lower subjective level of cognitive impairment associated with higher protein consumption (Yeh et al. 2022). In Fernando et al. (2018), high protein diets may prevent the accumulation of amyloid beta particles in the brain, particularly before the onset of cognitive decline and collecting issues in older adults (Fernando et al. 2018). During digestion, any dietary protein that is not absorbed is fermented by proteolytic bacteria, leading to the production of beneficial byproducts that affect both host functions and the microbiota's composition (Scott et al. 2013). The production of neurotoxic metabolites from eating animal proteins has also been linked in studies to an increased risk of Alzheimer's disease (Zhao et al. 2018). On the other hand, replacing animal proteins with plant-based proteins has been linked to enhanced brain health and lower dementia.

13.2.2 Dietary Fibre

Carbohydrates are typically classified into complex sugars and simple sugars. As a major source of microbiota-accessible carbohydrates, dietary fibre allows microorganisms to supply the host with carbon and energy. As a result, fibres are considered prebiotics, and the risk of Alzheimer's disease is higher when simple sugars are consumed, while greater fibre consumption reduces the risk of the condition (Dominika et al. 2011; Satokari 2020). Short-chain fatty acids are produced by gut bacteria when soluble fibre is consumed (Dhingra et al. 2012; Cuervo-Zanatta et al. 2023). Research recommends that consuming soluble fibre increases butyrate levels, reduces propionate production, increases cognitive performance, and decreases astrocyte activation in the Alzheimer's disease mouse model. These benefits are ascribed to dysbiosis of the gut microbiota (Syeda et al. 2018).

13.2.3 Dietary Fat

Dietary fats are mainly divided into unsaturated and saturated categories; consuming polyunsaturated fats, free fatty acids, and monounsaturated fats can affect several beneficial microorganisms, such as *Bifidobacterium* and *Prevotella* (David et al. 2014). Monounsaturated fatty acids and polyunsaturated fatty acids are linked to positive outcomes, with the prevention of neurodegenerative diseases and enhanced brain function (Field et al. 2001; Gardener et al. 2012; Matura et al. 2021). A diet high in trans and saturated fats is linked to an increase in proinflammatory bacteria, while a diet rich in polyunsaturated fatty acids and monounsaturated fatty acids supports the growth of short-chain fatty acid-producing bacteria. Human studies have demonstrated that omega-3 PUFAs can reduce the *Bacteroidetes:Firmicutes* ratio and elevate the abundance of short-chain fatty acids-producing bacterial genera, such as *Roseburia, Lachnospiraceae*, and *Bifidobacterium* (Balfegó et al. 2016).

13.2.4 Dietary Polyphenols

A polyphenol is a type of antioxidant micronutrient found in plants and in foods derived from them. These compounds are predominantly found in vegetables and fruits, including apples, cocoa, grapes, almonds, tomatoes, pomegranates, red onions, and oranges and are also present in drinks like wine, coffee, and green tea (Sadeghi Ekbatan et al. 2018). Studies show that polyphenols can change the gut microbiota's diversity and structure. In addition to reducing hazardous pathogens like *Clostridium* spp., *Staphylococcus aureus*, and *Salmonella typhimurium*, a diet high in polyphenols promotes the growth of good bacteria like *Lactobacillus* and *Bifidobacterium*. Furthermore, dietary polyphenols have been associated with the prevention of neurodegenerative diseases by interacting with the Advanced Glycation End Products-Receptor for Advanced Glycation Ed Products (AGEs—RAGE) axis and influencing the gut–brain–microbiota axis (Li et al. 2023).

13.2.5 Dietary Probiotics

The gut microbiota metabolises prebiotics, which are fibre compounds, whereas probiotics are live microbes that promote the host's health (Sanders et al. 2019). The ability of probiotics, such as *Bifidobacterium* and lactic acid bacteria, to reduce neuroinflammation has attracted considerable attention. Probiotic-associated bacteria can improve cognitive performance and help stop Alzheimer's disease-related memory loss. *Bifidobacterium* and *Lactobacillus* are among the most frequently used probiotic species (Hill et al. 2014). An 8-week course of *Bifidobacterium longum, Bifidobacterium infantis*, and *Bifidobacterium breve* led to modifications in the gastrointestinal microbiome and elevated levels of short-chain fatty acids in the serum and brain. Akbari et al. (2016) discovered that probiotic milk containing *Lactobacillus casei, Lactobacillus acidophilus, Lactobacillus fermentum,* and *Bifidobacterium bifidum* is linked to significant improvements in Mini-Mental State Examination scores, along with reductions in plasma and C-reactive protein levels and malondialdehyde (Akbari et al. 2016). In a study where MCI patients received *Bifidobacterium breve A1* for 16 weeks, notable improvements were observed in neuropsychological assessments, such as the Repeatable Battery for the Assessment of Neuropsychological Status and the Japanese version of the MCI Screen, compared to the placebo group. These findings suggest that prebiotics, synbiotics (combinations of probiotics and prebiotics), postbiotics (functional bioactive compounds), and probiotics can effectively alter Alzheimer's disease-related disease progression and neuropathology, offering a promising therapeutic strategy (Xiao et al. 2020).

13.3 PRESERVATION TECHNIQUES AND RETENTION OF MICRONUTRIENTS

Oxygen is essential for aerobic metabolism in brain and muscle cells, yet it also contributes to degenerative conditions due to oxygen-free radical's high reactivity. This is particularly true in tissues like the muscles and brain, which use significant amounts of oxygen. In order to combat this, the body has established a strong antioxidant defence mechanism, in which specific minerals and vitamins are essential (Butnariu 2012). Niacin, through NAD, is engaged in the regulation of intracellular calcium levels release. NAD is essential for nucleotide metabolism and the energy state of the cell, both of which are intimately related to calcium signalling. Given calcium's role in generating action potentials in neurons, niacin's involvement in brain functions is highly significant (Kennedy 2016). The palmitoylation of specific neuronal proteins, which depends on pantothenic acid, is necessary for neurotransmitter release at synapses, making it essential for information transmission in the mind (Bâ 2008).

Numerous preclinical studies have demonstrated that iron deficiency can modify the electrophysiological properties of neurotransmitter systems and neural circuits, which leads to the belief that iron is essential for synaptic function. The transcription of genes involved in synaptic plasticity is subsequently made possible by calcium signals activating signalling pathways (Muñoz and Humeres 2012).

A key role that magnesium plays in calcium and potassium transport is the active transport across membranes. Magnesium plays a vital role in the nervous system, facilitating neuromuscular coordination and optimising nerve transmission. By binding to the aspartate receptor and inhibiting its calcium channel, which needs to be blocked for glutamatergic excitatory transmission to take place, it also aids in preventing excessive excitement, which can result in cell death. Theoretically, low magnesium levels could increase glutamatergic neurotransmission, potentially causing oxidative stress and neuronal cell death (Kirkland, Sarlo, and Holton 2018). Certain neurons, referred to as "zinc-containing" neurons, have large amounts of zinc in their synaptic vesicles. These neurons use zinc to modulate neurotransmission mediated by both glutamate and gamma-aminobutyric acid, which in turn helps regulate synaptic excitability. Despite the fact that zinc is extremely cytotoxic and may cause more neuronal death in cases of temporary global ischemia and brain trauma, neurons have various homeostatic mechanisms that keep zinc levels at safe ranges (Sensi et al. 2009).

13.4 CONCLUSION

In conclusion, the relationship between gut health and overall well-being is increasingly acknowledged in scientific fields, with current research highlighting the complex role of gut bacteria in several physiological systems. The microbial ecosystem within the human gastrointestinal tract is vital for health, affecting a spectrum of functions from traditional food preservation methods to the gut–brain axis. Historical food processing techniques, such as fermentation and dehydration, have been utilised for centuries to ensure food safety and enhance nutritional content, often leveraging beneficial microorganisms. These continuing practices not only facilitate food preservation but also help sustain a balanced gut microbiome, encouraging the growth of probiotics that aid in digestion, fortify the immune system, and support mental health. Furthermore, dietary elements like probiotics, proteins, polyphenols, and fibre significantly influence gut microbiota, affecting both metabolic and cognitive functions. A nutrient-dense diet can enhance gut health, potentially reducing the risk of neurological conditions, inflammatory disorders, and other health concerns. The gut–brain axis, a critical communication pathway, underscores the necessity of a balanced microbiome for cognitive and emotional health. As scientific progress continues to uncover the importance of gut health, the integration of traditional food preservation techniques with modern dietary strategies offers a holistic approach to enhancing overall health. By fostering a symbiotic relationship with gut microbiota, improvements in digestion, immune resilience, and cognitive health can be realised, highlighting the profound impact of the gut on human health.

REFERENCES

Aaldijk, Emma, and Yannick Vermeiren. 2022. "The Role of Serotonin within the Microbiota-Gut-Brain Axis in the Development of Alzheimer's Disease: A Narrative Review." *Ageing Research Reviews* 75 (March): 101556. https://doi.org/10.1016/j.arr.2021.101556.

Akbari, Elmira, Zatollah Asemi, Reza Daneshvar Kakhaki, Fereshteh Bahmani, Ebrahim Kouchaki, Omid Reza Tamtaji, Gholam Ali Hamidi, and Mahmoud Salami. 2016. "Effect of Probiotic Supplementation on Cognitive Function and Metabolic Status in Alzheimer's Disease: A Randomized, Double-Blind and Controlled Trial." *Frontiers in Aging Neuroscience* 8 (November). https://doi.org/10.3389/fnagi.2016.00256.

Amit, Sadat Kamal, Md. Mezbah Uddin, Rizwanur Rahman, Syed Muhammad Rezwanul Islam, and Mohidus Samad Khan. 2017. "A Review on Mechanisms and Commercial Aspects of Food Preservation and Processing." *Agriculture & Food Security* 6 (1): 51. https://doi.org/10.1186/s40066-017-0130-8.

Ashie, Isaac N. A., James P. Smith, Benjamin K. Simpson, and Norman F. Haard. 1996. "Spoilage and Shelf-life Extension of Fresh Fish and Shellfish." *Critical Reviews in Food Science and Nutrition* 36 (1–2): 87–121. https://doi.org/10.1080/10408399609527720.

Askarova, Sholpan, Bauyrzhan Umbayev, Abdul-Razak Masoud, Aiym Kaiyrlykyzy, Yuliya Safarova, Andrey Tsoy, Farkhad Olzhayev, and Almagul Kushugulova. 2020. "The Links Between the Gut Microbiome, Aging, Modern Lifestyle and Alzheimer's Disease." *Frontiers in Cellular and Infection Microbiology* 10 (March). https://doi.org/10.3389/fcimb.2020.00104.

Bâ, Abdoulaye. 2008. "Metabolic and Structural Role of Thiamine in Nervous Tissues." *Cellular and Molecular Neurobiology* 28 (7): 923–31. https://doi.org/10.1007/s10571-008-9297-7.

Bäckhed, Fredrik, Claire M. Fraser, Yehuda Ringel, Mary Ellen Sanders, Ryan Balfour Sartor, Philip M. Sherman, James Versalovic, Vincent Young, and B. Brett Finlay. 2012. "Defining a Healthy Human Gut Microbiome: Current Concepts, Future Directions, and Clinical Applications." *Cell Host & Microbe* 12 (5): 611–22. https://doi.org/10.1016/j.chom.2012.10.012.

Baiardi, Donatella, Riccardo Puglisi, and Simona Scabrosetti. 2016. "Individual Attitudes on Food Quality and Safety: Empirical Evidence on EU Countries." *Food Quality and Preference* 49 (April): 70–4. https://doi.org/10.1016/j.foodqual.2015.11.011.

Balfegó, Mariona, Silvia Canivell, Felicia A. Hanzu, Aleix Sala-Vila, Margarita Martínez-Medina, Serafín Murillo, Teresa Mur et al. 2016. "Effects of Sardine-Enriched Diet on Metabolic Control, Inflammation and Gut Microbiota in Drug-Naïve Patients with Type 2 Diabetes: A Pilot Randomized Trial." *Lipids in Health and Disease* 15 (1): 78. https://doi.org/10.1186/s12944-016-0245-0.

Benlloch-Tinoco, María, Marta Igual, Dolores Rodrigo, and Nuria Martínez-Navarrete. 2015. "Superiority of Microwaves over Conventional Heating to Preserve Shelf-Life and Quality of Kiwifruit Puree." *Food Control* 50 (April): 620–29. https://doi.org/10.1016/j.foodcont.2014.10.006.

Butnariu, Monica. 2012. "The Oxygen Paradox." *Journal of Pharmacogenomics & Pharmacoproteomics* 3 (1). https://doi.org/10.4172/2153-0645.1000e104.

Clarke, Gerard, Kiran V. Sandhu, Brendan T. Griffin, Timothy G. Dinan, John F. Cryan, and Niall P. Hyland. 2019. "Gut Reactions: Breaking Down Xenobiotic–Microbiome Interactions." *Pharmacological Reviews* 71 (2): 198–224. https://doi.org/10.1124/pr.118.015768.

Cuervo-Zanatta, Daniel, Tauqeerunnisa Syeda, Vicente Sánchez-Valle, Mariangel Irene-Fierro, Pablo Torres-Aguilar, Mónica Adriana Torres-Ramos, Mineko Shibayama-Salas et al. 2023. "Dietary Fiber Modulates the Release of Gut Bacterial Products Preventing Cognitive Decline in an Alzheimer's Mouse Model." *Cellular and Molecular Neurobiology* 43 (4): 1595–1618. https://doi.org/10.1007/s10571-022-01268-7.

Dalile, Boushra, Lukas Van Oudenhove, Bram Vervliet, and Kristin Verbeke. 2019. "The Role of Short-Chain Fatty Acids in Microbiota–Gut–Brain Communication." *Nature Reviews Gastroenterology & Hepatology* 16 (8): 461–78. https://doi.org/10.1038/s41575-019-0157-3.

David, Lawrence A., Corinne F. Maurice, Rachel N. Carmody, David B. Gootenberg, Julie E. Button, Benjamin E. Wolfe, Alisha V. Ling et al. 2014. "Diet Rapidly and Reproducibly Alters the Human Gut Microbiome." *Nature* 505 (7484): 559–63. https://doi.org/10.1038/nature12820.

Deborah, Blum. 2012. "Food That Lasts Forever," *Time*, March 12, 2012.

Dhillon, Jaapna, Zhaoping Li, and Rudy M. Ortiz. 2019. "Almond Snacking for 8 Wk Increases Alpha-Diversity of the Gastrointestinal Microbiome and Decreases Bacteroides Fragilis Abundance Compared with an Isocaloric Snack in College Freshmen." *Current Developments in Nutrition* 3 (8): nzz079. https://doi.org/10.1093/cdn/nzz079.

Dhingra, Devinder, Mona Michael, Hradesh Rajput, and R. T. Patil. 2012. "Dietary Fibre in Foods: A Review." *Journal of Food Science and Technology* 49 (3): 255–66. https://doi.org/10.1007/s13197-011-0365-5.

Dominika, Świątecka, Narbad Arjan, Ridgway P. Karyn, and Kostyra Henryk. 2011. "The Study on the Impact of Glycated Pea Proteins on Human Intestinal Bacteria." *International Journal of Food Microbiology* 145 (1): 267–72. https://doi.org/10.1016/j.ijfoodmicro.2011.01.002.

Faruqui, Nairita Ahsan, Durdana Hossain Prium, Sadrina Afrin Mowna, Md. Asad Ullah, Yusha Araf, Bishajit Sarkar, Umme Salma Zohora, and Mohammad Shahedur Rahman. 2021. "Gut Microorganisms and Neurological Disease Perspectives." *Future Neurology* 16 (1). https://doi.org/10.2217/fnl-2020-0026.

Fernando, Warnakulasuriya Mithila Anoma Deepanie Binosha, Stephanie R. Rainey-Smith, Samantha L. Gardener, Victor L. Villemagne, Samantha C. Burnham, S. Lance Macaulay, Belinda M. Brown et al. 2018. “Associations of Dietary Protein and Fiber Intake with Brain and Blood Amyloid-β.” *Journal of Alzheimer's Disease* 61 (4): 1589–98. https://doi.org/10.3233/JAD-170742.

Field, Alison E., Eugenie H. Coakley, Aviva Must, Jennifer L. Spadano, Nan Laird, William H. Dietz, Eric Rimm, and Graham A. Colditz. 2001. “Impact of Overweight on the Risk of Developing Common Chronic Diseases During a 10-Year Period.” *Archives of Internal Medicine* 161 (13): 1581. https://doi.org/10.1001/archinte.161.13.1581.

Filippis, Francesca De, Edoardo Pasolli, Adrian Tett, Sonia Tarallo, Alessio Naccarati, Maria De Angelis, Erasmo Neviani et al. 2019. “Distinct Genetic and Functional Traits of Human Intestinal Prevotella Copri Strains Are Associated with Different Habitual Diets.” *Cell Host & Microbe* 25 (3): 444–53.e3. https://doi.org/10.1016/j.chom.2019.01.004.

González, Sonia, Nuria Salazar, Sergio Ruiz-Saavedra, María Gómez-Martín, Clara G. de los Reyes-Gavilán, and Miguel Gueimonde. 2020. “Long-Term Coffee Consumption Is Associated with Fecal Microbial Composition in Humans.” *Nutrients* 12 (5): 1287. https://doi.org/10.3390/nu12051287.

Govindarajan, Nambirajan, Roberto Carlos Agis-Balboa, Jonas Walter, Farahnaz Sananbenesi, and André Fischer. 2011. “Sodium Butyrate Improves Memory Function in an Alzheimer's Disease Mouse Model When Administered at an Advanced Stage of Disease Progression.” *Journal of Alzheimer's Disease* 26 (1): 187–97. https://doi.org/10.3233/JAD-2011-110080.

Grunert, Klaus G. 2005. “Food Quality and Safety: Consumer Perception and Demand.” *European Review of Agricultural Economics* 32 (3): 369–91. https://doi.org/10.1093/eurrag/jbi011.

Hill, Colin, Francisco Guarner, Gregor Reid, Glenn R. Gibson, Daniel J. Merenstein, Bruno Pot, Lorenzo Morelli et al. 2014. “The International Scientific Association for Probiotics and Prebiotics Consensus Statement on the Scope and Appropriate Use of the Term Probiotic.” *Nature Reviews Gastroenterology & Hepatology* 11 (8): 506–14. https://doi.org/10.1038/nrgastro.2014.66.

Johnson, Abigail J., Pajau Vangay, Gabriel A. Al-Ghalith, Benjamin M. Hillmann, Tonya L. Ward, Robin R. Shields-Cutler, Austin D. Kim et al. 2019. “Daily Sampling Reveals Personalized Diet-Microbiome Associations in Humans.” *Cell Host & Microbe* 25 (6): 789–802.e5. https://doi.org/10.1016/j.chom.2019.05.005.

Kaczmarek, Jennifer L., Xiaoji Liu, Craig S. Charron, Janet A. Novotny, Elizabeth H. Jeffery, Harold E. Seifried, Sharon A. Ross, Michael J. Miller, Kelly S. Swanson, and Hannah D. Holscher. 2019. “Broccoli Consumption Affects the Human Gastrointestinal Microbiota.” *The Journal of Nutritional Biochemistry* 63 (January): 27–34. https://doi.org/10.1016/j.jnutbio.2018.09.015.

Kennedy, David. 2016. “B Vitamins and the Brain: Mechanisms, Dose and Efficacy—A Review.” *Nutrients* 8 (2): 68. https://doi.org/10.3390/nu8020068.

Khanna, Sahil, Yoshiki Vazquez-Baeza, Antonio González, Sophie Weiss, Bradley Schmidt, David A. Muñiz-Pedrogo, John F. Rainey et al. 2017. “Changes in Microbial Ecology after Fecal Microbiota Transplantation for Recurrent C. Difficile Infection Affected by Underlying Inflammatory Bowel Disease.” *Microbiome* 5 (1): 55. https://doi.org/10.1186/s40168-017-0269-3.

Kim, Yeji, Sung Wook Hwang, Seungil Kim, Yong-Soo Lee, Tae-Young Kim, Su-Hyun Lee, Su Jung Kim, Hyun Ju Yoo, Eun Na Kim, and Mi-Na Kweon. 2020. “Dietary Cellulose Prevents Gut Inflammation by Modulating Lipid Metabolism and Gut Microbiota.” *Gut Microbes* 11 (4): 944–61. https://doi.org/10.1080/19490976.2020.1730149.

Kirkland, Anna E., Gabrielle L. Sarlo, and Kathleen F. Holton. 2018. “The Role of Magnesium in Neurological Disorders.” *Nutrients* 10 (6): 730. https://doi.org/10.3390/nu10060730.

Li, Yueqin, Yao Peng, Yingbin Shen, Yunzhen Zhang, Lianliang Liu, and Xinquan Yang. 2023. “Dietary Polyphenols: Regulate the Advanced Glycation End Products-RAGE Axis and the Microbiota-Gut-Brain Axis to Prevent Neurodegenerative Diseases.” *Critical Reviews in Food Science and Nutrition* 63 (29): 9816–42. https://doi.org/10.1080/10408398.2022.2076064.

Lianou, Alexandra, Efstathios Z. Panagou, and George-John E. Nychas. 2016. “Microbiological Spoilage of Foods and Beverages.” In *The Stability and Shelf Life of Food*, edited by P. Subramaniam and P. Wareing, 3–42. Elsevier. https://doi.org/10.1016/B978-0-08-100435-7.00001-0.

Magne, Fabien, Martin Gotteland, Lea Gauthier, Alejandra Zazueta, Susana Pesoa, Paola Navarrete, and Ramadass Balamurugan. 2020. “The Firmicutes/Bacteroidetes Ratio: A Relevant Marker of Gut Dysbiosis in Obese Patients?” *Nutrients* 12 (5): 1474. https://doi.org/10.3390/nu12051474.

Martínez, Inés, Catherine E. Muller, and Jens Walter. 2013. “Long-Term Temporal Analysis of the Human Fecal Microbiota Revealed a Stable Core of Dominant Bacterial Species.” *PLoS ONE* 8 (7): e69621. https://doi.org/10.1371/journal.pone.0069621.

Mascarello, Giulia, Anna Pinto, Nicoletta Parise, Stefania Crovato, and Licia Ravarotto. 2015. "The Perception of Food Quality Profiling Italian Consumers." *Appetite* 89 (June): 175–82. https://doi.org/10.1016/j.appet.2015.02.014.

Matura, Silke, David Prvulovic, Nina Mohadjer, Fabian Fusser, Viola Oertel, Andreas Reif, Johannes Pantel, and Tarik Karakaya. 2021. "Association of Dietary Fat Composition with Cognitive Performance and Brain Morphology in Cognitively Healthy Individuals." *Acta Neuropsychiatrica* 33 (3): 134–40. https://doi.org/10.1017/neu.2021.1.

Moore, Katie, Catherine F. Hughes, Mary Ward, Leane Hoey, and Helene McNulty. 2018. "Diet, Nutrition and the Ageing Brain: Current Evidence and New Directions." *Proceedings of the Nutrition Society* 77 (2): 152–63. https://doi.org/10.1017/S0029665117004177.

Muñoz, Pablo, and Alexis Humeres. 2012. "Iron Deficiency on Neuronal Function." *BioMetals* 25 (4): 825–35. https://doi.org/10.1007/s10534-012-9550-x.

Necidová, Lenka, Šárka Bursová, František Ježek, Danka Haruštiaková, Lenka Vorlová, and Jozef Golian. 2019. "Effect of Preservatives on the Shelf-Life and Sensory Characteristics of Pasteurized Liquid Whole Egg Stored at 4°C." *Poultry Science* 98 (11): 5940–48. https://doi.org/10.3382/ps/pez378.

Ogunrinola, Grace A., John O. Oyewale, Oyewumi O. Oshamika, and Grace I. Olasehinde. 2020. "The Human Microbiome and Its Impacts on Health." *International Journal of Microbiology* 2020 (June):1–7. https://doi.org/10.1155/2020/8045646.

Redondo-Useros, Noemí, Esther Nova, Natalia González-Zancada, Ligia E. Díaz, Sonia Gómez-Martínez, and Ascensión Marcos. 2020. "Microbiota and Lifestyle: A Special Focus on Diet." *Nutrients* 12 (6): 1776. https://doi.org/10.3390/nu12061776.

Rijswijk, Wendy van, and Lynn J. Frewer. 2008. "Consumer Perceptions of Food Quality and Safety and Their Relation to Traceability." *British Food Journal* 110 (10): 1034–46. https://doi.org/10.1108/00070700810906642.

Rinninella, Emanuele, Pauline Raoul, Marco Cintoni, Francesco Franceschi, Giacinto Abele Donato Miggiano, Antonio Gasbarrini, and Maria Cristina Mele. 2019. "What Is the Healthy Gut Microbiota Composition? A Changing Ecosystem across Age, Environment, Diet, and Diseases." *Microorganisms* 7 (1): 14. https://doi.org/10.3390/microorganisms7010014.

Rowland, Ian, Glenn Gibson, Almut Heinken, Karen Scott, Jonathan Swann, Ines Thiele, and Kieran Tuohy. 2018. "Gut Microbiota Functions: Metabolism of Nutrients and Other Food Components." *European Journal of Nutrition* 57 (1): 1–24. https://doi.org/10.1007/s00394-017-1445-8.

Rutsch, Andrina, Johan B. Kantsjö, and Francesca Ronchi. 2020. "The Gut-Brain Axis: How Microbiota and Host Inflammasome Influence Brain Physiology and Pathology." *Frontiers in Immunology* 11 (December). https://doi.org/10.3389/fimmu.2020.604179.

Sadeghi Ekbatan, Shima, Michele Iskandar, Lekha Sleno, Kebba Sabally, Joelle Khairallah, Satya Prakash, and Stan Kubow. 2018. "Absorption and Metabolism of Phenolics from Digests of Polyphenol-Rich Potato Extracts Using the Caco-2/HepG2 Co-Culture System." *Foods* 7 (1): 8. https://doi.org/10.3390/foods7010008.

Sajdakowska, Marta, Jerzy Gębski, Krystyna Gutkowska, and Sylwia Żakowska-Biemans. 2018. "Importance of Health Aspects in Polish Consumer Choices of Dairy Products." *Nutrients* 10 (8). https://doi.org/10.3390/nu10081007.

Samantha L., Yian Gu, Stephanie R. Rainey-Smith, Jennifer B. Keogh, Peter Marshall Clifton, Stacey L. Mathieson, Kevin Taddei et al. 2012. "Adherence to a Mediterranean Diet and Alzheimer's Disease Risk in an Australian Population." *Translational Psychiatry* 2 (10): e164. https://doi.org/10.1038/tp.2012.91.

Sanders, Mary Ellen, Daniel J. Merenstein, Gregor Reid, Glenn R. Gibson, and Robert A. Rastall. 2019. "Probiotics and Prebiotics in Intestinal Health and Disease: From Biology to the Clinic." *Nature Reviews Gastroenterology & Hepatology* 16 (10): 605–16. https://doi.org/10.1038/s41575-019-0173-3.

Santos, Susanne Fonseca, Hadassa Loth de Oliveira, Elizabeth Sumi Yamada, Bianca Cruz Neves, and Antonio Pereira. 2019. "The Gut and Parkinson's Disease—A Bidirectional Pathway." *Frontiers in Neurology* 10 (June). https://doi.org/10.3389/fneur.2019.00574.

Saresella, Marina, Elena Calabrese, Ivana Marventano, Federica Piancone, Andrea Gatti, Maria Gaetana Calvo, Raffaello Nemni, and Mario Clerici. 2010. "PD1 Negative and PD1 Positive CD4+ T Regulatory Cells in Mild Cognitive Impairment and Alzheimer's Disease." *Journal of Alzheimer's Disease* 21 (3): 927–38. https://doi.org/10.3233/JAD-2010-091696.

Satokari, Reetta. 2020. "High Intake of Sugar and the Balance between Pro- and Anti-Inflammatory Gut Bacteria." *Nutrients* 12 (5): 1348. https://doi.org/10.3390/nu12051348.

Scott, Karen P., Silvia W. Gratz, Paul O. Sheridan, Harry J. Flint, and Sylvia H. Duncan. 2013. "The Influence of Diet on the Gut Microbiota." *Pharmacological Research* 69 (1): 52–60. https://doi.org/10.1016/j.phrs.2012.10.020.

Sensi, Stefano L., Pierre Paoletti, Ashley I. Bush, and Israel Sekler. 2009. "Zinc in the Physiology and Pathology of the CNS." *Nature Reviews Neuroscience* 10 (11): 780–91. https://doi.org/10.1038/nrn2734.

Solfrizzi, Vincenzo, Carlo Custodero, Madia Lozupone, Bruno P. Imbimbo, Vincenzo Valiani, Pasquale Agosti, Andrea Schilardi et al. 2017. "Relationships of Dietary Patterns, Foods, and Micro- and Macronutrients with Alzheimer's Disease and Late-Life Cognitive Disorders: A Systematic Review." *Journal of Alzheimer's Disease* 59 (3): 815–49. https://doi.org/10.3233/JAD-170248.

Sun, Jing, Jingxuan Xu, Bo Yang, Keyang Chen, Yu Kong, Na Fang, Tianyu Gong, Fangyan Wang, Zongxin Ling, and Jiaming Liu. 2020. "Effect of *Clostridium butyricum* against Microglia-Mediated Neuroinflammation in Alzheimer's Disease via Regulating Gut Microbiota and Metabolites Butyrate." *Molecular Nutrition & Food Research* 64 (2). https://doi.org/10.1002/mnfr.201900636.

Sun, Yangbo, Buyun Liu, Linda G. Snetselaar, Robert B. Wallace, Aladdin H. Shadyab, Candyce H. Kroenke, Bernhard Haring et al. 2021. "Association of Major Dietary Protein Sources With All-Cause and Cause-Specific Mortality: Prospective Cohort Study." *Journal of the American Heart Association* 10 (5). https://doi.org/10.1161/JAHA.119.015553.

Sweeney, Timothy E., and John M. Morton. 2013. "The Human Gut Microbiome." *JAMA Surgery* 148 (6): 563. https://doi.org/10.1001/jamasurg.2013.5.

Syeda, Tauqeerunnisa, Mónica Sanchez-Tapia, Laura Pinedo-Vargas, Omar Granados, Daniel Cuervo-Zanatta, Eleazar Rojas-Santiago, Sof&a Díaz-Cintra, Nimbe Torres, and Claudia Perez-Cruz. 2018. "Bioactive Food Abates Metabolic and Synaptic Alterations by Modulation of Gut Microbiota in a Mouse Model of Alzheimer's Disease." *Journal of Alzheimer's Disease* 66 (4): 1657–82. https://doi.org/10.3233/JAD-180556.

Szablewski, Leszek. 2018. "Human Gut Microbiota in Health and Alzheimer's Disease." *Journal of Alzheimer's Disease* 62 (2): 549–60. https://doi.org/10.3233/JAD-170908.

Sze, Marc A., and Patrick D. Schloss. 2016. "Looking for a Signal in the Noise: Revisiting Obesity and the Microbiome." *mBio* 7 (4). https://doi.org/10.1128/mBio.01018-16.

Tang, Wei, Huifeng Zhu, Yanmei Feng, Rui Guo, and Dong Wan. 2020. "The Impact of Gut Microbiota Disorders on the Blood–Brain Barrier." *Infection and Drug Resistance* 13 (September): 3351–63. https://doi.org/10.2147/IDR.S254403.

Taylor, Bryn C., Franck Lejzerowicz, Marion Poirel, Justin P. Shaffer, Lingjing Jiang, Alexander Aksenov, Nicole Litwin et al. 2020. "Consumption of Fermented Foods Is Associated with Systematic Differences in the Gut Microbiome and Metabolome." *mSystems* 5 (2). https://doi.org/10.1128/mSystems.00901-19.

Thomas, Andrew Maltez, and Nicola Segata. 2019. "Multiple Levels of the Unknown in Microbiome Research." *BMC Biology* 17 (1): 48. https://doi.org/10.1186/s12915-019-0667-z.

Tojo, Rafael. 2014. "Intestinal Microbiota in Health and Disease: Role of Bifidobacteria in Gut Homeostasis." *World Journal of Gastroenterology* 20 (41): 15163. https://doi.org/10.3748/wjg.v20.i41.15163.

Wang, Baohong, Mingfei Yao, Longxian Lv, Zongxin Ling, and Lanjuan Li. 2017. "The Human Microbiota in Health and Disease." *Engineering* 3 (1): 71–82. https://doi.org/10.1016/J.ENG.2017.01.008.

Watson, Henry, Suparna Mitra, Fiona C. Croden, Morag Taylor, Henry M. Wood, Sarah L. Perry, Jade A. Spencer et al. 2018. "A Randomised Trial of the Effect of Omega-3 Polyunsaturated Fatty Acid Supplements on the Human Intestinal Microbiota." *Gut* 67 (11): 1974–83. https://doi.org/10.1136/gutjnl-2017-314968.

WHO. 2018. *Food Borne Disease Outbreak*. WHO, August 4, 2018.

Wu, Gary D., Jun Chen, Christian Hoffmann, Kyle Bittinger, Ying-Yu Chen, Sue A. Keilbaugh, Meenakshi Bewtra et al. 2011. "Linking Long-Term Dietary Patterns with Gut Microbial Enterotypes." *Science* 334 (6052): 105–8. https://doi.org/10.1126/science.1208344.

Wu, Gary D., Charlene Compher, Eric Z. Chen, Sarah A. Smith, Rachana D. Shah, Kyle Bittinger, Christel Chehoud et al. 2016. "Comparative Metabolomics in Vegans and Omnivores Reveal Constraints on Diet-Dependent Gut Microbiota Metabolite Production." *Gut* 65 (1): 63–72. https://doi.org/10.1136/gutjnl-2014-308209.

Xiao, Jinzhong, Noriko Katsumata, Francois Bernier, Kazuya Ohno, Yuki Yamauchi, Toshitaka Odamaki, Kenji Yoshikawa, Kumie Ito, and Toshiyuki Kaneko. 2020. "Probiotic Bifidobacterium Breve in Improving Cognitive Functions of Older Adults with Suspected Mild Cognitive Impairment: A Randomized, Double-Blind, Placebo-Controlled Trial." *Journal of Alzheimer's Disease* 77 (1): 139–47. https://doi.org/10.3233/JAD-200488.

Yatsunenko, Tanya, Federico E. Rey, Mark J. Manary, Indi Trehan, Maria Gloria Dominguez-Bello, Monica Contreras, Magda Magris et al. 2012. "Human Gut Microbiome Viewed across Age and Geography." *Nature* 486 (7402): 222–27. https://doi.org/10.1038/nature11053.

Yeh, Tian-Shin, Changzheng Yuan, Alberto Ascherio, Bernard A. Rosner, Deborah Blacker, and Walter C. Willett. 2022. "Long-Term Dietary Protein Intake and Subjective Cognitive Decline in US Men and Women." *The American Journal of Clinical Nutrition* 115 (1): 199–210. https://doi.org/10.1093/ajcn/nqab236.

Zhao, Jianfei, Xiaoya Zhang, Hongbin Liu, Michael A. Brown, and Shiyan Qiao. 2018. "Dietary Protein and Gut Microbiota Composition and Function." *Current Protein & Peptide Science* 20 (2): 145–54. https://doi.org/10.2174/1389203719666180514145437.

14 Evolution of Functional Foods for Brain Health

Practices and Interventions

Ruchi Jariwala, Niralee Patel, and Apurva Patel

14.1 INTRODUCTION

Functional foods are food products with health benefits beyond basic nutrition. Such functional foods include fortified, enhanced, or enriched foods; dietary supplements; and also, some conventional foods (Scrinis, 2008), which are consumed on a regular basis as part of a varied diet and designed to have a potentially targeted health benefit (Figure 14.1).

In Asian countries, the concept of functional foods is well rooted with the therapeutic effect of herbs containing foods and a balanced meal (Xu, 2001). In these countries, there is a long history of regular use of ginseng, turmeric, lemon grass, various spices, etc. in their diet, which is reported to have different health benefits (Xu, 2001). For the first time in Japan, the functional food concept was introduced as FOSHU (Foods for Specified Health Use). The concept of functional

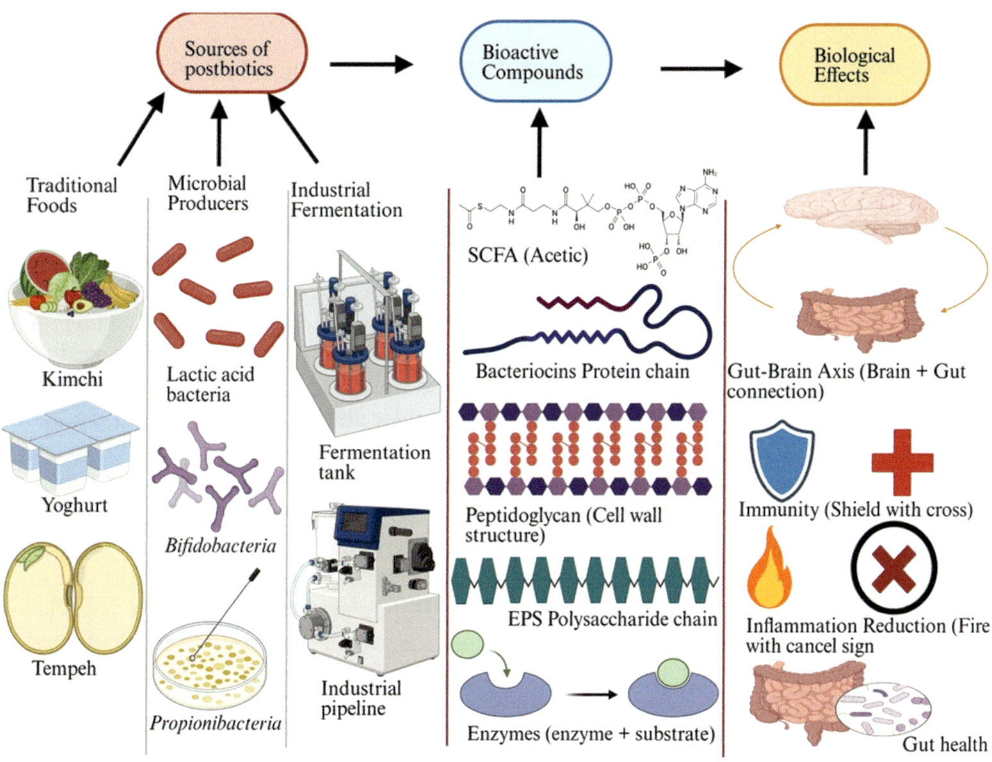

FIGURE 14.1 The role of food composition and postbiotics with brain function and health (Mafe et al. 2025). (Published by Wiley in Open Access under Creative Commons Attribution 4.0 International License.)

DOI: 10.1201/9781003610274-17

food has spread to North America and other countries. Although a minimal amount of biologically active compounds is present in this food, their benefits have been observed in many clinical studies. Regular consumption of certain animal products, fruits, and vegetables containing bioactive compounds has been associated with reducing the risk of cancer and many metabolic disorders by directly or indirectly influencing different molecular pathways (Shivanna, 2018; Bintanah et al., 2023). Such bioactive compounds include resveratrol, ellagic acid, quercetin, curcumin, sulforaphane, and epigallocatechin. Resveratrol, found in pomegranate and grapes, improves immunity, vascular functions, and the gut microbiota (Galiniak et al., 2019; Meng et al., 2020). Curcumin, a major bioactive compound of *Curcuma longa,* is well known for its anti-inflammatory, antioxidant, and antiviral properties (Shah et al., 2022). It is consumed as a spice in every culinary dish of Indian food. Sulforaphane from cruciferous plants protects from oxidative stress by activating the Nrf2-ARE pathway. It directly inhibits the binding of mutagenic agents to DNA and also modulates Phase I and Phase II xenobiotic-metabolizing enzymes (Sohel et al., 2018).

Food habits play an important role in the health of an individual, extending beyond basic nutrition to impact various bodily systems. The connection between the gastrointestinal tract and the central nervous system, known as the gut–brain axis, underscores the importance of dietary choices in maintaining brain health. Hence, designing functional foods to improve blood–brain barrier (BBB) dysfunction and brain function has gained much attention. As research in this field continues to expand, scientists are exploring new ways to incorporate functional foods into dietary recommendations for disease prevention and management. The mechanisms underlying interactions of functional foods with various body functions focusing on brain health are being actively explored. This emerging field holds the potential to revolutionize the approach with which many diseases can be prevented and the health of an individual can be maintained.

14.2 TYPES OF FUNCTIONAL FOODS

Functional foods can be classified into different categories depending mainly on their purpose of use and composition.

14.2.1 Conventional and Modified Functional Foods

Whole-food ingredients or natural foods without any modification are usually classified as conventional functional foods. Such foods are rich in antioxidants, phenolic compounds, vitamins, fibre, omega-3 fatty acids, and minerals. Few examples of such foods include herbs, nuts, fruits, vegetables, cereals, pulses, whole grains, and beans (Arshad et al., 2021). Conventional foods with organic characteristics with bioactive compounds free from any harmful substances such as pesticides, chemical fertilizers, and toxins (Vigar et al., 2019). These conventional foods provide many health benefits due to the presence of bioactive compounds and their nutrient-dense profiles. Usually recommended as a part of a balance diet, regular consumption of such food is also reported to reduce the risk of many chronic diseases. For example, in several epidemiological and experimental studies, consumption of cruciferous vegetables showed reduced risk of several types of cancer (Moriarty et al., 2007). Lycopene-rich tomato products have been reported to reduce the risk of prostate, ovarian, gastric, and pancreatic cancers (Kavanaugh et al., 2007). Consumption of citrus fruits may decrease the risk of stomach cancer (Bae et al., 2008). Dark chocolate may improve heart health while peanuts and tree nuts have been shown to reduce the risk of sudden death due to associated cardiac arrest (Faridi et al., 2008; Kris-Etherton et al., 2008). Cranberry juices have been associated with preventing urinary tract bacterial infection (Nowack, 2007).

Certain foods modified through enrichment, fortification, or enhancement by the addition of ingredients for specific health benefits are also included as functional foods. Such foods include folate-enriched bread for the proper growth and development of foetus, calcium-fortified packaged fruit juices for bone health, and bioactive compound-added foods (Sharma et al., 2024). Foods can also be modified by the biotechnological approach to enhance their nutritional value and health attributes (Damude and Kinney, 2008). A few examples include non-trans-fat oils or omega-3 fatty acid-enhanced fat oils (Dhiman et al., 2024).

14.2.2 Probiotics and Prebiotics as Functional Foods

Probiotics are the live microorganisms found in fermented foods and cultured milk, which provide health benefits upon consumption. Such microorganisms provide many health benefits to host such as improvement in gut health and immune system, alleviation of symptoms of lactose intolerance, and decreasing risk of colorectal cancer and irritable bowel syndrome (Jariwala et al., 2017). Recently, these microorganisms have been reported to modulate the gut–brain axis and improve cognition, anxiety, and depression (Shi et al., 2016). The microorganisms known for their probiotic properties usually belong to the lactic acid group of bacteria. On the other hand, the non-digestible foods that nurture the growth of such microorganisms in the gastrointestinal tract are known as prebiotics. Obesity is one of the major factors for physical and mental malfunctioning. Food composition may lead to an increase or a decrease in obesity (Figure 14.2). Prebiotics have been reported to possess antiobesity, anticancer, antimicrobial, antidiabetic, and anti-inflammatory properties (Kumari et al., 2024). The most common examples of prebiotics include different forms of oligofructose such as galacto-oligofructose, fructo-oligosaccharides, trans-galacto-oligosaccharides, and inulin.

14.2.3 Plant-Sourced Functional Foods

Functional foods based on plant sources usually involve fresh vegetables, fruits, spices, pulses, cereals, millets, and oilseeds. Such foods are rich in bioactive molecules such as vitamins, minerals, polyphenols, polysterols, and some omega-3 fatty acids (Nayak et al., 2021). These bioactive compounds possess antioxidant properties, which help in reducing the toxic effects of free radicals on the body. Free radicals or oxidant molecules are produced in living cells during metabolic

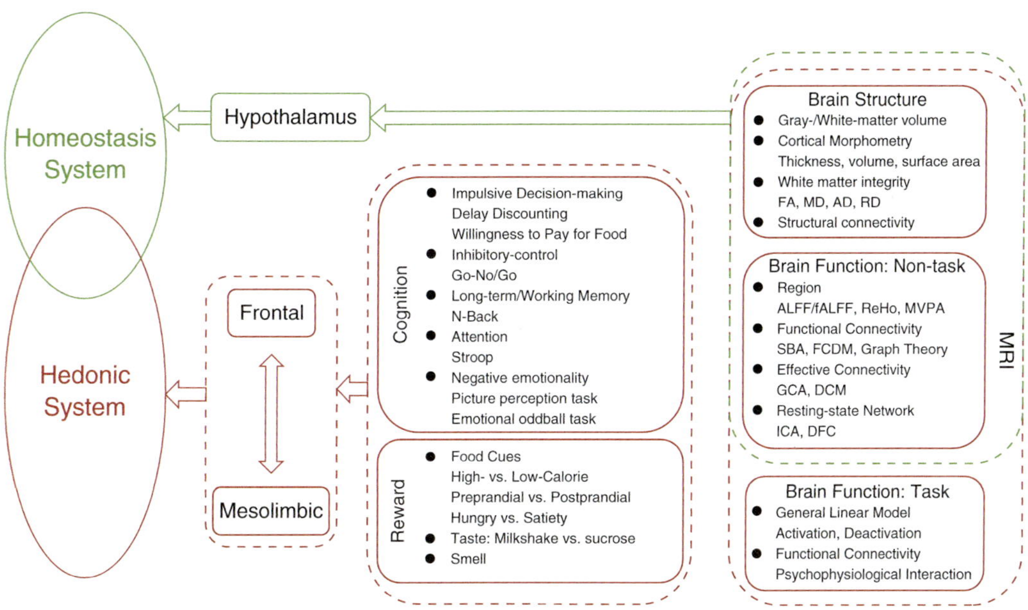

FIGURE 14.2 The association between obesity and brain activity with homeostasis and hedonic process (Li et al. 2023). (Published by Springer Nature in Open Access under Creative Commons Attribution 4.0 International License.)

TABLE 14.1
Functional Foods from Plant-Based Sources and Their Potential Health Benefits on Mental Health

Food Source	Bioactive Compound(s)	Health Benefits (Mental Health & Gut–Brain Axis)	References
Turmeric (*Curcuma longa*)	Curcumin	Antidepressant, anxiolytic, neuroprotective via gut microbiota modulation	Lopresti et al. (2014)
Flaxseeds	Alpha-linolenic acid (ALA), Lignans	Modulation of gut microbiota, improvement in stress resilience, neuroprotection	Davari et al. (2013)
Broccoli	Sulforaphane	Antidepressant and anxiolytic via antioxidant pathways and gut microbiome modulation	You et al. (2024), Sohel et al. (2018)
Green tea (*Camellia sinensis*)	Catechins (EGCG)	1. Anxiolytic, cognitive enhancement, modulates gut microbiota	Kuriyama et al. (2006)
Pomegranate	Polyphenols (punicalagins)	Neuroprotection, anti-anxiety via modulation of oxidative stress and microbiota	Bookheimer et al. (2013), Chen et al. (2023)
Berries (Blueberry)	Anthocyanins	Antioxidant, improves memory, reduces neuroinflammation, modulates gut bacteria	Barooah and Hazarika (2021), Parveen et al. (2022)

reactions. Such free radicals cause oxidation of molecules, affecting physiological reactions. These free radicals also cause mutations to DNA. Anti-oxidants such as free radicals present within these plant-derived functional foods protect against DNA damage. For example, leaf extract of *Moringa oleifera* has been reported to reduce free radicals and enhance cell viability by protecting H_2O_2-induced oxidative damage in human neuroblastoma cells. Moreover, these compounds present in the plant-derived foods are also known to have neuroprotective and anti-allergic properties (Das et al., 2016; Dubey et al., 2024; Table 14.1).

14.2.4 Animal-Origin Functional Foods

Animal origin foods with bioactive compounds include meat, poultry, fish, and milk and dairy products, which have health benefits. The bioactive compounds usually found in animal-origin products are noteworthy, especially short amino-acid sequences, omega-3 fatty acids, minerals, conjugated linoleic acid, etc. (Table 14.2). Bovine colostrum is rich in many bioactive compounds and hence considered a significant source of nutritional value. Bovine-derived products contain enzymes, proteins, growth factors, immunoglobulins, and nucleotides, which have demonstrated many therapeutic benefits in respiratory and gastrointestinal tract disorders in clinical and preclinical studies (Poonia and Shiva, 2022). Omega-3 fatty acids, proteins, and other similar compounds present in fish-based functional foods help in alleviating many health conditions and contribute to better mental health and immunity (Awuchi et al., 2022; Davari et al., 2013). A betaine derivative of B-hydroxybutyrate, L-cartine, is present in animal liver and animal-derived foods. It acts as a long-chain fatty acid carrier across the mitochondrial membrane becoming an essential nutrient in infants and humans with

TABLE 14.2
Animal-Based Functional Foods and Their Health Benefits

Food Source	Bioactive Compound(s)	Health Benefits (Mental Health & Gut-Brain Axis)	References
Yogurt (Probiotic)	*Lactobacillus* spp., *Bifidobacterium* spp.	Modulates gut–brain axis, reduces anxiety and depression symptoms, enhances mood	Wallace and Milev 2017)
Milk	Lactoferrin, CLA	Anxiolytic, stress-resilience effects via microbiome modulation	Wang et al. (2023)
Fish (Salmon, Mackerel)	Omega-3 fatty acids (EPA, DHA)	Antidepressant, improves cognitive function, modulates gut–brain signalling	Grosso et al. (2014)
Colostrum	Immunoglobulins, lactoperoxidase, growth factor	Supports gut barrier integrity, reduces neuroinflammation, mood stabilizer	Playford et al. (2000)
Honey	Phenolic acids, flavonoids	Neuroprotective, anxiolytic, modulates gut microbiome composition	Erejuwa et al. (2014)
Cheese (probiotic varieties)	*Lactobacillus* spp., bioactive peptides	Improves gut microbiota diversity, reduces anxiety and stress-like behaviour	Lozano et al. (2024)

high energy requirements (Rebouche and Seim, 1998). Phylloquinone and menaquinones found in many animal foods scavenge free radicals and deficiency of it is associated with ageing, cardiovascular disorders, and bone health (Zittermann, 2001; Koivu-Tikkanen et al., 2000).

14.3 MECHANISM OF ACTION OF FUNCTIONAL FOODS FOR MENTAL HEALTH

The mechanism by which functional foods benefit mental health involves addressing the ongoing dysfunction and loss of nerve structures, which gradually impair cognitive and motor skills, including memory, learning, decision-making, balance, movement, speech, respiration, and cardiac function. Advances in health care and hygiene have increased life expectancy of people around the world with a proportionate increase in the prevalence of neurodegenerative diseases in the ageing population. Age-related disorders such as Alzheimer's disease (AD) and Parkinson's disease are normally detected late, which restricts the efficacy of the treatment options. Biochemical constituents of functional foods are known to impart health benefits and to decrease the risk of several major diseases, including noncommunicable diseases and neurodegenerative disorders.

14.3.1 Neuroprotection

The functional foods contain bioactive compounds that have the potential to enhance mental health through various mechanisms. Polyphenols, flavonoids, and anthocyanins modulate signalling pathways related to neurotransmission, survival, inflammation, and neuroplasticity, thereby providing neuroprotective activity (Singh et al., 2025). The specific mechanisms by which they exhibit neuroprotection in many neurodegenerative diseases differ greatly. Anthocyanins possess antioxidant and anti-inflammatory properties that reduce oxidative stress, modulate neuroprotection signalling pathways, and inhibit neuroinflammation (Subash et al., 2014). Damage to mitochondrial DNA is a precursor to the onset of multiple sclerosis. This antioxidant facilitates the transport of fatty acid molecules for beta-oxidation in mitochondria, thus protecting tissues from ionizing radiation by chemically quenching ROS and biologically inducing TFs (NrF2) to promote the expression of genes, primarily phase II detoxifier genes. Neuroprotective effect is also exhibited by other bioactive compounds. For example, L-carnitine is oxidized to produce energy by transferring long-chain fatty acids into mitochondria and also provides protection against inflammation. Another compound with significant neuroprotective potential is caffeic acid, which is found in common berries. Caffeic acid stimulates guanine triphosphatase activity and reduces oxidative stress. Moreover, it also possesses anti-inflammatory properties through the inhibition of cyclooxygenase. This compound provides anti-inflammatory effects, improves performance in the Morris Water Maze, and prevents neurochemical changes. Furthermore, it leads to a reduction in brain-derived neurotrophic factor, interleukin-1, tumour markers, and Aβ levels. These effects collectively contribute to improvements in memory, motor skills, and cognition (Erdogan et al., 2024; Hein et al., 2019). Ellagic acid, a compound present in strawberries, has been linked to a reduction in the ratio of insoluble amyloid beta 42/40, a factor associated with AD pathology (Xiong et al., 2024; Zota et al., 2024). Quercetin, the most studied polyphenol, exhibits a neuroprotective effect by suppressing caspase activity, regulating glutathione levels, inducing Nrf2-ARE and paraxonase 2, decreasing intracellular reactive oxygen species, and promoting autophagy. Curcumin from turmeric has been shown to possess neuroprotective effects in rats by modifying PI3K/AKT signalling pathways, thus providing protection against sciatic nerve injury (Sang et al., 2018). Moreover, it has the potential to reverse the effects of dementia on memory by modulating BDNF and ERK protein expression (Liu et al., 2014). In addition to plant-based polyphenols, certain phytochemicals that mimic human hormones also demonstrate neuroprotective effects. Estrogen, a phytochemical found in flaxseeds, garlic, sesame seeds, and peaches, activates rapid signalling pathways such as PI3K and MAPK (Singh et al., 2025). These pathways are essential for neuroprotection (Bryant et al., 2006). Flavonoids and their metabolites may exert neuroprotection through mechanisms beyond conventional antioxidant activities, such as interacting with cell signalling cascades and influencing gene expression (Youdim et al., 2002). These compounds show promise in maintaining cognitive function and preventing age-related neurological deficits, highlighting the potential of dietary interventions in neuroprotection (Costa et al., 2016). The neuroprotective mechanisms of flavonoids and their metabolites extend beyond their antioxidant properties, highlighting the complexity of their interactions with cellular processes.

14.3.2 Neurotransmitters Regulation

The bioactive compounds present in the functional foods promote brain health by regulating neurotransmitters. These foods have the potential to prevent brain and psychiatric disorders by influencing cellular processes such as synaptic transmission, neurogenesis, and synaptogenesis (Reza-Zaldívar and Jacobo-Velázquez, 2023; Gasmi et al., 2023). Certain nutrients, such as flavonoids, omega-3 fatty acids, and probiotics, can modulate neurotransmitter/neuropeptide regulation through which they influence BBB function (Aslam et al., 2020; Scuto et al., 2024). The peroxisome-proliferator-activated receptor pathway could potentially link functional foods to mood improvement by regulating the biosynthesis of allopregnanolone and brain-derived neurotrophic factor (Matrisciano and Pinna, 2020; Playford et al., 2000). Plant polyphenols have shown neuroprotective properties by modulating neurotransmitter metabolism and receptor function, offering potential therapeutic applications in many neurodegenerative disorders (Rebas et al., 2020). These compounds can activate the Nrf2 antioxidant pathway, protecting against oxidative stress and neurotoxicity (Scuto et al., 2024). Ω-3 fatty acid such as alpha-linolenic acid (ALA) and Ω-6 fatty acid like linoleic acid (LA) from walnuts enhance cognitive functions by producing pro- and anti-inflammatory signalling molecules like prostaglandins. Such prostaglandins are derived from metabolism of ALA and LA into lipid messengers like eicosapentaenoic acid (EPA), docosahexaenoic acid (DHA), or arachidonic acid (Das, 2006; Carey et al., 2013). Thus, functional foods containing bioactive compounds offer promising potential for promoting brain health, preventing neurological disorders, and modulating neurotransmitter function through various cellular and molecular mechanisms, highlighting their importance in neuroprotection and potential therapeutic applications.

14.3.3 Gut Microbiome Regulation

Fermented foods, considered functional due to their potential health benefits, contain probiotics, prebiotics, and bioactive components that can modify the gut microbiota and influence host nutrition (Schneider et al., 2024). The gut microbiota plays a crucial role in the bidirectional communication between the gut and the brain, known as the gut microbiota-brain axis. This axis has significant effects on mental health and brain function. Alterations in gut microbiota composition have been associated with anxiety-like and depressive-like behaviours. Stress has been linked to changes in gut microbiota composition (O'Mahony et al., 2009; Collins et al., 2012). Several studies have reported altered compositions of intestinal microbiota in ASD patients, including higher levels of certain bacterial species and lower levels of others. Certain probiotics have been shown to affect brain activity regions that control central processing of emotion and sensation (Wang et al., 2011; De Angelis et al., 2013). The gut microbiota can communicate with the brain via the vagus nerve. The enteric nervous system (ENS) also plays a crucial role in this communication, with gut microbiota regulating ENS integrity through TLR-2 signalling (Bravo et al., 2011). Certain probiotics have been shown to normalize immune responses, affecting the release of cytokines like IL-6 and altering corticotrophin-releasing factor mRNA levels in the amygdala (Desbonnet et al., 2010). Certain strains of *Lactobacillus, Saccharomyces, Escherichia,* and *Bifidobacterium* play a crucial role in the production of neurotransmitters such as GABA, norepinephrine, serotonin, and dopamine (Barrett et al., 2012; Dicks, 2023; Ragavan and Hemalatha, 2024). Short-chain fatty acids (SCFAs) produced by such strains can influence brain function by activation of G-protein-coupled receptors and many other mechanisms (Carretta et al., 2021; Kasubuchi et al., 2015). Gut microbiota can influence tryptophan metabolism, which is crucial for serotonin production and thus affects mood and behaviour (Hou et al., 2023; Roth et al., 2021). The intricate relationship between fermented foods, gut microbiota, and the gut–brain axis highlights the potential of probiotic-rich foods to influence mental health and brain function through various mechanisms, including neurotransmitter production, immune modulation, and metabolite signalling.

14.4 CURRENT PRACTICES AND INTERVENTIONS

Dietary interventions, including specific supplementations and diet plans, have emerged as a promising strategy for managing various disorders by modulating the microbiota–gut–brain axis. This axis plays a crucial role in influencing brain health and cognitive functions. Research has shown that incorporating nutraceuticals and functional foods into regular diets can significantly impact this axis. Mediterranean-like diets and fermented foods, in particular, have demonstrated benefits for symptoms of depression and immune function in healthy individuals (Roy et al., 2024). The gut microbiota produces a range of compounds, including cytokines, neurotransmitters, and microbial metabolites, which can affect brain function through immune, endocrine, and neural pathways (Roy et al., 2024; Ribeiro et al., 2022). This diet is also associated with greater microbial diversity of health-promoting bacterial taxa and better metabolomic profiles such as reductions in bile acids, branched chain fatty acids (BCFAs), and trimethylamine N-oxide, and increased levels of SCFAs.

Traditional fermented foods consumed in different cultures offer potential cost-effective alternatives to commercial probiotic supplements. These foods are often prepared using seasonal raw materials from local agriculture, making them accessible and culturally relevant (Dahiya and Nigam, 2022a). While animal studies have provided valuable insights into the effects of nutritional interventions on the microbiota–gut–brain axis (Berding et al., 2022), there is a need for more clinical trials to establish direct evidence-based recommendations for cognitive and emotional functioning in humans (Schneider et al., 2024; Sharma et al., 2024).

Current practices and interventions for functional foods targeting the gut–brain axis increasingly focus on personalized nutrition and combined approaches. Researchers are recognizing the importance of tailoring dietary interventions based on an individual's unique gut microbiota composition, as baseline microbial profiles can significantly influence the efficacy of functional foods (Donoso et al., 2023; Zhou et al., 2024). Personalized nutrition approaches are emerging, recognizing that individual responses to dietary interventions may vary. For example, probiotic supplementation has shown cognitive benefits in healthy older individuals and patients with mild cognitive impairment and AD, but results are mixed in younger, healthier individuals (Fekete et al., 2024). Personalized approaches considering individual metabolic and microbiome profiles are emphasized for optimizing health outcomes. Combined dietary-probiotic approaches, such as incorporating probiotics into non-dairy substrates like grains, cereals, vegetables, and fruits, are being explored to cater to diverse dietary preferences and needs. By integrating specific probiotic strains with carefully selected functional foods, researchers aim to create synergistic effects that may more effectively modulate the gut microbiome and, consequently, impact brain function and behaviour (Dahiya and Nigam, 2022b). Combined dietary-probiotic approaches, such as incorporating probiotics into non-dairy substrates like grains, cereals, vegetables, and fruits, are being explored to cater to diverse dietary preferences and needs. However, it is important to note that not all probiotic, prebiotic, or dairy-rich diet-based treatments exhibit psychobiotic effects on the central nervous system. Factors such as probiotic composition, quantity, and study duration significantly influence outcomes (Zagórska et al., 2020). These combined interventions hold promise for addressing individual variability in response to functional foods and potentially improving outcomes in gut–brain axis modulation. Interestingly, traditional fermented foods, while different from commercial probiotic preparations, may also offer potential biotic benefits in nutrition (Dahiya and Nigam, 2022a, Collins et al., 2012). Advanced computational tools, in vitro techniques, and in vivo studies are being employed to better understand and predict personalized responses to dietary, prebiotic, and probiotic interventions (Gibbons et al., 2022). Future research should focus on comprehensive characterization of participants, including dietary patterns, gastrointestinal phenotype, and gut microbiota composition, to identify potential responders to interventions and develop more effective personalized nutrition strategies (Roy, 2024; Ceccon et al., 2024).

14.5 CONCLUSION

Functional foods impact the gut-brain axis, potentially improving mental health and well-being. Bioactive compounds, probiotics, and prebiotics can modulate neurotransmitters, provide neuroprotection, and influence gut microbiome composition, affecting cognitive function, mood, and neurological disorders. Personalized nutrition approaches based on individual microbiome profiles and metabolic characteristics are emerging. Combined interventions, like integrating probiotics with specific functional foods, show promise for enhanced gut–brain axis modulation. Further, research is needed to establish evidence-based recommendations, focusing on comprehensive participant characterization. Advanced computational tools, in vitro techniques, and in vivo studies will be crucial for predicting and optimizing individual responses to functional food interventions, potentially revolutionizing nutrition's role in mental health.

REFERENCES

Arshad, M. S., Khalid, W., Ahmad, R. S., Khan, M. K., Ahmad, M. H., Safdar, S., & Kousar, S. et al. 2021. Functional foods and human health: an overview. In *Functional Foods: Phytochemicals and Health Promoting Potential*, edited by M. S. Arshad and M. H. Ahmad. IntechOpen. https://doi.org/10.5772/intechopen.99000.

Aslam, H., Green, J., Jacka, F. N., Collier, F., Berk, M., Pasco, J. & Dawson, S. L. 2020. Fermented foods, the gut and mental health: a mechanistic overview with implications for depression and anxiety. *Nutritional Neuroscience* 23, 659–671.

Awuchi, C. G., Chukwu, C. N., Iyiola, A. O., Noreen, S., Morya, S., Adeleye, A. O., Twinomuhwezi, H., Leicht, K., Mitaki, N. B., & Okpala, C. O. R. 2022. Bioactive compounds and therapeutics from fish: revisiting their suitability in functional foods to enhance human wellbeing. *BioMed Research International* 2022, 3661866. https://doi.org/10.1155/2022/3661866

Bae, J. M., Lee, E. J., & Guyatt, G. J. 2008. Citrus fruit intake and stomach cancer risk: a quantitative systematic review. *Gastric Cancer* 11, 23–32.

Barooah, M. & Hazarika, D. J. 2021. Roles of Functional Foods in Neuroprotection. *Antioxidants and Functional Foods for Neurodegenerative Disorders*. CRC Press.

Barrett, E., Ross, R., O' Toole, P. W., Fitzgerald, G. F. & Stanton, C. 2012. γ-Aminobutyric acid production by culturable bacteria from the human intestine. *Journal of Applied Microbiology* 113, 411–417.

Berding, K., Kaulmann, D. & Cryan, J. F. 2022. Food-gut microbiota interactions. *Probiotics*. Elsevier.

Bintanah, S., Ulvie, Y. N. S., Kusuma, H. S., Jauharany, F. F. & Sulistyaningrum, H. 2023. Functional food innovation for the treatment of metabolic syndrome. In *Food Sustainability, Environmental Awareness, and Adaptation and Mitigation Strategies for Developing Countries*, edited by M. D. Rahman, H. S. Jauharany, and I. P. Sulistyaningrum. IGI Global. https://doi.org/10.4018/978-1-6684-5629-3.ch002.

Bookheimer, S. Y., Renner, B. A., Ekstrom, A., Li, Z., Henning, S. M., Brown, J. A., Jones, M., Moody, T., & Small, G. W. 2013. Pomegranate juice augments memory and fmri activity in middle-aged and older adults with mild memory complaints. *Evidence-based Complementary and Alternative Medicine* 2013, 946298.

Bravo, J. A., Forsythe, P., Chew, M. V., Escaravage, E., Savignac, H. M., Dinan, T. G., Bienenstock, J., & Cryan, J. F. 2011. Ingestion of Lactobacillus strain regulates emotional behavior and central GABA receptor expression in a mouse via the vagus nerve. *Proceedings of the National Academy of Sciences of the United States of America* 108, 16050–16055.

Bryant, D. N., Sheldahl, L. C., Marriott, L. K., Shapiro, R. A. & Dorsa, D. M. 2006. Multiple pathways transmit neuroprotective effects of gonadal steroids. *Endocrinology* 29, 199–207.

Carey, A. N., Fisher, D. R., Joseph, J. A., & Shukitt-Hale, B. 2013. The ability of walnut extract and fatty acids to protect against the deleterious effects of oxidative stress and inflammation in hippocampal cells. *Journal of Nutritional Neuroscience* 16, 13–20.

Carretta, M. D., Quiroga, J., López, R., Hidalgo, M. A., & Burgos, R. A. 2021. Participation of short-chain fatty acids and their receptors in gut inflammation and colon cancer. *Frontiers in Immunology* 12, 662739.

Ceccon, M., Kantsjö, J. B., & Ronchi, F. 2024. Personalized paths: unlocking Alzheimer's via the gut-brain axis. *Journal of Veterinary Medicine* 40, 194–209.

Chen, P., Chen, F., Lei, J., & Zhou, B. 2023. Pomegranate polyphenol punicalagin improves learning memory deficits, redox homeostasis, and neuroinflammation in aging mice. *Journal of Pharmaceutical Research* 37, 3655–3674.

Collins, S. M., Surette, M., & Bercik, P. 2012. The interplay between the intestinal microbiota and the brain. *Nature Reviews Microbiology* 10, 735–742.

Costa, L. G., Garrick, J. M., Roquè, P. J., & Pellacani, C. 2016. Mechanisms of neuroprotection by quercetin: counteracting oxidative stress and more. *Oxidative Medicine and Cellular Longevity* 2016, 2986796. https://doi.org/10.1155/2016/2986796.

Dahiya, D., & Nigam, P. S. 2022a. Probiotics, prebiotics, synbiotics, and fermented foods as potential biotics in nutrition improving health via microbiome-gut-brain axis. *Journal of Food* 8, 303.

Dahiya, D., & Nigam, P. S. 2022b. Use of characterized microorganisms in fermentation of non-dairy-based substrates to produce probiotic food for gut-health and nutrition. *Journal of Food* 9, 1.

Damude, H. G., & Kinney, A. J. 2008. Engineering oilseeds to produce nutritional fatty acids. *Journal of Plant Physiology* 132, 1–10.

Das, R., Biswas, S., & Banerjee, E. R. 2016. Nutraceutical-prophylactic and therapeutic role of functional food in health. *Journal of Nutraceuticals and Functional Foods* 6, 1–17.

Das, U. N. 2006. Essential fatty acids-a review. *Current Pharmaceutical Biotechnology* 7, 467–482.

Davari, S., Talaei, S. A. & Alaei, H. 2013. Probiotics treatment improves diabetes-induced impairment of synaptic activity and cognitive function: behavioral and electrophysiological proofs for microbiome–gut–brain axis. *Journal of Neuroimmunology* 240, 287–296.

De Angelis, M., Piccolo, M., Vannini, L., Siragusa, S., De Giacomo, A., Serrazanetti, D. I., Cristofori, F., Guerzoni, M. E., Gobbetti, M., & Francavilla, R. 2013. Fecal microbiota and metabolome of children with autism and pervasive developmental disorder not otherwise specified. *PLoS ONE* 8, e76993.

Desbonnet, L., Garrett, L., Clarke, G., Kiely, B., Cryan, J. F., & Dinan, T. G. 2010. Effects of the probiotic Bifidobacterium infantis in the maternal separation model of depression. *Neuroscience* 170, 1179–1188.

Dhiman, A., Chopra, R., Singh, P. K., Homroy, S., Chand, M., & Talwar, B. 2024. Amelioration of nutritional properties of bakery fat using omega-3 fatty acid-rich edible oils: a review. *Journal of the Science of Food and Agriculture* 104, 3175–3184.

Dicks, L. M. T. 2023. Our mental health is determined by an intrinsic interplay between the central nervous system, enteric nerves, and gut microbiota. *International Journal of Molecular Sciences* 25, 38.

Donoso, F., Cryan, J. F., Olavarría-Ramírez, L., Nolan, Y. M., & Clarke, G. 2023. Inflammation, lifestyle factors, and the microbiome-gut-brain axis: relevance to depression and antidepressant action. *Clinical Pharmacology & Therapeutics* 113, 246–259.

Dubey, R. K., Shukla, S., Shukla, V., & Singh, S. 2024. Sea buckthorn: a potential dietary supplement with multifaceted therapeutic activities. *International Phytomedicine* 2, 681–687.

Erdogan, M., Colak, T., Ceylan, F. S., Kir, H. M., Kurnaz, S., Ozsoy, O. D., & Sahin, Z. 2024. Neuroprotective activity of strawberry tree (*Arbutus unedo* L.) against formaldehyde-induced oxidative stress in the rat hippocampus. *International Journal of Morphology* 42(1).

Erejuwa, O. O., Sulaiman, S. A., & Wahab, M. S. A.. 2014. Effects of honey and its mechanisms of action on the development and progression of cancer. *Molecules* 19, 2497–2522.

Faridi, Z., Njike, V. Y., Dutta, S., Ali, A., & Katz, D. L. 2008. Acute dark chocolate and cocoa ingestion and endothelial function: a randomized controlled crossover trial1. *The American Journal of Clinical Nutrition* 88, 58–63.

Fekete, M., Lehoczki, A., Major, D., Fazekas-Pongor, V., Csípő, T., Tarantini, S., Csizmadia, Z., & Varga, J. T. 2024. Exploring the influence of gut–brain axis modulation on cognitive health: a comprehensive review of prebiotics, probiotics, and symbiotics. *Nutrients* 16, 789.

Galiniak, S., Aebisher, D., & Bartusik-Aebisher, D. 2019. Health benefits of resveratrol administration. *Acta Biochimica Polonica* 66, 13–21.

Gasmi, A., Nasreen, A., Menzel, A., Gasmi Benahmed, A., Pivina, L., Noor, S., Peana, M., Chirumbolo, S., & Bjørklund, G. 2022. Neurotransmitters regulation and food intake: the role of dietary sources in neurotransmission. *Molecules* 28, 210.

Gibbons, S. M., Gurry, T., Lampe, J. W., Chakrabarti, A., Dam, V., Everard, A., Goas, A., Gross, G., Kleerebezem, M., & Lane, J. 2022. Perspective: leveraging the gut microbiota to predict personalized responses to dietary, prebiotic, and probiotic interventions. *Advances in Nutrition* 13, 1450–1461.

Grosso, G., Galvano, F., Marventano, S., Malaguarnera, M., Bucolo, C., Drago, F., & Caraci, F. 2014. Omega-3 fatty acids and depression: scientific evidence and biological mechanisms. *Oxidative Medicine and Cellular Longevity* 2014, 313570.

Hein, S., Whyte, A. R., Wood, E., Rodriguez-Mateos, A., & Williams, C. M. 2019. Systematic review of the effects of blueberry on cognitive performance as we age. *The Journals of Gerontology: Series A* 74, 984–995.

Hou, Y., Li, J., & Ying, S. 2023. Tryptophan metabolism and gut microbiota: a novel regulatory axis integrating the microbiome, immunity, and cancer. *Molecules* 13, 1166.

Jariwala, R., Mandal, H., & Bagchi, T. 2017. Indigenous lactobacilli strains of food and human sources reverse enteropathogenic E. coli O26: H11-induced damage in intestinal epithelial cell lines: effect on redistribution of tight junction proteins. *Microbiological Research* 163, 1263–1272.

Kasubuchi, M., Hasegawa, S., Hiramatsu, T., Ichimura, A., & Kimura, I. 2015. Dietary gut microbial metabolites, short-chain fatty acids, and host metabolic regulation. *Nutrients* 7, 2839–2849.

Kavanaugh, C. J., Trumbo, P. R., & Ellwood, K. C. 2007. The US Food and Drug Administration's evidence-based review for qualified health claims: tomatoes, lycopene, and cancer. *Journal of the National Cancer Institute* 99, 1074–1085.

Koivu-Tikkanen, T. J., Ollilainen, V., & Piironen, V. I. 2000. Determination of phylloquinone and menaquinones in animal products with fluorescence detection after postcolumn reduction with metallic zinc. *Journal of Agricultural and Food Chemistry* 48, 6325–6331.

Kris-Etherton, P. M., Hu, F. B., Ros, E., & Sabaté, J.. 2008. The role of tree nuts and peanuts in the prevention of coronary heart disease: multiple potential mechanisms. *The Journal of Nutrition* 138, 1746S–1751S.

Kumari, A., K. G., R., Sudhakaran, V. A., Warrier, A. S., & Singh, N. K. 2024. Unveiling the health benefits of prebiotics: a comprehensive review. *International Journal of Microbiology* 64, 376–388.

Kuriyama, S., Hozawa, A., Ohmori, K., Shimazu, T., Matsui, T., Ebihara, S., Awata, S., Nagatomi, R., Arai, H., & Tsuji, I. 2006. Green tea consumption and cognitive function: a cross-sectional study from the Tsurugaya Project. *American Journal of Clinical Nutrition* 83, 355–361.

Li, G., Hu, Y., Zhang, W., Wang, J., Ji, W., Manza, P., Volkow, N. D., Zhang, Y., & Wang, G.-J. 2023. Brain functional and structural magnetic resonance imaging of obesity and weight loss interventions. *Molecular Psychiatry* 28 (March). https://doi.org/10.1038/s41380-023-02025-y.

Liu, D., Wang, Z., Gao, Z., Xie, K., Zhang, Q., Jiang, H. & Pang, Q. 2014. Effects of curcumin on learning and memory deficits, BDNF, and ERK protein expression in rats exposed to chronic unpredictable stress. *American Journal of Clinical Nutrition* 271, 116–121.

Lopresti, A. L., Maes, M., Maker, G. L., Hood, S. D. & Drummond, P. D. 2014. Curcumin for the treatment of major depression: a randomised, double-blind, placebo controlled study. *Behavioural Brain Research* 167, 368–375.

Lozano, J., Fabius, S., Fernández-Ciganda, S., Urbanavicius, J., Piccini, C., Scorza, C. & Zunino, P. 2024. Beneficial effect of GABA-producing Lactiplantibacillus strain LPB145 isolated from cheese starters evaluated in anxiety-and depression-like behaviours in rats. *Journal of Affective Disorders* 1, 1–15.
Mafe, A. N., Sharifi-Rad, J., Calina, D., Ogunyemi, A. J., & Tubi, A. O. 2025. Postbiotics in functional foods: microbial derivatives shaping health, immunity and next-generation nutrition. *Food Frontiers* 7 (1). https://doi.org/10.1002/fft2.70205.
Matrisciano, F. & Pinna, G. 2020. PPAR and functional foods: rationale for natural neurosteroid-based interventions for postpartum depression. *Behavioral Microbiology* 12, 100222.
Meng, X., Zhou, J., Zhao, C.-N., Gan, R.-Y. & Li, H.-B. 2020. Health benefits and molecular mechanisms of resveratrol: a narrative review. *Neurobiology of Stress* 9, 340.
Moriarty, R. M., Naithani, R. & Surve, B. 2007. Organosulfur compounds in cancer chemoprevention. *Mini-Reviews in Medicinal Chemistry* 7, 827–838.
Nayak, S. N., Aravind, B., Malavalli, S. S., Sukanth, B., Poornima, R., Bharati, P., Hefferon, K., Kole, C. & Puppala, N. 2021. Omics technologies to enhance plant based functional foods: an overview. *Frontiers in Genetics* 12, 742095.
Nowack, R. 2007. Cranberry juice: a well-characterized folk-remedy against bacterial urinary tract infection. *Wiener Medizinische Wochenschrift* 157, 325–330.
O' Mahony, S. M., Marchesi, J. R., Scully, P., Codling, C., Ceolho, A.-M., Quigley, E. M., Cryan, J. F. & Dinan, T. G. 2009. Early life stress alters behavior, immunity, and microbiota in rats: implications for irritable bowel syndrome and psychiatric illnesses. *Biological Psychiatry* 65, 263–267.
Parveen, S., Bhat, I. U. H. & Bhat, R. 2022. Berries as functional foods: molecular mechanisms promoting human health. In *Molecular Mechanisms of Functional Food*, edited by R. Bhat, 596–621. Wiley.
Playford, R. J., Macdonald, C. E. & Johnson, W. S. 2000. Colostrum and milk-derived peptide growth factors for the treatment of gastrointestinal disorders. *The American Journal of Clinical Nutrition* 72, 5–14.
Poonia, A., & Shiva, P. 2022. Bioactive compounds, nutritional profile and health benefits of colostrum: a review. *Journal of Food Production and Nutrition* 4, 26.
Ragavan, M. L., & Hemalatha, S. 2024. The functional roles of short chain fatty acids as postbiotics in human gut: future perspectives. *Food Science and Biotechnology* 33, 275–285.
Rebas, E., Rzajew, J., Radzik, T. & Zylinska, L. 2020. Neuroprotective polyphenols: a modulatory action on neurotransmitter pathways. *Current Neuropharmacology* 18, 431–445.
Rebouche, C. J. & Seim, H. 1998. Carnitine metabolism and its regulation in microorganisms and mammals. *Annual Review of Nutrition* 18, 39–61.
Reza-Zaldívar, E. E. & Jacobo-Velázquez, D. A. 2023. Comprehensive review of nutraceuticals against cognitive decline associated with Alzheimer's disease. *Antioxidants* 8, 35499–35522.
Ribeiro, G., Ferri, A., Clarke, G., & Cryan, J. F. 2022. Diet and the microbiota–gut–brain-axis: a primer for clinical nutrition. *Current Opinion in Clinical Nutrition and Metabolic Care* 25, 443–450.
Roth, W., Zadeh, K., Vekariya, R., Ge, Y. & Mohamadzadeh, M. 2021. Tryptophan metabolism and gut-brain homeostasis. *International Journal of Molecular Sciences* 22, 2973.
Roy, N. 2024. Diet, Gut Physiology, and the Microbiota-Gut-Brain Axis. *Proceedings of the Nutrition Society* 83, E13.
Roy, S., Nisha, P. & Chakraborty, R. 2024. *Traditional Foods: The Reinvented Superfoods*, Springer.
Sang, Q., Sun, D., Chen, Z., & Zhao, W. 2018. NGF and PI3K/Akt signaling participate in the ventral motor neuronal protection of curcumin in sciatic nerve injury rat models. *Biomedicine & Pharmacotherapy* 103, 1146–1153.
Schneider, E., Balasubramanian, R., Ferri, A., Cotter, P. D., Clarke, G., & Cryan, J. F. 2024. Fibre & fermented foods: differential effects on the microbiota-gut-brain axis. *Proceedings of the Nutrition Society*. 1–16.
Scrinis, G. 2008. Functional foods or functionally marketed foods? A critique of, and alternatives to, the category of 'functional foods'. *Public Health Nutrition* 11, 541–545.
Scuto, M., Majzúnová, M., Torcitto, G., Antonuzzo, S., Rampulla, F., Di Fatta, E. & Trovato Salinaro, A. 2024. Functional food nutrients, redox resilience signaling and neurosteroids for brain health. *International Journal of Molecular Sciences* 25, 12155.
Shah, J. M., Murad, W., Mubin, S., Ullah, O., Rehman, N. U., & Rahman, M. H. 2022. Multiple health benefits of curcumin and its therapeutic potential. *Environmental Science and Pollution Research* 29, 43732–43744.
Sharma, M., Vidhya, C., Ojha, K., Yashwanth, B., Singh, B., Gupta, S. & Pandey, S. K. 2024. The role of functional foods and nutraceuticals in disease prevention and health promotion. *European Journal of Nutrition & Food Safety* 16, 61–83.
Shi, L. H., Balakrishnan, K., Thiagarajah, K., Ismail, N. I. M. & Yin, O. S. 2016. Beneficial properties of probiotics. *Malaysian Journal of Life Sciences Research* 27, 73.

Shivanna, K. R. 2018. HY Mohan Ram (1930–2018). *Current Science* 115(6), 1005–1006.

Singh, L., Wani, A. W., Shaifali, Sadawarti, R. K., Kaur, H., Bashir, O., Majeed, J., Mirza, A. A., Sharma, J., & Ali, S. 2025. A comprehensive review on neurotrophic receptors and their implications in brain health: exploring the neuroprotective potential of berries. *Journal of Berry Research* 15, 48–64.

Sohel, M. M. H., Amin, A., Prastowo, S., Linares-Otoya, L., Hoelker, M., Schellander, K., & Tesfaye, D. 2018. Sulforaphane protects granulosa cells against oxidative stress via activation of NRF2-ARE pathway. *Cell and Tissue Research* 374, 629–641.

Subash, S., Essa, M. M., Al-Adawi, S., Memon, M. A., Manivasagam, T., & Akbar, M. 2014. Neuroprotective effects of berry fruits on neurodegenerative diseases. *Neural Regeneration Research* 9, 1557–1566.

Vigar, V., Myers, S., Oliver, C., Arellano, J., Robinson, S., & Leifert, C. 2019. A systematic review of organic versus conventional food consumption: is there a measurable benefit on human health? *Nutrients* 12, 7.

Wallace, C. J. K., & Milev, R. 2017. The effects of probiotics on depressive symptoms in humans: a systematic review. *Annals of General Psychiatry* 16, 1–10.

Wang, L., Christophersen, C. T., Sorich, M. J., Gerber, J. P., Angley, M. T., & Conlon, M. A. 2011. Low relative abundances of the mucolytic bacterium Akkermansia muciniphila and Bifidobacterium spp. in feces of children with autism. *Applied and Environmental Microbiology* 77, 6718–6721.

Wang, W., Cheng, Z., Wang, X., An, Q., Huang, K., Dai, Y., Meng, Q., & Zhang, Y. 2023. Lactoferrin deficiency during lactation increases the risk of depressive-like behavior in adult mice. *Brain Behavior* 1, 242.

Xiong, H.-Y., Hendrix, J., Schabrun, S., Wyns, A., Van Campenhout, J., Nijs, J., & Polli, A. 2024. The role of the brain-derived neurotrophic factor in chronic pain: links to central sensitization and neuroinflammation. *Brain Sciences* 14, 71.

Xu, Y. J. 2001. Perspectives on the 21st century development of functional foods: bridging Chinese medicated diet and functional foods. *International Journal of Food Science & Technology* 36, 229–242.

You, M., Chen, N., Yang, Y., Cheng, L., He, H., Cai, Y., Liu, Y., Liu, H., & Hong, G. 2024. The gut microbiota–brain axis in neurological disorders. *Medicine* 5, e656.

Youdim, K. A., Spencer, J. P. E., Schroeter, H., & Rice-Evans, C. 2002. Dietary flavonoids as potential neuroprotectants. *Biological Chemistry* 383(3–4), 503–519.

Zagórska, A., Marcinkowska, M., Jamrozik, M., Wiśniowska, B., & Paśko, P. 2020. From probiotics to psychobiotics–the gut-brain axis in psychiatric disorders. *Biomedicine & Pharmacotherapy* 11, 717–732.

Zhou, P., Chen, C., Patil, S., & Dong, S. 2024. Unveiling the therapeutic symphony of probiotics, prebiotics, and postbiotics in gut-immune harmony. *Frontiers in Immunology* 11, 1355542.

Zittermann, A. 2001. Effects of vitamin K on calcium and bone metabolism. *Current Opinion in Clinical Nutrition and Metabolic Care* 4, 483–487.

Zota, I., Chanomidou, K., Gravanis, A., & Charalampopoulos, I. 2024. Stimulating myelin restoration with BDNF: a promising therapeutic approach for Alzheimer's disease. *Frontiers in Cellular Neuroscience* 18, 1422130.

15 Adaptations in Lifestyles, Herbal Care, Physical and Psychosocial Therapies

Monica Saini, Preeti Solanki, Kanwarbir Singh Gill, and Jaibeer Singh Chhina

15.1 INTRODUCTION

The gut–brain axis (GBA) is a significant bidirectional communication system, which intensely impacts mood, cognition, and immune functioning, particularly in geriatric inhabitants. With progressing age, gut microbial range decreases, leading to intensified inflammation and cognitive disorders. Though, lifestyle variations, herbal preparations, and therapeutic involvements can reinstate microbial balance, decrease inflammation, and revitalize mental health. A stable gut microbiota decreases neuroinflammation and sustains brain strength by producing neurotransmitters, for instance, serotonin and gamma-aminobutyric acid (GABA). Nutritional supplements including prebiotics, probiotics, and symbiotics improve gastrointestinal (GI) utility and advance cognitive progressions comprising attention, insight, and memory (Fekete et al. 2024). Dietary preparations abundant in fibre, chiefly plant-based and Mediterranean nourishments, encourage microbial multiplicity and improve gut–brain communication, ensuring decreased complete inflammation and mental health decline (Patil and Mehdi 2025). Cryan et al. (2019) established that probiotic supplements in the elderly escalate microbial range, positively impacting neurotransmitter levels, thus increasing cognitive functioning and maintenance of mood.

This underlines the GBA's crucial role in geriatric mental well-being. Likewise, Ghosh et al. (2020) stated that adherence to a Mediterranean diet slows cognitive deterioration by modifying gut microbiota make-up. Furthermore, mindfulness-based stress reduction (MBSR) enriches gut health by lessening stress-related GI symptoms and improving microbial profiles in older adults while highlighting psychosocial treatments' significance in GBA variation (Kuo et al. 2017). These results support a holistic method merging diet, lifestyle, and psychosocial involvements to enhance mental and physical well-being in ageing inhabitants via GBA improvement (Figure 15.1).

Lifestyle modifications, herbal medications, and therapies play vital roles in dealing with mental health, especially in geriatric and psychiatric care. These interventions decrease symptoms of psychiatric disorders, improve cognition, and upgrade quality of life more efficiently than only pharmacotherapy (Sharma et al. 2024; Nguyen and Lavretsky 2020). Planned routines, workouts, sleep hygiene, and social connections reduce depression and anxiety in older individuals and patients with psychiatric disorders. Exercise improves neuroplasticity and self-worth, whereas social engagements shield loneliness—a crucial geriatric risk (Sharma et al. 2024). Herbal medications such as St. John's wort and Ginkgo biloba modify neurotransmitters and inflammation with lesser side effects than Selective serotonin reuptake inhibitors (SSRIs) in the weak elderly (Nguyen and Lavretsky 2020). Mind-body treatments (yoga, meditation, music/art) further lessen stress and cognitive deterioration. They advance emotional regulation, adaptation, and adherence in long-lasting illness (Nguyen and Lavretsky 2020).

Dysbiosis triggers depression, anxiety, and cognitive disorders through inflammation and reduced serotonin production. Lifestyle variations promote microbial variety and short-chain fatty acids (SCFAs),

DOI: 10.1201/9781003610274-18

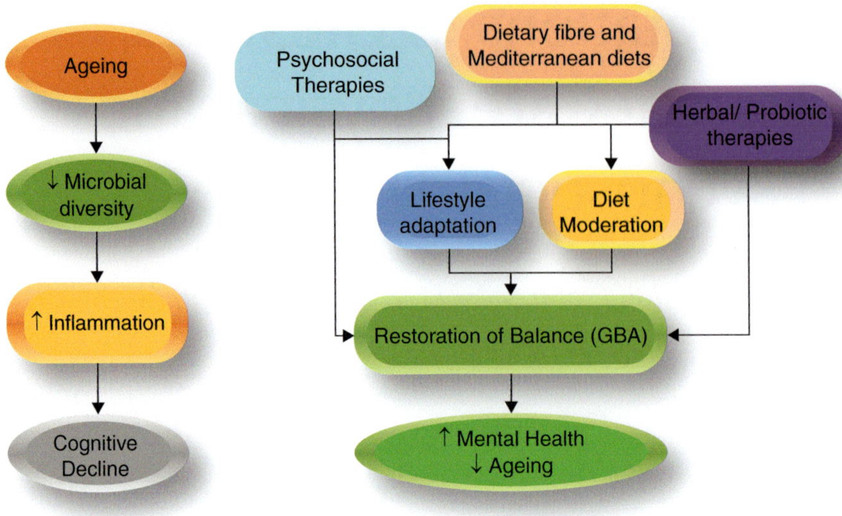

FIGURE 15.1 The role of lifestyle modifications, herbal remedies, and therapies in mental well-being.

dropping gut permeability (Sharma et al. 2024). Herbals with polyphenols reform microbiota, improving vagus nerve signalling (Pferschy-Wenzig et al. 2022; Guan et al. 2024). Therapies reduce hypothalamic–pituitary–adrenal (HPA) axis movement, stabilizing microbiome balance in some way (Nguyen and Lavretsky 2020). Thus, multimodal methods pointing to gut–brain dysbiosis suggest holistic support for dysbiosis-related conditions in vulnerable people (Sharma et al. 2024). Natural therapeutics have shown a significant effect on healthcare in the following ways:

- **Integrated natural approaches and therapies**: Diet, food, probiotics, exercise, and mind-body treatments restore gut microbiota equilibrium, decreasing inflammation and improving neurotransmitter production for improved geriatric mental well-being (Ugwu et al. 2025; Cryan et al. 2019).
- **Therapeutic advantages**: Probiotics, herbal preparations (e.g., Ginkgo biloba), and yoga improve depression and anxiety in elderly by controlling the HPA axis and SCFAs through gut–brain signalling (Nguyen and Lavretsky 2020; O'Riordan et al. 2025).
- **Geriatric/psychiatric efficacy**: These interventions develop cognition, mood, and resistance in ageing populations, outpacing isolated pharmacotherapy by dealing with dysbiosis-linked comorbidities (Ugwu et al. 2025).
- **Gut–brain mechanism**: Natural remedies encourage microbial diversity, vagus nerve action, and anti-inflammatory metabolites, connecting gut repair to psychiatric condition relief (Cryan et al. 2019) (Table 15.1).

15.2 LIFESTYLE ADAPTATIONS FOR MENTAL AND PHYSICAL WELL-BEING

15.2.1 Diet and Nutrition

Role of a balanced diet in preservation of mental and physical well-being: A balanced diet enriched with nutrients such as fibre, polyphenols, omega-3s, and vitamin B complex reduces inflammation, supports gut microbiota, enhances neurotransmitter function, and helps sustain physical and mental well-being. Nutritional deficiency can contribute to mood symptoms and gradual

TABLE 15.1
Effectiveness of Natural and Therapeutic Interventions and Gut–Brain Dysbiosis

Condition	Lifestyle/Herbal Remedies/ Therapy efficiency	Natural/Therapeutic Efficiency	Gut–Brain Dysbiosis Link
Depression	Decreased symptoms via diet/workout/meditation (Sharma et al. 2024; Nguyen and Lavretsky 2020)	Probiotics/diet lessen symptoms through microbiota modulation (Ugwu et al. 2025; Nguyen and Lavretsky, 2020)	Altered microbiota, reduced SCFAs/inflammation (Chen et al. 2022); low SCFAs, inflammation elevate mood disorders (Chen et al. 2022)
Anxiety	Reduced via yoga/prayer/community/social support (Sharma et al. 2024; Nguyen and Lavretsky, 2020)	Yoga/meditation bring down stress over vagus nerve activation (O'Riordan et al. 2025)	HPA dysregulation, lessened GABA producers (Pferschy-Wenzig et al. 2022); dysbiosis disturbs GABA/serotonin production (Cryan et al. 2019)
Cognitive decline	Enhanced through Mediterranean diet/Ginkgo/yoga (Nguyen and Lavretsky 2020)	Herbal supplements (Ginkgo) increase memory/function (Nguyen and Lavretsky 2020)	Dysbiosis associates with amyloid/inflammation (Chunduri et al. 2022); microbial loss relates with amyloid build-up (Ugwu et al. 2025)
Stress disorders	Improved coping by managing routine/time (Sharma et al. 2024)	Lifestyle incorporation improves social/cognitive outcomes (Chen et al. 2022)	Permeability grows, cytokine elevation (Guan et al. 2024); barrier dysfunction rises translocation (O'Riordan et al. 2025)

TABLE 15.2
Effects of Dietary Components on Physical and Mental Well-Being

Nutrient/Diet Component	Physical Well-Being Benefits	Mental Well-Being Benefits
Fibre & microbiota	Gut health, immunity enhancement (Grajek et al. 2022)	Anxiety relief and stress modulation (Gheonea et al. 2023)
Omega-3 fatty acids	Cardiovascular protection, reduced chronic disease risk (Firth et al. 2019)	Mood enhancement, reduction in low mood symptoms (Grajek et al. 2022)
Polyphenols/antioxidants	Frailty prevention, inflammation control (Gheonea et al. 2023)	Cognitive improvement, well-being improvement (Muscaritoli 2021)
Vitamin B & micronutrients	Metabolic stability (Muscaritoli 2021)	Neurotransmitter function (Firth et al. 2019)

cognitive decline. Nutrient-rich food such as whole grains, vegetables, fermented sources, and fruits can help in the prevention of mental health disorders in an individual (Grajek et al. 2022; Firth et al. 2019). Microbiota in the GI tract is promoted by mechanisms like the GBA, where fibre-producing SCFAs decrease stress and improve recovery. Polyphenols and antioxidants reduce oxidative stress linked to mental decline and physical vulnerability, while omega-3s s tabilize mood and cardiovascular health. Evidence also shows that dietary improvements lower low mood symptoms and enhance quality of life (Firth et al. 2019; Gheonea et al. 2023).

Mediterranean-style diet patterns reduce dementia risk and help in metabolic resilience (Grajek et al. 2022; Muscaritoli 2021). Adopting such diets fosters holistic preventive care (Table 15.2).

15.2.1.1 GBA Support Through Dietary Modification

- **GBA modulation**: Diet balanced with nutrients influences microbiota, producing metabolites that regulate anxiety, mood, and cognition via bidirectional signalling (Patil and Mehdi 2025).
- **Mediterranean diet enrichment**: Enhances neuroprotection and decreases neurodegeneration risk through anti-inflammatory nutrients and gut health support (Picone et al. 2024).
- **Nutritional psychiatry**: Omega-3s, zinc, and folate from whole foods improve brain function, promote emotional resilience, and stabilize mood (Lachance and Ramsey 2015).
- **Nutrient effects on cognition**: Flavonoids & Omega-3s boost synaptic plasticity, neuroprotection against oxidative stress, leading to improvement in cognition (Gómez-Pinilla 2008).
- **Antioxidants in diet**: Micronutrients and polyphenols fight inflammation, supporting mental well-being along with cognitive longevity (Hunt, Pontifex, and Vauzour 2024).

15.2.2 Physical Activity and Exercise

Through intertwined mechanisms (both physiological and psychological), regular exercise regulates mood, mitigates stress, and improves cognitive function (Callaghan 2004; Mahindru, Patil, and Agrawal 2023). Downregulation of the HPA axis, through aerobic and resistance activities, improves sleep and lowers overall cortisol levels (Callaghan 2004; Hossain et al. 2024). Physical activity also enhances monoamines, brain-derived neurotrophic factor (BDNF), and endorphins, supporting positive affect, self-esteem, and resilience in people with and without depressive/anxiety disorders (Pujari 2024; Hossain et al. 2024). It is suggested that even a total of 150 minutes of moderate-intensity physical activity per week presents certain mental health benefits, such as decreased incidence of clinical depression and better mental health outcomes (Mahindru, Patil, and Agrawal 2023; Román et al. 2022).

By promoting hippocampal neurogenesis, regulated cerebral blood flow, and synaptic plasticity, regular physical exercise leads to improved cognitive functions such as better attention and concentration, memory, and executive functions (Callaghan 2004; Liu and Nusslock 2018). Altogether, evidence-based studies point towards physical activity as a beneficial non-pharmacological treatment modality that suggestively regulates mood and decreases stress along with safeguarding cognitive functions (Mahindru, Patil, and Agrawal 2023) (Table 15.3).

Regular walking, by decreasing cortisol and through endorphins release, leads to a reduction in stress and helps regulate mood. Especially in elderly, it also boosts cognitive function by improving brain blood flow and neuroplasticity (Han, Kim, and Kim 2021; Xu et al. 2024). These advantages make regular walking a convenient mediation for emotional well-being and anxiety (Xu et al. 2024) (Tables 15.4 and 15.5).

TABLE 15.3
Relationship of Regular Physical Activity with Mental Health

Exercise Aspect	Mental Health Outcome	Key Mechanism/Notes
Long-term active lifestyle	Lower incidence of anxiety and depressive disorders (Mahindru, Patil, and Agrawal 2023)	Preventive effect through the whole lifespan
Moderate aerobic activity	Reduction in anxiety and stress (Callaghan 2004)	HPA-axis regulation, decreased cortisol
Regular aerobic/resistance	Regulated mood (Hossain et al. 2024; Mahindru, Patil, and Agrawal 2023)	Monoamines, self-efficacy, Endorphins
Sustained training (weeks–months)	Improved cognitive functions—executive functions and attention/concentration (Pujari 2024; Liu and Nusslock 2018)	Hippocampal neurogenesis, cerebral blood flow, buffers stress-induced damage

TABLE 15.4
Relationship of Regular Walking with Mental Health

Walking Aspect	Mental Health Outcome	Key Mechanism/Notes
Moderate leisure walking	Overall reduction in stress (Han, Kim, and Kim 2021)	Downregulation of cortisol, relaxation in elderly population
Structured walking programs	Decreased incidence of depression/anxiety, improvement in mood (Xu et al. 2024)	Endorphins, comparable to active controls, <3-month time period optimal
Daily/vigorous walking	Improved cognitive functioning (Han, Kim, and Kim 2021)	Enhanced health prospects, neuroprotection vs. light walking

TABLE 15.5
Influence of Physical Workout on Stress Resilience and Neuroplasticity

Activity Type	Neuroplasticity Impact	Stress Resilience Impact
Aerobic workouts	Cognitive reserve enhancement, neurogenesis (Neumann et al. 2021)	Better emotional regulation, lower reactivity (Kvam et al. 2016)
Structured programs	Increased grey matter preservation, neural efficiency (Neumann et al. 2021)	Guard psychopathology onset (Kvam et al. 2016)
Walking (moderate)	Increased hippocampal volume, BDNF, synaptic plasticity (Arida and Teixeira-Machado 2021)	Decreased cortisol, HPA regulation (Hossain et al. 2024)

15.2.3 Sleeping Habits

Good sleep habits also help in the reduction of mood symptoms and anxiety and promote mental health (Alanazi et al. 2023; Scott et al. 2021). Relaxing routines, constant sleeping schedules, and screen time limitations—along with sleep hygiene practices—help in improving sleep quality, eventually lowering mental health symptoms (Scott et al. 2021). Better sleep quality also strengthens cardiovascular health, metabolic regulation, and immune function, lowering the risks of obesity and chronic diseases (S. Sharma and Kavuru 2010). Overall, sleep hygiene assimilates preventive care, generating health benefits for comprehensive well-being in all populations (Scott et al. 2021).

Strategies to improve sleep in older individuals and patients with mental health disorders include sleep hygiene, evaluating psychological and behavioural causes of sleep disturbances, cognitive behavioural therapy for insomnia (CBT-I), and various sleep restriction techniques (Gupta et al. 2017; Suzuki, Miyamoto, and Hirata 2017). In elderly, considering the risk-benefit ratio, pharmacological treatments are generally approved only for severe cases (Gupta et al. 2017). Sound sleep promotes stable mood, better cognitive function, and also mitigates the risk of relapse in older individuals with mental health disorders (Fu and Li 2025) (Table 15.6).

15.2.4 Lifestyle Modifications to Manage Stress

Relaxation techniques such as self-guided relaxation and Jacobson's progressive muscle relaxation, promotion of structured routines, and physical activity along with active social engagement help in promoting well-being and reducing psychological distress (Sharma et al. 2024; Akarsu et al. 2025). Key strategies in lifestyle management include the establishment of daily structured routines with management of time and along with regular exercise and other behavioural activation strategies (Sharma et al. 2024; Braun, Foreyt, and Johnston 2016). HPA axis dysregulation and stress-induced inflammation can be addressed by limiting screen media use and dietary

TABLE 15.6
Relationship of Mental Health and Sleep in Elderly Psychiatric Patients

Sleep Strategy	Mental Health Outcome	Mechanism/Notes
Behavioural therapies/cognitive behavioural therapy	Better mood regulation, improvement in insomnia symptoms (Gupta et al. 2017)	Behavioural changes targeting sleep patterns
Circadian rhythm regulation	Improved cognition, relapse prevention (Fu and Li 2025)	Light management and chronotherapy
Judicious pharmacotherapy	Symptom relief in resistant insomnia (Gupta et al. 2017)	Monitoring for interactions and side effects and avoiding polypharmacy
Sleep hygiene modification	Better quality of sleep, decreased anxiety symptoms (Wang et al. 2023)	Lifestyle modifications

improvements (Akarsu et al. 2025; Morgan et al. 2014). Nature and relaxation techniques can be incorporated to reduce stress and anxiety and improve mental well-being, as mentioned below:

- **Spending time with nature**: Nature exposure decreases cortisol and improves overall mood by regulating dopamine and serotonin release subsequently reducing anxiety symptoms (Jimenez et al. 2021; Moncrieff et al. 2022).
- **Deep breathing for anxiety relief**: Breathing exercises also help in regulating the nervous system by activating the parasympathetic system (Bentley et al. 2023).
- **Other relaxation techniques**: Progressive muscle relaxation, self-guided relaxation, and guided imagery can help in mind-body relaxation, thus helping in stress management (Toussaint et al. 2021).
- **Combined effects of nature and relaxation**: Integrating nature exposure with relaxation practices enhances mental well-being and also helps in getting rid of depressive and anxiety symptoms (Jimenez et al. 2021; Owens and Bunce 2023).

15.3 HERBAL REMEDIES FOR GUT–BRAIN HEALTH

Modulation of GBA via dietary herbs listed in Table 15.7 has shown significant ameliorative effects on ageing, cognitive functions, and gut microbiota.

The herbs discussed have successfully demonstrated favourable safety profiles and efficacy for stress reduction, sleep improvement, and cognitive support in the elderly, primarily via anti-inflammatory, antioxidant, and GABAergic mechanisms, and are easily available as dietary and prescription supplements in the market. Older adults require standardized extracts at conservative doses due to age-related interactions with SSRIs or antipsychotics such as increased bleeding risk with *Ginkgo biloba* extract, or sedation potentiation with valerian root extract, not to forget the polypharmacy concerns. Clinical monitoring and herb–drug interaction screening improve safe integration of geriatric with psychiatric care.

15.4 PHYSICAL THERAPIES FOR HEALTH AND WELL-BEING

Physical therapies as a part of lifestyle modifications offer consolidative methods to improve mental and physical welfare in the ageing and psychiatric persons via modulating the GBA, decreasing inflammation, and enhancing neuroplasticity. Yoga and meditation improve both physical and cognitive flexibility; massage therapy alleviates stress; acupuncture practices balance Qi for pain respite; physiotherapy conserves mobility; and all have been successfully tailored for older adults, fostering physical and emotional resilience.

TABLE 15.7
Herbal Remedies for Gut–Brain Health

Herb	Medicinal Characteristics (Physiological Effects)	Effects on Mental Health	Effects on Gut Health	References
Ashwagandha (*Withania somnifera*)	Reduces cortisol, modulates HPA axis, anti-inflammatory via NF-κB inhibition, antioxidant (Nrf2 activation), GABAergic enhancement.	Decreases stress, anxiety, depression; improves sleep, cognition, and mood.	Supports microbiome balance, reduces inflammation; potential prebiotic effects.	Jamnekar et al. (2025), Mikulska et al. (2023), Panossian and Efferth (2022)
Bacopa monnieri	Neuroprotective, antioxidant, enhances synaptic plasticity.	Improves memory, cognition; anxiolytic, antidepressant.	Reduces gut inflammation, supports microbiota diversity.	Wightman et al. 2023)
Chamomile (*Matricaria chamomilla*)	Anti-anxiolytic, mild sedative, anti-spasmodic.	Reduces anxiety and promotes relaxation/ sleep.	Soothes GI tract, anti-inflammatory; potential prebiotic effects.	Kazemi et al. (2024)
Fennel (*Foeniculum vulgare*)	Carminative, anti-spasmodic, antioxidant.	Indirect anxiety relief via better digestion.	Enhances digestion, reduces bloating; prebiotic potential.	Rafieian et al. (2023)
Garlic (*Allium sativum*)	Antimicrobial, prebiotic, antioxidant, anti-inflammatory.	Indirect mood improvement via reduced inflammation.	Promotes probiotic growth, improves microbiota diversity.	Chen et al. (2020), Mony et al. (2025)
Ginger (*Zingiber officinale*)	Anti-inflammatory (6-shogaol), anti-nausea, Nrf2 activator.	Reduces stress-related anxiety via gut–brain axis.	Synergistic with turmeric; enhances SCFA, reduces gut inflammation.	Zahedi et al. (2024)
Ginkgo biloba	Antioxidant, improves cerebral blood flow, neuroprotective.	Enhances memory, cognition; reduces neuropsychiatric symptoms.	Supports gut barrier integrity, anti-inflammatory.	Obrenovich et al. (2022), Zhu et al. (2024)
Ginseng (*Panax ginseng*)	Adaptogenic, boosts energy, immunomodulatory, antioxidant, "stress vaccine" effect.	Improves mood, reduces anxiety/ depression symptoms; enhances cognitive function.	Promotes beneficial microbiota shifts, anti-inflammatory in gut mucosa.	Panossian and Efferth (2022)
Lavender (*Lavandula angustifolia*)	Aromatherapy anxiolytic, GABAergic, anti-inflammatory.	Decreases anxiety, improves sleep quality.	Indirect via stress reduction; calms gut motility.	Manzoor et al. (2025)
Peppermint (*Mentha piperita*)	Antispasmodic, carminative, relaxes GI smooth muscle.	Reduces stress-induced IBS symptoms affecting mood.	Improves IBS, motility; supports gut microbiota balance.	Rasooli et al. (2010), Sarikhani et al. (2021)
Rhodiola (*Rhodiola rosea*)	Adaptogenic, enhances stress resilience, modulates the HPA axis, antioxidant, anti-fatigue.	Reduces stress and anxiety; improves fatigue and cognitive performance under stress.	Modulates gut microbiota, enhances SCFA production indirectly via stress reduction.	Burns (2022), Panossian and Efferth (2022)

(*Continued*)

TABLE 15.7 (*Continued*)
Herbal Remedies for Gut–Brain Health

Herb	Medicinal Characteristics (Physiological Effects)	Effects on Mental Health	Effects on Gut Health	References
Turmeric (*Curcuma longa*)	Curcumin: anti-inflammatory, antioxidant, modulates microbiota.	Improves depression/anxiety via gut–brain modulation.	Increases beneficial bacteria, reduces dysbiosis/inflammation.	Enayati et al. (2023), Nunes et al. (2024)
Valerian root (*Valeriana officinalis*)	Sedative, GABA_A receptor modulator, reduces sleep latency.	Improves sleep quality, reduces insomnia and anxiety.	Mild gut relaxant; supports microbiome via reduced stress on gut–brain axis.	Gutiérrez-Romero et al. (2024)

15.4.1 Yoga and Meditation

Yoga improves physical and mental flexibility, strength, and clarity through the stimulation of vagal tone, a decrease in cortisol (and hence stress), and promoting a parasympathetic dominance. In a study conducted in Arhatic Yoga retreats, it was observed that beta-diversity shifted rapidly ($p<0.05$), enriching *Akkermansia* and SCFA producers (Swarup et al. 2025). Mindfulness meditation decreases anxiety and depression while enhancing cognition through modulation of gut microbiota and HPA axis (Soni 2025). For elderly/psychiatric patients, chair yoga, gentle pranayama (e.g., Nadi Shodhana), and short-session-tailored interventions improve adherence and mitigate frailty risks.

15.4.2 Massage Therapy and Acupuncture

Massage therapy decreases physical tension and stress stimulation of vagal afferents, encouraging parasympathetic activation and moderating the GBA, as evidenced in the study for irritable bowel syndrome with constipation, where one-finger meditation massage targeted abdominal acupoints and helped to regulate 5-HT and substance P levels. Hence, as a complementary therapy, massages and acupuncture techniques can alleviate visceral hypersensitivity, increase bowel motility, and improve quality of life for anxiety and depression patients, outperforming the use of probiotics alone in relieving symptoms (Zeng et al. 2023). Complementing the GBA effects via acupuncture methods are analogous in supporting its role in psychiatric care (Shi et al. 2024).

Acupuncture poises mental and physical health by decreasing stress and increasing Qi flow via restoration of the microbiota–gut–brain axis (MGBA), ameliorating dysbiosis, and discouraging the progression of amnestic mild cognitive impairment (aMCI) towards Alzheimer's. In aMCI subjects, acupuncture at GV20/ST36/SP6 acupoints augments advantageous taxa (*Lactobacillus*, *Bifidobacterium*), increases SCFAs, reinforces intestinal barriers (ZO-1/occludin upregulation), and decreases NF-κB inflammation, thereby diminishing Aβ deposition and cognitive deterioration (Shi et al. 2024; Xing, Hu, and Bao 2025). Multi-omics analyses further confirm neurotransmitter modulation and neuroprotection, thus proving acupuncture as a multi-target therapy in cases of geriatric stress and neurodegeneration (Yin et al. 2024).

15.4.3 Physiotherapy

Physiotherapy is vital for older adults, preserving mobility and independence via structured rehabilitation that counteracts sarcopenia, improves gait, and decreases fall risks. Schrenk et al. (2023) outlined a protocol for online guided training in elderly patients targeting cognition and GBA interactions. Physical exercise enhanced hippocampal volume, neurogenesis, and microbiota diversity (increased *Akkermansia* taxa and metabolites—SCFAs), mitigating AD pathology via GBA modulation. By increasing BDNF, decreasing neuroinflammation, and alleviating depression via endorphin release and social engagement, physiotherapy nurtures mental health resilience, with improved cognitive scores and functional outcomes in frail older adults (Sanchez-Martinez et al. 2024; Schrenk et al. 2023).

15.5 PSYCHOSOCIAL THERAPIES AND MENTAL HEALTH SUPPORT

Psychosocial therapies play a vital function in the management of mental health, particularly amongst the elderly, who frequently face psychiatric conditions and GBA dysregulation. This segment features various psychosocial interventions that are efficient in this population, comprising cognitive behavioural therapy (CBT), MBSR, social support networks, art and music therapy, and animal-assisted therapy (AAT) (Figure 15.2).

15.5.1 Cognitive Behavioural Therapy (CBT)

CBT in managing stress, anxiety, and depression: CBT is an extensively used psychotherapy concentrating on classifying and altering dysfunctional thoughts as well as behaviours. Research by Hofmann et al. (2012) confirms its efficiency among mental health problems, as it was observed that CBT significantly lessens anxiety and stress signs, increasing coping techniques across people. Cuijpers et al. (2014) established CBT's strong efficacy in decreasing depressive signs and preventing deterioration. Gould, Coulson, and Howard (2012) stated that CBT personalized for elderly people effectively lessens anxiety symptoms and increases quality of life.

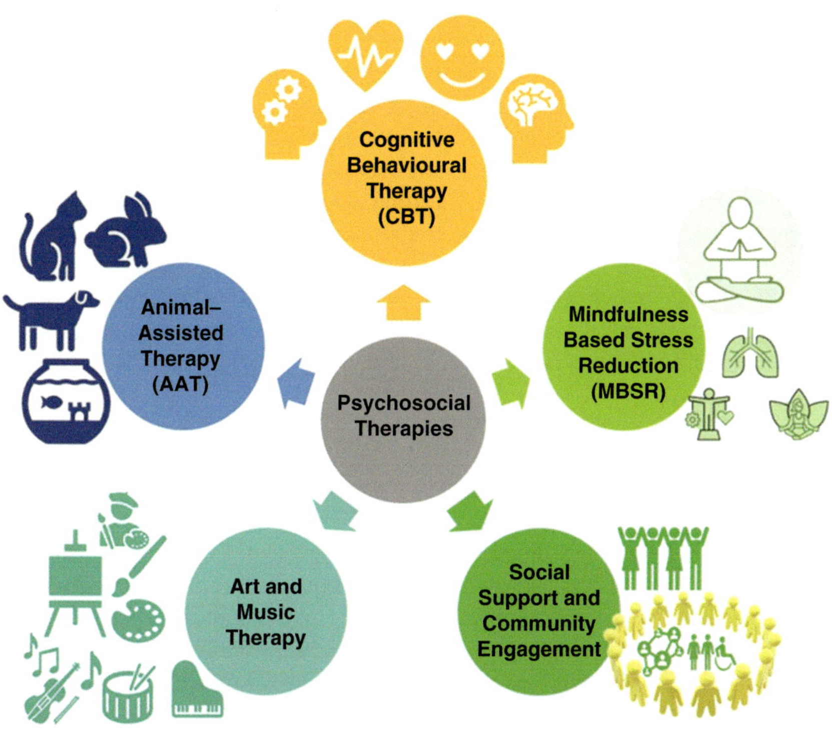

FIGURE 15.2 The psychosocial therapies for improving physical and mental well-being in older adults.

TABLE 15.8
Conditions, CBT Benefit and Gut–Brain Dysbiosis

Condition	Indication of CBT Benefit (GI/Psych)	Evidence of Gut–Brain Dysbiosis
Irritable bowel syndrome (IBS)	CBT enhanced signs significantly vs. education controls with long-term effects (3–6 months) (Lackner et al. 2018)	IBS shows changed microbiota, permeability and visceral hypersensitivity as per brain–gut dysregulation characteristics (Mayer 2000; Keefer, Palsson, and Pandolfino 2018)
Functional GI disorders	CBT decreases acuteness, catastrophizing and incapacity among functional GI disorders (Keefer, Palsson, and Pandolfino 2018).	Functional GI conditions are measured brain–gut–microbiome disorders through immune activation and barrier dysfunction (Mayer 2000; Loh et al. 2024)
Major depressive disorder (MDD)	CBT is recommended treatment; lessens rumination and tension, enhancing mood (Dipalma et al. 2025)	Depression illustrates altered microbiota, enlarged permeability, and inflammatory signalling involving microbiota (Loh et al. 2024)
Anxiety disorders	CBT consistently decreases worry, autonomic arousal in addition to avoidance (Dipalma et al. 2025)	Anxiety disorders display stress-sensitive barrier variations and microbiota alterations (Loh et al. 2024)
Fibromyalgia and related syndromes	CBT expands pain coping, mood and disability often with exercise/education (Dipalma et al. 2025)	Changed microbiota and reactions to probiotics/prebiotics confirm dysbiosis-linked model (Dipalma et al. 2025)

CBT and GBA: In irritable bowel syndrome, both clinic-based and minimal-contact CBT considerably enhanced GI symptoms, with constant benefits at 6 months (Lackner et al. 2018). CBT and gut-directed hypnosis are preferred "brain–gut therapies" directing visceral sensitivity, mobility, pain, and anxiety connected to GI indicators (Riehl 2018; Keefer, Palsson, and Pandolfino 2018). Mechanistic assessments recommend that gut-focused CBT can modify gut microbiome make-up and function, affecting the brain–gut–microbiome axis ahead of symptom relief (Jacobs et al. 2021). CBT protocols comprise psychoeducation on the brain–gut axis, initiate self-monitoring, easing, mindfulness, cognitive reformation, and exposure-based evasion reduction (Lackner et al. 2018; Riehl 2018). These elements diminish HPA-axis stimulation, sympathetic activation, increase parasympathetic/vagal tone, and enhance coping—controllers of intestinal flexibility, absorptivity, and immune activity (Mayer 2000; Loh et al. 2024). Likewise, in psychiatric and pain conditions associated with dysbiosis alike major depression and fibromyalgia, CBT-induced stress physiology and behavioural variations may regularize gut microbial bionetworks, refining mood, cognition, and pain (Dipalma et al. 2025; Loh et al. 2024) (Table 15.8).

15.5.2 Mindfulness-Based Stress Reduction (MBSR)

MBSR is an organized program imparting mindfulness meditation to decrease stress and enhance mental well-being. Keng, Smoski, and Robins (2011) highlighted incorporating mindfulness into daily habits to improve health and practical stress management. Progressive muscle relaxation, breathing, and guided imagery concentrate attention on body parts providing and elevating relaxation. Toussaint et al. (2021) recorded physiological and psychological relief and improvements in getting relaxed after such training. Arch and Craske (2006) established that mindful breathing declines negative feelings and progresses emotional regulation (Figure 15.3).

Enhanced cognitive functioning and mental health: Goyal et al. (2014) found that MBSR is capable of lowering stress and anxiety in people. Kabat-Zinn (2003) observed that MBSR resolves anxiety and depression by means of emotional regulation besides resilience. Zeidan et al. (2010)

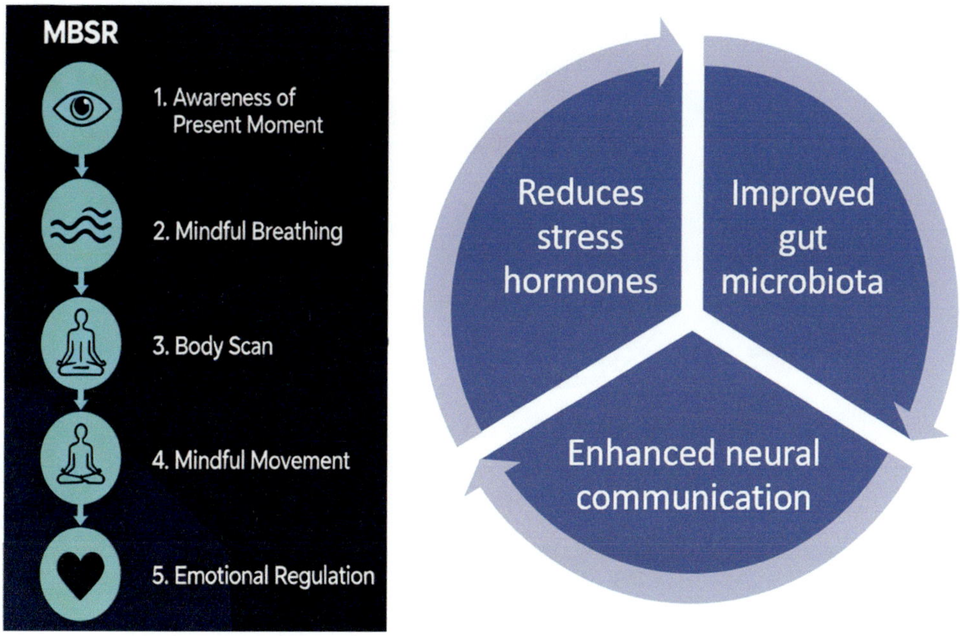

FIGURE 15.3 Mindfulness-based stress reduction (MBSR) and its beneficial effects.

established that short-term mindfulness training improves attention and working memory, helping stress management.

Physical well-being benefits: In a study by Black and Slavich (2016), it was observed that mindfulness meditation modifies inflammation markers, resistance, and ageing biology. Mindful breathing exercises and guided imagery of body scans decrease negative belief cycles, recovering emotional control along with gut health (Lenze et al. 2014).

Geriatric population: Keng, Smoski, and Robins (2011) stated enhanced quality of life and minimized anxiety and depressive indications in older adults after MBSR. In another research study, it was found that MBSR declined depression and increased emotion regulation with better sleep quality amongst depressed elderly persons, supporting its practice by caregivers in the nursing homes (Javadzade et al. 2024).

15.5.3 Social Support and Community Engagement

Social support associations are dynamic for mental well-being and emotional security. Everard et al. (2000) presented that engaging in group activities definitely affects physical health positively independent of demographic features. It was observed that offering ≥100 hours per annum reduced mortality and physical constraints while refining mood and determination in life (Kim et al. 2020) (Figure 15.4).

Emotional regulation and coping mechanism: Cohen and Wills (1985) associated strong social networks with reduced stress, apprehension, and depression. Elderly persons with healthy social ties show better emotional regulation and improved gut microbiota (Santini et al. 2015). Research study by Berkman et al. (2000) exhibited that social networks enhance psychological resilience and lessen distress.

Increased endurance and help in chronic illness: Holt-Lunstad, Smith, and Layton (2010) explained that individuals with robust social networks have greater survival rates. Similarly, it was noticed in another study that social support nurtures psychological resilience in elderly managing chronic illnesses (Lin et al. 2025).

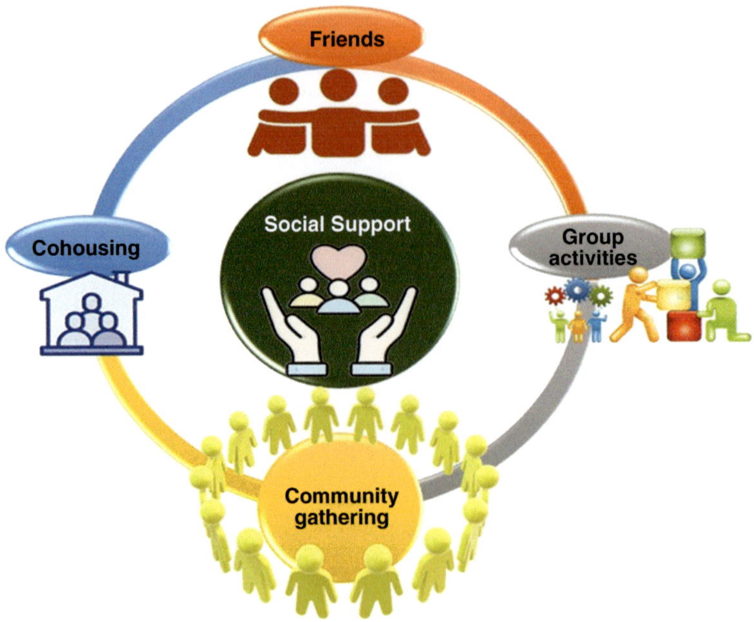

FIGURE 15.4 Social support and community engagement methods.

TABLE 15.9
Efficiency of Animal-Assisted Therapy (AAT) and Association with GBA

Condition	AAT Efficiency	Gut–Brain Dysbiosis Connection
Depression	Decreased symptoms, improved mood, reduction in treatment-resistant cases (Pandey et al. 2024; Pereira & Fonte, 2018)	Dysbiosis related to neurotransmitter changes and inflammation (Kelly et al., 2016)
Anxiety	Reduced anxiety, improved coping, decreased suffering (Pandey et al. 2024; Lundqvist et al., 2017)	Dysbiosis disturbs stress reaction via the HPA axis (Kelly et al., 2016)
Schizophrenia	Improved symptoms, better social functioning and quality of life (Pandey et al. 2024; Chen et al., 2022)	Dysbiosis related with inflammation and acuteness of symptom (Kelly et al., 2016)

15.5.4 Animal-Assisted Therapy (AAT)

AAT supports emotional regulation, security, and social engagement while reducing negative emotions such as anxiety and depression. Barker and Dawson (1998) described that AAT fosters social communications that lead to reducing the feelings of isolation.

Decreased anxiety and depression: AAT progresses emotional conditions by developing comfort and safety, lessening symptoms in psychological disorders associated with gut–brain dysbiosis (Pandey et al. 2024). Gee et al. (2025) and Travers et al. (2013) observed that AAT aids in anxiety, sadness, and dementia care.

Aid in dementia care: A study by Peluso et al. (2018) exhibited that AAT lessens aggression and anxiety, refining quality of life and social competence in dementia patients (Table 15.9).

15.5.5 Art and Music Therapy

Art and music therapy helps in making better and improved mood, understanding, cognition, social communication, and quality of life in geriatric and psychiatric patients. Group music therapy lessens depression and isolation in dementia (Sukumar et al. 2024; Shukla et al. 2022). Art therapy nurtures emotional expression and communication, managing stress reduction (Yu et al. 2022). Neuroimaging displays that these therapies produce neuroplasticity in brain regions involved in emotion regulation, adjustment, and attention (Yu et al. 2022; Shukla et al. 2022). Positive effects comprise decreased chronic stress, enhanced sleep, and better social engagement, maintaining GBA strength indirectly.

15.6 HOLISTIC INTEGRATION OF THERAPIES FOR OPTIMAL HEALTH

The integration of physical, herbal, and psychosocial therapies nurtures optimal health in geriatric and psychiatric patient care through addressing the unified physical, mental, and social aspects, as evidenced by sustainable lifestyles that mediate stress and depression via knowledge of physical therapies (Akarsu et al. 2025). Physical activities such as yoga and exercise improve hippocampal neurogenesis, BDNF expression, and brain elasticity against neurodegeneration via GBA modulation (Arida and Teixeira-Machado 2021; Loh et al. 2024). Herbal remedies like ashwagandha and curcumin complement physical therapies by decreasing cortisol, inflammation, and GBA dysbiosis while promoting microbiota diversity and neurotransmitter modulation (Chandrasekhar, Kapoor, and Anishetty 2012; Enayati et al. 2023). Psychosocial interventions, including mindfulness and social support, alleviate stress via emotional regulation and increased immunity, further amplifying therapy efficacy (Arch and Craske 2006; Black and Slavich 2016).

This multimodal synergy is well in line with the biopsychosocial model, wherein lifestyle factors, i.e., diet, exercise, and sleep hygiene, work together to enhance mental well-being and overall functional performance (Alanazi et al. 2023; Grajek et al. 2022). For example, polyphenols and adaptogens in the Mediterranean diet modulate MGBA signalling, decreasing frailty and ameliorating cognitive decline in older adults (Ghosh et al. 2020; Jamnekar et al. 2025). Physical education mediates the connection between healthy lifestyles and decreased anxiety/depression, accentuating tailored exercise programs (Akarsu et al. 2025). Psychosocial elements—social integration & MBSR—further potentiate the outcomes by decreasing inflammation and increasing emotional regulation (Berkman et al. 2000; Grossman et al. 2004).

In elderly patients, this combination alleviates polypharmacy risks; herbs like valerian help in promoting sleep without sedative-like interactions, whereas physiotherapy conserves mobility, indirectly enhancing mood via endorphins (Bent et al. 2006; Callaghan 2004). Mind-body practices like MBSR and breathing exercises/yoga regulate the HPA axis, while complementing herbal therapies' anti-inflammatory activity for holistic GBA optimization (Bentley et al. 2023; Kucukosmanoglu et al. 2024). Randomized trials show that combined interventions produce greater remission rates for depression compared to singular therapy (Cuijpers et al. 2014; Kvam et al. 2016). Although challenges do exist like herb–drug interactions (e.g., ginkgo with anticoagulants), it requires personalized plans (De Souza Silva et al. 2014).

Hence, holistic amalgamation helps in patient-centred care by employing synergies for resilience. Future models need to prioritize multidisciplinary teams integrating nutrition, exercise, herbs, and psychosocial support to improve quality of life in ageing populations (Nguyen and Lavretsky 2020; Sharma et al. 2024).

15.7 CONCLUSION

Lifestyle adaptations (exercise, diet, sleep hygiene), herbal care (ashwagandha, gingko, turmeric, etc.), and therapies (yoga, meditation, massage, acupuncture, physiotherapy) can synergistically improve mental and physical health in the elderly by modulating the GBA, decreasing inflammation, and improving neuroplasticity. They encourage microbiota eubiosis, BDNF upregulation, and resilience to stress, and better remission of depression/anxiety than isolated. Future trends should emphasize precision interventions via microbiome-multi-omics profiling, AI-guided personalization, and scalable digital platforms for geriatric psychiatry, combining nutraceuticals with mind–body practices amid increasing multimorbidity.

DECLARATION OF GENERATIVE AI AND AI-ASSISTED TECHNOLOGIES IN THE WRITING PROCESS

During the preparation of this work, the author(s) used AI assistance to support language refinement, structural editing, and generation of alternative phrasings for sections of the text. The tool was not used for data analysis, interpretation of results, or the creation or verification of scientific claims or references. All AI-generated suggestions were critically reviewed, revised, and verified by the author(s), who take full responsibility for the accuracy, integrity, and originality of the content.

REFERENCES

Akarsu, Mehmet, İsmail İlbak, Zeliha Çavuşoğlu, Ratko Pavlović, Ana Maria Vulpe, Adina Camelia Şlicaru, Nicolae Lucian Voinea, and Cristina Ioana Alexe. 2025. "The Relationship between a Sustainable Healthy Lifestyle and Depression, Stress, and Anxiety: A Structural Model on the Mediating Role of Physical Literacy." *Healthcare* 13 (14): 1646. doi:10.3390/healthcare13141646.

Alanazi, Eman M., Abeer Mohammed M. Alanazi, Asmaa Hamed Albuhairy, and Alshaymaa Akram A. Alanazi. 2023. "Sleep Hygiene Practices and Its Impact on Mental Health and Functional Performance among Adults in Tabuk City: A Cross-Sectional Study." *Cureus* 15 (3): e36221. doi:10.7759/cureus.36221.

Arch, Joanna J., and Michelle G. Craske. 2006. "Mechanisms of Mindfulness: Emotion Regulation Following a Focused Breathing Induction." *Behaviour Research and Therapy* 44 (12): 1849–1858. doi:10.1016/j.brat.2005.12.007.

Arida, Ricardo Mario and Lavinia Teixeira-Machado. 2021. "The Contribution of Physical Exercise to Brain Resilience." *Frontiers in Behavioral Neuroscience* 14 (January): 626769. doi:10.3389/fnbeh.2020.626769.

Barker, Sandra B., and Kathryn S. Dawson. 1998. "The Effects of Animal-Assisted Therapy on Anxiety Ratings of Hospitalized Psychiatric Patients." *Psychiatric Services* 49 (6): 797–801. doi:10.1176/ps.49.6.797.

Bent, Stephen, Amy Padula, Dan Moore, Michael Patterson, and Wolf Mehling. 2006. "Valerian for Sleep: A Systematic Review and Meta-Analysis." *The American Journal of Medicine* 119 (12): 1005–1012. doi:10.1016/j.amjmed.2006.02.026.

Bentley, Tanya G. K., Gina D'Andrea-Penna, Marina Rakic, Nick Arce, Michelle LaFaille, Rachel Berman, Katie Cooley, and Preston Sprimont. 2023. "Breathing Practices for Stress and Anxiety Reduction: Conceptual Framework of Implementation Guidelines Based on a Systematic Review of the Published Literature." *Brain Sciences* 13 (12): 1612. doi:10.3390/brainsci13121612.

Berkman, Lisa F., Thomas Glass, Ian Brissette, and Teresa E. Seeman. 2000. "From Social Integration to Health: Durkheim in the New Millennium." *Social Science & Medicine* 51 (6): 843–857. doi:10.1016/s0277-9536(00)00065-4.

Black, David S., and George M. Slavich. 2016. "Mindfulness Meditation and the Immune System: A Systematic Review of Randomized Controlled Trials." *Annals of the New York Academy of Sciences* 1373 (1): 13–24. doi:10.1111/nyas.12998.

Braun, Cameron, John P. Foreyt, and Craig A. Johnston. 2016. "Stress." *American Journal of Lifestyle Medicine* 10 (4): 235–238. doi:10.1177/1559827616642400.

Burns, Joshua. 2022. "Common Herbs for Stress: The Science and Strategy of a Botanical Medicine Approach to Self-care." *Journal of Interprofessional Education & Practice* 30 (December): 100592. doi:10.1016/j.xjep.2022.100592.

Callaghan, P. 2004. "Exercise: A Neglected Intervention in Mental Health Care?" *Journal of Psychiatric and Mental Health Nursing* 11 (4): 476–483. doi:10.1111/j.1365-2850.2004.00751.x.

Chandrasekhar, K., Jyoti Kapoor, and Sridhar Anishetty. 2012. "A Prospective, Randomized Double-Blind, Placebo-Controlled Study of Safety and Efficacy of a High-Concentration Full-Spectrum Extract of Ashwagandha Root in Reducing Stress and Anxiety in Adults." *Indian Journal of Psychological Medicine* 34 (3): 255–262. doi:10.4103/0253-7176.106022.

Chen, Keyu, Yasushi Nakasone, Kun Xie, Kozue Sakao, and De-Xing Hou. 2020. "Modulation of Allicin-Free Garlic on Gut Microbiome." *Molecules* 25 (3): 682. doi:10.3390/molecules25030682.

Chen, Pan, Ling Zhang, Yuan Feng, Yu-Fei Liu, Tong Leong Si, Zhaohui Su, Teris Cheung, Gabor S. Ungvari, Chee H. Ng, and Yu-Tao Xiang. 2022. "Brain-Gut Axis and Psychiatric Disorders: A Perspective from Bibliometric and Visual Analysis." *Frontiers in Immunology* 13 (November): 1047007. doi:10.3389/fimmu.2022.1047007.

Chunduri, Alisha, S. Deepak Mohan Reddy, M. Jahanavi, and C. Nagendranatha Reddy. 2022. "Gut–Brain Axis, Neurodegeneration and Mental Health: A Personalized Medicine Perspective." *Indian Journal of Microbiology* 62 (4): 505–515. doi:10.1007/s12088-022-01033-w.

Cohen, Sheldon, and Thomas A. Wills. 1985. "Stress, Social Support, and the Buffering Hypothesis." *Psychological Bulletin* 98 (2): 310–357. doi:10.1037/0033-2909.98.2.310.

Cryan, John F., Kenneth J. O'Riordan, Caitlin S. M. Cowan, Kiran V. Sandhu, Thomaz F. S. Bastiaanssen, Marcus Boehme, Martin G. Codagnone et al. 2019. "The Microbiota-Gut-Brain Axis." *Physiological Reviews* 99 (4): 1877–2013. doi:10.1152/physrev.00018.2018.

Cuijpers, Pim, Eirini Karyotaki, Erica Weitz, Gerhard Andersson, Steven D. Hollon, and Annemieke Van Straten. 2014. "The Effects of Psychotherapies for Major Depression in Adults on Remission, Recovery and Improvement: A Meta-Analysis." *Journal of Affective Disorders* 159 (February): 118–126. doi:10.1016/j.jad.2014.02.026.

De Souza Silva, José Edson, Carlos Adriano Santos Souza, Tamara Bispo Da Silva, Isla Alcântara Gomes, Giselle De Carvalho Brito, Adriano Antunes De Souza Araújo, Divaldo Pereira De Lyra-Júnior, Wellington Barros Da Silva, and Francilene Amaral Da Silva. 2014. "Use of Herbal Medicines by Elderly Patients: A Systematic Review." *Archives of Gerontology and Geriatrics* 59 (2): 227–233. doi:10.1016/j.archger.2014.06.002.

Dipalma, Gianna, Grazia Marinelli, Laura Ferrante, Angela Di Noia, Claudio Carone, Valeria Colonna, Pierluigi Marotti et al. 2025. "Modulating the Gut Microbiota to Target Neuroinflammation, Cognition and Mood: A Systematic Review of Human Studies with Relevance to Fibromyalgia." *Nutrients* 17 (14): 2261. doi:10.3390/nu17142261.

Enayati, Ayesheh, Aida Soghi, Alexandra E. Butler, Manfredi Rizzo, and Amirhossein Sahebkar. 2023. "The Effect of Curcumin on the Gut-Brain Axis: Therapeutic Implications." *Journal of Neurogastroenterology and Motility* 29 (4): 409–418. doi:10.5056/jnm23065.

Everard, K. M., H. W. Lach, E. B. Fisher, and M. C. Baum. 2000. "Relationship of Activity and Social Support to the Functional Health of Older Adults." *The Journals of Gerontology Series B* 55 (4): S208–S212. doi:10.1093/geronb/55.4.s208.

Fekete, Mónika, Andrea Lehoczki, Dávid Major, Vince Fazekas-Pongor, Tamás Csípő, Stefano Tarantini, Zoltán Csizmadia, and János Tamás Varga. 2024. "Exploring the Influence of Gut–Brain Axis Modulation on Cognitive Health: A Comprehensive Review of Prebiotics, Probiotics, and Symbiotics." *Nutrients* 16 (6): 789. doi:10.3390/nu16060789.

Firth, Joseph, Wolfgang Marx, Sarah Dash, Rebekah Carney, Scott B. Teasdale, Marco Solmi, Brendon Stubbs et al. 2019. "The Effects of Dietary Improvement on Symptoms of Depression and Anxiety: A Meta-Analysis of Randomized Controlled Trials." *Psychosomatic Medicine* 81 (3): 265–280. doi:10.1097/psy.0000000000000673.

Fu, Wangsheng, and Wei Li. 2025. "Sleep Quality, Mental Health, and Cognitive Function Among Older Adults in Chinese Communities: A Cross-Sectional Study." *Frontiers in Public Health* 13 (October): 1592886. doi:10.3389/fpubh.2025.1592886.

Gee, Nancy R., Lisa Townsend, Erika Friedmann, Sandra Barker, and Megan Mueller. 2025. "A Pilot Randomized Controlled Trial to Examine the Impact of a Therapy Dog Intervention on Depression, Mood, and Anxiety in Hospitalized Older Adults." *Healthcare* 13 (15): 1819. doi:10.3390/healthcare13151819.

Gheonea, Theodora Claudia, Carmen-Nicoleta Oancea, Magdalena Mititelu, Elena Carmen Lupu, Corina-Bianca Ioniță-Mîndrican, and Ion Rogoveanu. 2023. "Nutrition and Mental Well-Being: Exploring Connections and Holistic Approaches." *Journal of Clinical Medicine* 12 (22): 7180. doi:10.3390/jcm12227180.

Ghosh, Tarini Shankar, Simone Rampelli, Ian B. Jeffery, Aurelia Santoro, Marta Neto, Miriam Capri, Enrico Giampieri et al. 2020. "Mediterranean Diet Intervention Alters the Gut Microbiome in Older People Reducing Frailty and Improving Health Status: The NU-AGE 1-Year Dietary Intervention across Five European Countries." *Gut* 69 (7): 1218–1228. doi:10.1136/gutjnl-2019-319654.

Gómez-Pinilla, Fernando. 2008. "Brain Foods: The Effects of Nutrients on Brain Function." *Nature Reviews. Neuroscience* 9 (7): 568–578. doi:10.1038/nrn2421.

Gould, Rebecca L., Mark C. Coulson, and Robert J. Howard. 2012. "Cognitive Behavioral Therapy for Depression in Older People: A Meta-Analysis and Meta-Regression of Randomized Controlled Trials." *Journal of the American Geriatrics Society* 60 (10): 1817–1830. doi:10.1111/j.1532-5415.2012.04166.x.

Goyal, Madhav, Sonal Singh, Erica M. S. Sibinga, Neda F. Gould, Anastasia Rowland-Seymour, Ritu Sharma, Zackary Berger et al. 2014. "Meditation Programs for Psychological Stress and Well-Being." *JAMA Internal Medicine* 174 (3): 357. doi:10.1001/jamainternmed.2013.13018.

Grajek, Mateusz, Karolina Krupa-Kotara, Agnieszka Białek-Dratwa, Karolina Sobczyk, Martina Grot, Oskar Kowalski, and Wiktoria Staśkiewicz. 2022. "Nutrition and Mental Health: A Review of Current Knowledge about the Impact of Diet on Mental Health." *Frontiers in Nutrition* 9 (August): 943998. doi:10.3389/fnut.2022.943998.

Grossman, Paul, Ludger Niemann, Stefan Schmidt, and Harald Walach. 2004. "Mindfulness-Based Stress Reduction and Health Benefits." *Journal of Psychosomatic Research* 57 (1): 35–43. doi:10.1016/s0022-3999(03)00573-7.

Guan, Yueyue, Guohua Tang, Lei Li, Jianzhong Shu, Yuhua Zhao, Li Huang, and Jun Tang. 2024. "Herbal Medicine and Gut Microbiota: Exploring Untapped Therapeutic Potential in Neurodegenerative Disease Management." *Archives of Pharmacal Research* 47 (2): 146–164. doi:10.1007/s12272-023-01484-9.

Gupta, Ravi, Sourav Das, Kishore Gujar, Kk Mishra, Navendu Gaur, and Abdul Majid. 2017. "Clinical Practice Guidelines for Sleep Disorders." *Indian Journal of Psychiatry* 59 (5): 116. doi:10.4103/0019-5545.196978.

Gutiérrez-Romero, Sebastián Antonio, Erika Sofía Torres-Narváez, Adrián Camilo Zamora-Gómez, Silvana Castillo-Castillo, Angela Liliana Latorre-Velásquez, Carolina Betancourt-Villamizar, and Carlos O. Mendivil. 2024. "Effect of a Nutraceutical Combination on Sleep Quality Among People With Impaired Sleep: A Randomised, Placebo-controlled Trial." *Scientific Reports* 14 (1): 8062. doi:10.1038/s41598-24-58661-z.

Han, Areum, Junhyoung Kim, and Jaehyun Kim. 2021. "A Study of Leisure Walking Intensity Levels on Mental Health and Health Perception of Older Adults." *Gerontology and Geriatric Medicine* 7 (January): 2333721421999316. doi:10.1177/2333721421999316.

Hofmann, Stefan G., Anu Asnaani, Imke J. J. Vonk, Alice T. Sawyer, and Angela Fang. 2012. "The Efficacy of Cognitive Behavioral Therapy: A Review of Meta-Analyses." *Cognitive Therapy and Research* 36 (5): 427–440. doi:10.1007/s10608-012-9476-1.

Holt-Lunstad, Julianne, Timothy B. Smith, and J. Bradley Layton. 2010. "Social Relationships and Mortality Risk: A Meta-Analytic Review." *PLoS Medicine* 7 (7): e1000316. doi:10.1371/journal.pmed.1000316.

Hossain, Md Najmul, Jaeeun Lee, Hongseok Choi, Yi-Sub Kwak, Jongnam Kim, Jaeeun Lee, Hongseok Choi, Yi-Sub Kwak, and Jongnam Kim. 2024. "The Impact of Exercise on Depression: How Moving Makes Your Brain and Body Feel Better." *Physical Activity and Nutrition* 28 (2): 43–51. doi:10.20463/pan.2024.0015.

Hunt, Thomas, Matthew G. Pontifex, and David Vauzour. 2024. "(Poly)Phenols and Brain Health—Beyond Their Antioxidant Capacity." *FEBS Letters* 598 (24): 2949–2962. doi:10.1002/1873-3468.14988.

Jacobs, Jonathan P., Arpana Gupta, Ravi R. Bhatt, Jacob Brawer, Kan Gao, Kirsten Tillisch, Venu Lagishetty et al. 2021. "Cognitive Behavioral Therapy for Irritable Bowel Syndrome Induces Bidirectional Alterations in the Brain-Gut-Microbiome Axis Associated with Gastrointestinal Symptom Improvement." *Microbiome* 9 (1): 236. doi:10.1186/s40168-021-01188-6.

Jamnekar, Pallavi Prabhakar, Tejaswini Jeevan Dehankar, Rupali V. Bedre, Binthuja G. Dharan, Bharat Agravat, and Harsha Agravat. 2025. "Ashwagandha as an Adaptogenic Herb: A Comprehensive Review of Immunological and Neurological Effects." *Cureus*, November. doi:10.7759/cureus.96183.

Javadzade, Nima, Sayed Vahid Esmaeili, Victoria Omranifard, and Fatemeh Zargar. 2024. "Effect of Mindfulness-Based Stress Reduction (MBSR) Program on Depression, Emotion Regulation, and Sleep Problems: A Randomized Controlled Trial Study on Depressed Elderly." *BMC Public Health* 24 (1): 271. doi:10.1186/s12889-024-17759-9.

Jimenez, Marcia P., Nicole V. DeVille, Elise G. Elliott, Jessica E. Schiff, Grete E. Wilt, Jaime E. Hart, and Peter James. 2021. "Associations Between Nature Exposure and Health: A Review of the Evidence." *International Journal of Environmental Research and Public Health* 18 (9): 4790. doi:10.3390/ijerph18094790.

Kabat-Zinn, J. 2003. "Mindfulness-Based Stress Reduction (MBSR)." *Constructivism in the Human Sciences* 8 (2): 73–107.

Kazemi, Asma, Sara Shojaei-Zarghani, Parham Eskandarzadeh, and Mohammad Hashem Hashempur. 2024. "Effects of Chamomile (Matricaria Chamomilla L.) on Sleep: A Systematic Review and Meta-analysis of Clinical Trials." *Complementary Therapies in Medicine* 84 (August): 103071. doi:10.1016/j.ctim.2024.103071.

Keefer, Laurie, Olafur S. Palsson, and John E. Pandolfino. 2018. "Best Practice Update: Incorporating Psychogastroenterology into Management of Digestive Disorders." *Gastroenterology* 154 (5): 1249–1257. doi:10.1053/j.gastro.2018.01.045.

Kelly, John R., Yuliya Borre, Ciaran O' Brien, Elaine Patterson, Sahar El Aidy, Jennifer Deane, Paul J. Kennedy, et al. 2016. "Transferring the Blues: Depression-Associated Gut Microbiota Induces Neurobehavioural Changes in the Rat." *Journal of Psychiatric Research* 82 (July): 109–118. doi:10.1016/j.jpsychires.2016.07.019.

Keng, Shian-Ling, Moria J. Smoski, and Clive J. Robins. 2011. "Effects of Mindfulness on Psychological Health: A Review of Empirical Studies." *Clinical Psychology Review* 31 (6): 1041–1056. doi:10.1016/j.cpr.2011.04.006.

Kim, Eric S., Ashley V. Whillans, Matthew T. Lee, Ying Chen, and Tyler J. VanderWeele. 2020. "Volunteering and Subsequent Health and Well-Being in Older Adults: An Outcome-Wide Longitudinal Approach." *American Journal of Preventive Medicine* 59 (2): 176–186. doi:10.1016/j.amepre.2020.03.004.

Kucukosmanoglu, Hazal Sarak, Holger Cramer, Rahele Tavakoly, Alina Moosburner, and Mirela-Ioana Bilc. 2024. "Mind-Body Medicine in the Treatment of Depression: A Narrative Review of Efficacy, Safety and Mechanisms." *Current Psychiatry Reports* 26 (12): 729–740. doi:10.1007/s11920-024-01548-7.

Kuo, Braden, Manoj Bhasin, Jolene Jacquart, Matthew A. Scult, Lauren Slipp, Eric Isaac Kagan Riklin, Veronique Lepoutre et al. 2017. "Correction: Genomic and Clinical Effects Associated with a Relaxation Response Mind-Body Intervention in Patients with Irritable Bowel Syndrome and Inflammatory Bowel Disease." *PLoS ONE* 12 (2): e0172872. doi:10.1371/journal.pone.0172872.

Kvam, Siri, Catrine Lykkedrang Kleppe, Inger Hilde Nordhus, and Anders Hovland. 2016. "Exercise as a Treatment for Depression: A Meta-Analysis." *Journal of Affective Disorders* 202 (May): 67–86. doi:10.1016/j.jad.2016.03.063.

Lachance, Laura, and Drew Ramsey. 2015. "Food, Mood and Brain Health: Implications for the Modern Clinician." *Missouri Medicine* 112 (2): 111–115.

Lackner, Jeffrey M., James Jaccard, Laurie Keefer, Darren M. Brenner, Rebecca S. Firth, Gregory D. Gudleski, Frank A. Hamilton et al. 2018. "Improvement in Gastrointestinal Symptoms after Cognitive Behavior Therapy for Refractory Irritable Bowel Syndrome." *Gastroenterology* 155 (1): 47–57. doi:10.1053/j.gastro.2018.03.063.

Lenze, Eric J., Steven Hickman, Tamara Hershey, Leah Wendleton, Khanh Ly, David Dixon, Peter Doré, and Julie Loebach Wetherell. 2014. "Mindfulness-based Stress Reduction for Older Adults with Worry Symptoms and Co-occurring Cognitive Dysfunction." *International Journal of Geriatric Psychiatry* 29 (10): 991–1000. doi:10.1002/gps.4086.

Lin, Chunni, Xiayi Zhu, Xiaohui Wang, Lingyue Wang, Ying Wu, Xiang Hu, Jing Wen, and Li Cong. 2025. "The Impact of Perceived Social Support on Chronic Disease Self-Management among Older Inpatients in China: The Chain-Mediating Roles of Psychological Resilience and Health Empowerment." *BMC Geriatrics* 25 (1): 284. doi:10.1186/s12877-025-05902-z.

Liu, Patrick Z., and Robin Nusslock. 2018. "Exercise-Mediated Neurogenesis in the Hippocampus via BDNF." *Frontiers in Neuroscience* 12 (February): 52. doi:10.3389/fnins.2018.00052.

Loh, Jian Sheng, Wen Qi Mak, Li Kar Stella Tan, Chu Xin Ng, Hong Hao Chan, Shiau Hueh Yeow, Jhi Biau Foo, Yong Sze Ong, Chee Wun How, and Kooi Yeong Khaw. 2024. "Microbiota–Gut–Brain Axis and Its Therapeutic Applications in Neurodegenerative Diseases." *Signal Transduction and Targeted Therapy* 9 (1): 37. doi:10.1038/s41392-024-01743-1.

Lundqvist, Martina, Per Carlsson, Rune Sjödahl, Elvar Theodorsson, and Lars-Åke Levin. 2017. "Patient Benefit of Dog-Assisted Interventions in Health Care: A Systematic Review." *BMC Complementary and Alternative Medicine* 17 (1): 358. doi:10.1186/s12906-017-1844-7.

Mahindru, Aditya, Pradeep Patil, and Varun Agrawal. 2023. "Role of Physical Activity on Mental Health and Well-Being: A Review." *Cureus* 15 (1): e33475. doi:10.7759/cureus.33475.

Manzoor, Sana, Allah Rakha, Hina Rasheed, Zuhaib F. Bhat, Muhammad Shaffay Ali Khan, Gholamreza Abdi, and Rana Muhammad Aadil. 2025. "A Comprehensive Review on Anxiolytic Effect of Lavandula Angustifolia Mill. In Clinical Studies." *Food Science & Nutrition* 13 (9): e70993. doi:10.1002/fsn3.70993.

Mayer, E. A. 2000. "The Neurobiology of Stress and Gastrointestinal Disease." *Gut* 47 (6): 861–869. doi:10.1136/gut.47.6.861.

Mikulska, Paulina, Marta Malinowska, Miłosz Ignacyk, Paweł Szustowski, Joanna Nowak, Karolina Pesta, Monika Szeląg et al. 2023. "Ashwagandha (Withania Somnifera)—Current Research on the Health-Promoting Activities: A Narrative Review." *Pharmaceutics* 15 (4): 1057. doi:10.3390/pharmaceutics15041057.

Moncrieff, Joanna, Ruth E. Cooper, Tom Stockmann, Simone Amendola, Michael P. Hengartner, and Mark A. Horowitz. 2022. "The Serotonin Theory of Depression: A Systematic Umbrella Review of the Evidence." *Molecular Psychiatry* 28 (8): 3243–3256. doi:10.1038/s41380-022-01661-0.

Mony, Tamanna, Marcus Jackson, Amitai Zuckerman, Wenyan Yu, Thao Nguyen, Ashley Balderrama, Runting Li, Grace Sun, Jiankun Cui, and Zezong Gu. 2025. "Supplementation of Aged Garlic Extract Attenuates

Age-associated Memory Impairment and Cognitive Decline: Involvement of Molecular Pathways in the Cortex and Hippocampus." *Biomedical Reports* 24 (1): 1–13. doi:10.3892/br.2025.2075.

Morgan, Nani, Michael R. Irwin, Mei Chung, and Chenchen Wang. 2014. "The Effects of Mind-Body Therapies on the Immune System: Meta-Analysis." *PLoS ONE* 9 (7): e100903. doi:10.1371/journal.pone.0100903.

Muscaritoli, Maurizio. 2021. "The Impact of Nutrients On Mental Health and Well-Being: Insights from the Literature." *Frontiers in Nutrition* 8 (March): 656290. doi:10.3389/fnut.2021.656290.

Neumann, R. J., K. F. Ahrens, B. Kollmann, N. Goldbach, A. Chmitorz, D. Weichert, C. J. Fiebach et al. 2021. "The Impact of Physical Fitness on Resilience to Modern Life Stress and the Mediating Role of General Self-Efficacy." *European Archives of Psychiatry and Clinical Neuroscience* 272 (4): 679–692. doi:10.1007/s00406-021-01338-9.

Nguyen, Sarah A., and Helen Lavretsky. 2020. "Emerging Complementary and Integrative Therapies for Geriatric Mental Health." *Current Treatment Options in Psychiatry* 7 (4): 447–470. doi:10.1007/s40501-020-00229-5.

Nunes, Yandra Cervelim, Nathalia M. Mendes, Enzo Pereira De Lima, Amanda Chabrour Chehadi, Caroline Barbalho Lamas, Jesselina F. S. Haber, Manoela Dos Santos Bueno et al. 2024. "Curcumin: A Golden Approach to Healthy Aging: A Systematic Review of the Evidence." *Nutrients* 16 (16): 2721. doi:10.3390/nu16162721.

O'Riordan, Kenneth J., Gerard M. Moloney, Lily Keane, Gerard Clarke, and John F. Cryan. 2025. "The Gut Microbiota-Immune-Brain Axis: Therapeutic Implications." *Cell Reports Medicine* 6 (3): 101982. doi:10.1016/j.xcrm.2025.101982.

Obrenovich, Mark, Sandeep Kumar Singh, Yi Li, George Perry, Bushra Siddiqui, Waqas Haq, and V. Prakash Reddy. 2022. "Natural Product Co-Metabolism and the Microbiota–Gut–Brain Axis in Age-Related Diseases." *Life* 13 (1): 41. doi:10.3390/life13010041.

Owens, Matthew, and Hannah Bunce. 2023. "The Effect of Brief Exposure to Virtual Nature on Mental Wellbeing in Adolescents." *Scientific Reports* 13 (1): 17769. doi:10.1038/s41598-023-44717-z.

Pandey, Ramendra Pati, None Himanshu, None Gunjan, Riya Mukherjee, and Chung-Ming Chang. 2024. "The Role of Animal-Assisted Therapy in Enhancing Patients' Well-Being: Systematic Study of the Qualitative and Quantitative Evidence." *JMIRx Med* 5 (March): e51787. doi:10.2196/51787.

Panossian, Alexander, and Thomas Efferth. 2022. "Network Pharmacology of Adaptogens in the Assessment of Their Pleiotropic Therapeutic Activity." *Pharmaceuticals* 15 (9): 1051. doi:10.3390/ph15091051.

Patil, Shradha, and Sayed Farhan S. Mehdi. 2025. "The Gut-Brain Axis and Mental Health: How Diet Shapes Our Cognitive and Emotional Well-Being." *Cureus* 17 (7): e88420. doi:10.7759/cureus.88420.

Peluso, Silvio, Anna De Rosa, Natascia De Lucia, Antonella Antenora, Maddalena Illario, Marcello Esposito, and Giuseppe De Michele. 2018. "Animal-Assisted Therapy in Elderly Patients: Evidence and Controversies in Dementia and Psychiatric Disorders and Future Perspectives in Other Neurological Diseases." *Journal of Geriatric Psychiatry and Neurology* 31 (3): 149–157. doi:10.1177/0891988718774634.

Pereira, Jorge Mota, and Daniela Fonte. 2018. "Pets Enhance Antidepressant Pharmacotherapy Effects in Patients with Treatment Resistant Major Depressive Disorder." *Journal of Psychiatric Research* 104 (July): 108–113. doi:10.1016/j.jpsychires.2018.07.004.

Pferschy-Wenzig, Eva-Maria, Manuela R. Pausan, Karin Ardjomand-Woelkart, Stefanie Röck, Ramy M. Ammar, Olaf Kelber, Christine Moissl-Eichinger, and Rudolf Bauer. 2022. "Medicinal Plants and Their Impact on the Gut Microbiome in Mental Health: A Systematic Review." *Nutrients* 14 (10): 2111. doi:10.3390/nu14102111.

Picone, Pasquale, Antonella Girgenti, Miriam Buttacavoli, and Domenico Nuzzo. 2024. "Enriching the Mediterranean Diet Could Nourish the Brain More Effectively." *Frontiers in Nutrition* 11 (November): 1489489. doi:10.3389/fnut.2024.1489489.

Pujari, Venkateswar. 2024. "Moving to Improve Mental Health - The Role of Exercise in Cognitive Function: A Narrative Review." *Journal of Pharmacy and Bioallied Sciences* 16 (Suppl 1): S26–S30. doi:10.4103/jpbs.jpbs_614_23.

Rafieian, Fatemeh, Reza Amani, Atefe Rezaei, Aslı Can Karaça, and Seid Mahdi Jafari. 2023. "Exploring Fennel (Foeniculum Vulgare): Composition, Functional Properties, Potential Health Benefits, and Safety." *Critical Reviews in Food Science and Nutrition* 64 (20): 6924–6941. doi:10.1080/10408398.2023.2176817.

Rasooli, Iraj, Parviz Owlia, Massoud Taghizadeh, Shakiba DarvishAlipoor Astaneh, and Seyedeh Maryam Sharafi. 2010. "Protective Effects of Bioactive Phytochemicals From Mentha Piperita With Multiple Health Potentials." *Pharmacognosy Magazine* 6 (23): 147. doi:10.4103/0973-1296.66926.

Riehl, Megan E. 2018. "The Emerging Role of Brain-Gut Therapies for Irritable Bowel Syndrome." *Gastroenterol Hepatol (N Y)* 14(7) (July): 436–438.

Román, Julie E. Ibáñez, Ola Ekholm, Maria Holst Algren, Ai Koyanagi, Sarah Stewart-Brown, Eric E. Hall, Brendon Stubbs, Vibeke Koushede, Lau Caspar Thygesen, and Ziggi Ivan Santini. 2022. "Mental Wellbeing and Physical Activity Levels: A Prospective Cohort Study." *Mental Health and Physical Activity* 24 (November): 100498. doi:10.1016/j.mhpa.2022.100498.

Sanchez-Martinez, Javier, Patricio Solis-Urra, Jorge Olivares-Arancibia, and Julio Plaza-Diaz. 2024. "Physical Exercise and Mechanism Related to Alzheimer's Disease: Is Gut–Brain Axis Involved?" *Brain Sciences* 14 (10): 974. doi:10.3390/brainsci14100974.

Santini, Ziggi Ivan, Ai Koyanagi, Stefanos Tyrovolas, Catherine Mason, and Josep Maria Haro. 2015. "The Association between Social Relationships and Depression: A Systematic Review." *Journal of Affective Disorders* 175 (January): 53–65. doi:10.1016/j.jad.2014.12.049.

Sarikhani, Manizheh, Mahla Deylam, Effat Alizadeh, Marzie Hejazy, Saeideh Alizadeh-Salteh, Hassan Moeini, and Masoumeh Firouzamandi. 2021. "Anti-aging Effects of Peppermint (Mentha Piperita L.) and Shirazi Thyme (Zataria Multiflora Boiss.) Plant Extracts." *Food Bioscience* 41 (February): 100930. doi:10.1016/j.fbio.2021.100930.

Schrenk, Simon J., Stefan Brodoehl, Stefano Flor, Christiane Frahm, Christian Gaser, Rami Abou Hamdan, Marco Herbsleb et al. 2023. "Impact of an Online Guided Physical Activity Training on Cognition and Gut-Brain Axis Interactions in Older Adults: Protocol of a Randomized Controlled Trial." *Frontiers in Aging Neuroscience* 15 (September): 1254194. doi:10.3389/fnagi.2023.1254194.

Scott, Alexander J., Thomas L. Webb, Marrissa Martyn-St James, Georgina Rowse, and Scott Weich. 2021. "Improving Sleep Quality Leads to Better Mental Health: A Meta-Analysis of Randomised Controlled Trials." *Sleep Medicine Reviews* 60 (September): 101556. doi:10.1016/j.smrv.2021.101556.

Sharma, Indira, Arun V. Marwale, Roop Sidana, and Ishwar D. Gupta. 2024. "Lifestyle Modification for Mental Health and Well-Being." *Indian Journal of Psychiatry* 66 (3): 219–234. doi:10.4103/indianjpsychiatry.indianjpsychiatry_39_24.

Sharma, Sunil, and Mani Kavuru. 2010. "Sleep and Metabolism: An Overview." *International Journal of Endocrinology* 2010 (January): 1–12. doi:10.1155/2010/270832.

Shi, Jiatian, Xinyue Zhang, Jianhua Chen, Ruishi Shen, Huashun Cui, and Huangan Wu. 2024. "Acupuncture and Moxibustion Therapy for Cognitive Impairment: The Microbiome–gut–brain Axis and Its Role." *Frontiers in Neuroscience* 17 (January): 1275860. doi:10.3389/fnins.2023.1275860.

Shukla, Apoorva, Sonali G. Choudhari, Abhay M. Gaidhane, and Zahiruddin Quazi Syed. 2022. "Role of Art Therapy in the Promotion of Mental Health: A Critical Review." *Cureus* 14 (8): e28026. doi:10.7759/cureus.28026.

Soni, Kiran. 2025. "The Gut-Brain Axis and the Role of Yoga and Pranayama: A Holistic Perspective." *Journal of Emerging Technologies and Innovative Research* 12 (5). https://www.jetir.org/papers/JETIR2505105.pdf.

Sukumar, M., Suresh Kumar Thanneeru, Roshan Fakirchand Sutar, Anjan Sahu, and Amit Agrawal. 2024. "Effect of Music Therapy on Quality of Life in Geriatric Population: A Systematic Review and Meta-Analysis." *Indian Journal of Palliative Care* 30 (4): 296–303. doi:10.25259/ijpc_23_2024.

Suzuki, Keisuke, Masayuki Miyamoto, and Koichi Hirata. 2017. "Sleep Disorders in the Elderly: Diagnosis and Management." *Journal of General and Family Medicine* 18 (2): 61–71. doi:10.1002/jgf2.27.

Swarup, Sanjay, Abhishek Gupta, Marianne Chung, Vaishnavi Radhakrishnan, Valerie Davis, Michael D. J. Lynch, Trevor C. Charles, Jiujun Cheng, and Glenn Mendoza. 2025. "Rapid Shift of Gut Microbiome and Enrichment of Beneficial Microbes During Arhatic Yoga Meditation Retreat in a Single-arm Pilot Study." *BMC Complementary Medicine and Therapies* 25 (1): 51. doi:10.1186/s12906-025-04783-4.

Toussaint, Loren, Quang Anh Nguyen, Claire Roettger, Kiara Dixon, Martin Offenbächer, Niko Kohls, Jameson Hirsch, and Fuschia Sirois. 2021. "Effectiveness of Progressive Muscle Relaxation, Deep Breathing, and Guided Imagery in Promoting Psychological and Physiological States of Relaxation." *Evidence-Based Complementary and Alternative Medicine* 2021 (July): 1–8. doi:10.1155/2021/5924040.

Travers, Catherine, Jacqueline Perkins, Jacqui Rand, Helen Bartlett, and John Morton. 2013. "An Evaluation of Dog-Assisted Therapy for Residents of Aged Care Facilities with Dementia." *Anthrozoös* 26 (2): 213–225. doi:10.2752/175303713x13636846944169.

Ugwu, Okechukwu Paul-Chima, Michael Ben Okon, Esther Ugo Alum, Chinyere Nneoma Ugwu, Emeka Godson Anyanwu, Basajja Mariam, Fabian Chukwudi Ogenyi et al. 2025. "Unveiling the Therapeutic Potential of the Gut Microbiota–Brain Axis: Novel Insights and Clinical Applications in Neurological Disorders." *Medicine* 104 (30): e43542. doi:10.1097/md.0000000000043542.

Wang, Yingting, Zixuan Guang, Jinjing Zhang, Lixin Han, Rongqiang Zhang, Yichun Chen, Qi Chen et al. 2023. "Effect of Sleep Quality on Anxiety and Depression Symptoms among College Students in China's Xizang Region: The Mediating Effect of Cognitive Emotion Regulation." *Behavioral Sciences* 13 (10): 861. doi:10.3390/bs13100861.

Wightman, Emma, Julie Khan, Ellen Smith, Vivien Rolfe, Darren Smith, Greg Young, William Cheung, and David Kennedy. 2023. "Chronic Supplementation of a Multi-ingredient Herbal Supplement Increases Speed of Cognitive Task Performance Alongside Changes in the Urinary Metabolism of Dopamine and the Gut Microbiome in Cognitively Intact Older Adults Experiencing Subjective Memory Decline: A Randomized, Placebo Controlled, Parallel Groups Investigation." *Frontiers in Nutrition* 10 (October): 1257516. doi:10.3389/fnut.2023.1257516.

Xing, Hang, Wen-Long Hu, and Sheng-Yong Bao. 2025. "Acupuncture Modulates the Microbiota-gut-brain Axis: A Novel Therapeutic Strategy for Amnestic Mild Cognitive Impairment." *Frontiers in Neuroscience* 19 (November). doi:10.3389/fnins.2025.1710066.

Xu, Zijun, Xiaoxiang Zheng, Hanyue Ding, Dexing Zhang, Peter Man-Hin Cheung, Zuyao Yang, King Wa Tam et al. 2024. "The Effect of Walking on Depressive and Anxiety Symptoms: Systematic Review and Meta-Analysis." *JMIR Public Health and Surveillance* 10 (July): e48355. doi:10.2196/48355.

Yin, Zi-Han, Qiong-Nan Bao, Ya-Qin Li, Yi-Wei Liu, Zi-Qi Wang, Fang Ye, Xia He et al. 2024. "Discovery of the Microbiota-gut-brain Axis Mechanisms of Acupuncture for Amnestic Mild Cognitive Impairment Based on Multi-omics Analyses: A Pilot Study." *Complementary Therapies in Medicine* 88 (December): 103118. doi:10.1016/j.ctim.2024.103118.

Yu, Ai-Ling, Shu-Fen Lo, Pei-Yu Chen, and Shiou-Fang Lu. 2022. "Effects of Group Music Intervention on Depression for Elderly People in Nursing Homes." *International Journal of Environmental Research and Public Health* 19 (15): 9291. doi:10.3390/ijerph19159291.

Zahedi, Elham, Fatemeh Mokhtari Naseri, Elham Zamani, Marjan Nikbakhtzadeh, Tayebeh Rastegar, Ashkan Sanaeirad, and Seyed Shahabeddin Sadr. 2024. "Ginger Extract Improves Cognitive Dysfunction via Modulation of Gut Microbiota-Derived Short-Chain Fatty Acids in D-Galactose/Ovariectomy-Induced Alzheimer-Like Disease." *Molecular Neurobiology* 62 (4): 5095–5108. doi:10.1007/s12035-024-04583-w.

Zeidan, Fadel, Susan K. Johnson, Bruce J. Diamond, Zhanna David, and Paula Goolkasian. 2010. "Mindfulness Meditation Improves Cognition: Evidence of Brief Mental Training." *Consciousness and Cognition* 19 (2): 597–605. doi:10.1016/j.concog.2010.03.014.

Zeng, Xiayang, Jingjing He, Xiaoyu Li, Peng Chen, Jinhong Zuo, Xinlei Cai, Zhenyu Fan, and Jianpeng Qu. 2023. "Clinical Efficacy of One-finger Meditation Massage on IBS-C Based on the 'Gut-brain Axis' Theory: Study Protocol for a Randomized Controlled Trial." *BMC Complementary Medicine and Therapies* 23 (1): 185. doi:10.1186/s12906-023-04019-3.

Zhu, Cheng, Jie Liu, Jixin Lin, Jiaxi Xu, and Enyan Yu. 2024. "Investigating the Effects of Ginkgo Biloba Leaf Extract on Cognitive Function in Alzheimer's Disease." *CNS Neuroscience & Therapeutics* 30 (9): e14914. doi:10.1111/cns.14914.

Index

B

E

H

I

M

S

T

U

V

W

Y

For Product Safety Concerns and Information please contact our EU representative GPSR@taylorandfrancis.com Taylor & Francis Verlag GmbH, Kaufingerstraße 24, 80331 München, Germany